Genetically Modified Foods

Mahgoub, Sala E. O.
(Salah Eldin Osman)
Genetically

Genetically Modified Foods

BASICS, APPLICATIONS, AND CONTROVERSY

Salah E. O. Mahgoub

CRC Press is an imprint of the
Taylor & Francis Group, an **informa** business

CRC Press
Taylor & Francis Group
6000 Broken Sound Parkway NW, Suite 300
Boca Raton, FL 33487-2742

© 2015 by Taylor & Francis Group, LLC
CRC Press is an imprint of Taylor & Francis Group, an Informa business

No claim to original U.S. Government works

Printed on acid-free paper
Version Date: 20160328

International Standard Book Number-13: 978-1-4822-4281-2 (Hardback)

This book contains information obtained from authentic and highly regarded sources. Reasonable efforts have been made to publish reliable data and information, but the author and publisher cannot assume responsibility for the validity of all materials or the consequences of their use. The authors and publishers have attempted to trace the copyright holders of all material reproduced in this publication and apologize to copyright holders if permission to publish in this form has not been obtained. If any copyright material has not been acknowledged please write and let us know so we may rectify in any future reprint.

Except as permitted under U.S. Copyright Law, no part of this book may be reprinted, reproduced, transmitted, or utilized in any form by any electronic, mechanical, or other means, now known or hereafter invented, including photocopying, microfilming, and recording, or in any information storage or retrieval system, without written permission from the publishers.

For permission to photocopy or use material electronically from this work, please access www.copyright.com (http://www.copyright.com/) or contact the Copyright Clearance Center, Inc. (CCC), 222 Rosewood Drive, Danvers, MA 01923, 978-750-8400. CCC is a not-for-profit organization that provides licenses and registration for a variety of users. For organizations that have been granted a photocopy license by the CCC, a separate system of payment has been arranged.

Trademark Notice: Product or corporate names may be trademarks or registered trademarks, and are used only for identification and explanation without intent to infringe.

Library of Congress Cataloging-in-Publication Data

Mahgoub, Salah E. O. (Salah Eldin Osman)
  Genetically modified foods : basics, applications, and controversy / author, Salah E.O. Mahgoub.
        pages cm
   Includes bibliographical references and index.
   ISBN 978-1-4822-4281-2 (alk. paper)
   1. Food--Biotechnology. 2. Genetically modified foods. 3. Transgenic organisms. I. Title.

TP248.65.F66M34 2015
664--dc23   2015002730

Visit the Taylor & Francis Web site at
http://www.taylorandfrancis.com

and the CRC Press Web site at
http://www.crcpress.com

I would like to dedicate this work to my family. To my sincere, wonderful wife, Sumia, for her unlimited support, encouragement, and patience during my involvement in the manuscript preparation. To my son, Amar, and my daughters, Sali, Eman, and Umnia, for their continued support and follow-up on a daily basis. They have all created a conducive environment that enabled me to research and collect relevant information for the book. To all of them, I say thank you, and may Allah bless them all.

I present this work to all readers with interest in obtaining new information on contemporary issues that affect their lives.

CONTENTS

1 Introduction — 1
    1.1 Agricultural Practices — 2
    1.2 The World Food Prize — 3
    1.3 Advances — 5
    1.4 GM Technology — 6
    1.5 GM Controversy — 8
    1.6 Overview — 14
    References — 16
    Suggested References — 18
    Websites — 18

2 The Basics — 23
    2.1 Biotechnology — 23
        2.1.1 What Is Biotechnology? — 23
        2.1.2 Genetic Material (DNA) — 26
        2.1.3 GM and GE — 29
        2.1.4 GMOs, GM Crops, and GM Foods — 32
            2.1.4.1 Genetically Modified Organisms — 32
            2.1.4.2 Genetically Modified Crops — 32
            2.1.4.3 GM Foods — 33
    2.2 Biotechnology—Past, Present, and Future — 33
        2.2.1 Historical Developments in Biotechnology — 33
        2.2.2 Biotechnology Timeline — 36
        2.2.3 The Future of Biotechnology — 45
    2.3 Types of Biotechnology — 49
        2.3.1 Red Biotechnology — 49
        2.3.2 White Biotechnology — 49
        2.3.3 Green Biotechnology — 50
        2.3.4 Gray Biotechnology — 50
        2.3.5 Blue Biotechnology — 51
        2.3.6 Bioeconomy — 51

2.4	Position Statements on Biotechnology		51
References			53
Suggested References			58

3 Applications of Genetic Modification at the Laboratory and Greenhouse Levels — 59

- 3.1 Genetic Modification of Plants — 59
- 3.2 Nongenetic Engineering Techniques (Conventional Biotechnology) — 60
    - 3.2.1 Simple Selection — 60
    - 3.2.2 Marker-Assisted Selection — 60
    - 3.2.3 Crossing — 61
    - 3.2.4 Interspecies Crossing — 62
    - 3.2.5 Embryo Rescue — 62
    - 3.2.6 Somatic Hybridization — 63
    - 3.2.7 Somaclonal Variation — 63
    - 3.2.8 Mutation Breeding — 64
    - 3.2.9 Plant TC and Micropropagation — 65
- 3.3 Unintended Effects from Conventional Biotechnology — 69
    - 3.3.1 Mutations — 70
- 3.4 GE Techniques — 70
    - 3.4.1 Overview — 70
    - 3.4.2 From the Laboratory to the Greenhouse — 72
    - 3.4.3 Techniques of GE — 72
        - 3.4.3.1 Microbial Vectors — 74
        - 3.4.3.2 Electroporation — 77
        - 3.4.3.3 Microinjection — 78
        - 3.4.3.4 Microprojectile Bombardment (Biolistics) — 79
        - 3.4.3.5 Gene Splicing — 80
        - 3.4.3.6 Gene Silencing — 81
        - 3.4.3.7 Calcium Phosphate Precipitation — 83
        - 3.4.3.8 Lipofection — 83
    - 3.4.4 The Steps Followed in GE — 83
    - 3.4.5 Methods for Detection, Identification, and Quantification of GMOs in Food and Feed — 88
        - 3.4.5.1 Introduction — 88
        - 3.4.5.2 Sampling — 90
        - 3.4.5.3 GMO Testing Methods — 91
        - 3.4.5.4 Validation and Standardization of Methods — 96

		3.4.5.5	Reference Material for GMO Testing	97
		3.4.5.6	The Future of GMO Testing	97
		3.4.5.7	ISO Standards for Detecting GMOs in Food	98
3.5	Unintended Effects of GE			100
References				103
Suggested References				106

4 Applications of Genetic Modification at the Field and Commercial Levels — 109

- 4.1 Introduction — 109
- 4.2 Plant Biotechnology — 110
  - 4.2.1 Development of GM Crop Varieties with Improved Resistance to Herbicides Used for Weed Control — 112
  - 4.2.2 Development of GM Crop Varieties with Improved Resistance to Pests — 114
  - 4.2.3 Development of GM Crop Varieties with Improved Resistance to Bacterial, Fungal, and Viral Diseases — 116
    - 4.2.3.1 Transgenic Crop Varieties Resistant to Bacterial Diseases — 117
    - 4.2.3.2 Transgenic Crop Varieties Resistant to Fungal Diseases — 117
    - 4.2.3.3 Transgenic Crop Varieties Resistant to Viral Diseases — 117
  - 4.2.4 Development of GM Crop Varieties with Improved Resistance or Tolerance to Abiotic Factors — 118
  - 4.2.5 Development of GM Crop Varieties with Improved Nutritional Quality Traits — 119
  - 4.2.6 Weighing Risks and Benefits of Applying GE to Crops — 121
  - 4.2.7 Production of Biofertilizers — 122
  - 4.2.8 Global Status of GM Crops — 126
  - 4.2.9 GM Crops/Foods in Developing Countries with Special Reference to Africa — 130
- 4.3 Animal Biotechnology — 135
  - 4.3.1 Animal Genetic Modification Techniques — 135
    - 4.3.1.1 Techniques That Do Not Use GE — 135
    - 4.3.1.2 Techniques That Involve GE — 136
  - 4.3.2 Biotechnology for Animal Feed Production — 138
- 4.4 Food Biotechnology — 138

	4.4.1	Introduction—Genetically Modified Foods	138
	4.4.2	Composition of GE Foods	140
		4.4.2.1 Intended Changes in Composition	140
		4.4.2.2 Unintended Changes in Composition	141
	4.4.3	Biotechnology in Food Processing and the Uses of GMOs in the Food Industry	141
		4.4.3.1 GM Crops (Products) Used as Food	141
		4.4.3.2 Food Companies Using GMO Ingredients	145
		4.4.3.3 Examples of GM Foods	145
		4.4.3.4 Processed Foods and Ingredients Based on GM Products	149
		4.4.3.5 Foods Produced from Animals Consuming GM Crops or Treated with Bovine Growth Hormone (BGH)	150
		4.4.3.6 Non-GMO Companies and Food Products	151
		4.4.3.7 Non-GMO Seed Companies	156
		4.4.3.8 Non-GMO Project Verified	156
	4.4.4	Biotechnology in Functional Foods and Nutraceuticals	156
		4.4.4.1 Production of Plant-Based Functional Foods	159
		4.4.4.2 Production of Animal-Based Functional Foods	159
		4.4.4.3 Application of Biotechnology for Production of Foods and Food Ingredients	160
	4.4.5	The Future of GM Food	166
4.5	Biotechnology for Biofuel Production	170
References	172
Suggested References	180

5 Laws, Regulations, and Labeling for GM Foods — 183

5.1	Introduction	183
5.2	General Aspects of GM Food Regulations and Labeling	185
	5.2.1 The Principle of Substantial Equivalence	185
	5.2.2 GM Food Labeling	187
	5.2.3 Issues with GM Food Labeling	187
	5.2.4 Pro- and Anti-Labeling Arguments	189
	5.2.4.1 Pro-Labeling Arguments	189
	5.2.4.2 Anti-Labeling Arguments	190
	5.2.5 Global GM Food Labeling	191
5.3	GM Foods Regulations and Labeling in Selected Countries	193
	5.3.1 Introduction	193

		5.3.2	Australia–New Zealand	195
		5.3.3	Canada	197
			5.3.3.1 Regulations	197
			5.3.3.2 Labeling	198
		5.3.4	The European Union	199
		5.3.5	Japan	202
		5.3.6	South Africa	204
		5.3.7	United States	205
	5.4	Consumer Perspectives on GM Food Labeling		210
	5.5	Intellectual Property (IP) and Patents		211
	References			215
	Suggested References			220

6 GM Foods or Not? The Controversy — 221

- 6.1 Introduction — 221
- 6.2 Issues of Concern and Controversy — 226
  - 6.2.1 Concerns about Food Safety and Human Health — 226
  - 6.2.2 Environmental Concerns — 228
  - 6.2.3 Ethical Concerns — 230
  - 6.2.4 Regulatory and Legal Concerns — 230
  - 6.2.5 Economic Concerns — 230
  - 6.2.6 Animal Health and Welfare — 231
  - 6.2.7 Consumer Choice — 231
  - 6.2.8 Concerns about Bias in Scientific Publishing — 232
- 6.3 Proponents — 232
  - 6.3.1 Who Are the Proponents? — 232
    - 6.3.1.1 International Bodies and Governments — 232
    - 6.3.1.2 Biotech, Agrochemical, and Associated Companies — 233
    - 6.3.1.3 Pro-GM Lobby Groups — 236
    - 6.3.1.4 Global Network of Pro-Corporate Activists — 241
  - 6.3.2 Arguments in Favor of GM Foods — 241
    - 6.3.2.1 Benefits for Farmers — 242
    - 6.3.2.2 Benefits for Consumers — 243
    - 6.3.2.3 Benefits for the Environment — 246
    - 6.3.2.4 Benefits for the Economy — 249
    - 6.3.2.5 Other Points in Favor of GM Foods — 250
- 6.4 Opponents — 251
  - 6.4.1 Introduction — 251
  - 6.4.2 Who Are the Opponents? — 252

		6.4.2.1	United States	252
		6.4.2.2	Europe	256
		6.4.2.3	Canada, Australia, and New Zealand	258
		6.4.2.4	Africa	258
		6.4.2.5	Central and South America	260
		6.4.2.6	Asia Pacific	260
	6.4.3	Popular Protests against GMOs and GM Foods		260
	6.4.4	Vandalism and Threats by Anti-GMO Activists		261
	6.4.5	Arguments against GM Foods		262
		6.4.5.1	Better Alternatives to GM Technology	263
		6.4.5.2	Hazards and Risks of GM	264
		6.4.5.3	Impact on the Environment, the Ecosystem, and Farming	270
		6.4.5.4	Ethical and Moral Objections	275
		6.4.5.5	GM Technology Creates Socioeconomic and Political Threats	275
6.5	And the Debate Continues!			277
References				278
Suggested References				285

7 Consumer Issues — 295

7.1	Introduction	295
7.2	Consumer Rights	298
7.3	Studies on Consumers' Perceptions, Attitudes, and Preferences for GM Foods	300
7.4	Impact of Moving from Including GM Components to Non-GM Components on Food Sales	312
7.5	Roles of Mass Media	313
References		314
Suggested References		319

Glossary	323
Index	383

1
Introduction

Food is one of the basic needs of life. Throughout history, and ever since the time that man has used agriculture to grow plants and domesticate and raise animals for consumption, man has been trying to secure the availability of food and improve its quality by making it safer and healthier. The World Food Summit of 1996 (FAO 1996) defined food security as existing "when all people at all times have access to sufficient, safe, nutritious food to maintain a healthy and active life." Food security is used as an indicator of food sufficiency to everybody. Food security at different levels, that is, global, regional, national, and household, needs to be achieved and sustained. Global food security requires (1) sufficient and sustainable food availability, (2) economic and physical access to food by all, and (3) an adequate nutritional value of the diets that people consume. One of the main goals of governments around the world is to make available reliable sources of safe and sustainable food. Consumers' interest and welfare need to be a high priority of governments. There also needs to be a culture of openness on science, food, and farming.

Henry Kissinger, the former U.S. Secretary of State, pointed out that whoever controls food sources controls the people (Lovel and Mager 2008). To some extent, this is true because food is a basic need of life without which people cannot survive. It can be used as a very effective weapon to control people and nations and to dictate to them whatever the food controller wishes. In fact, many needy nations have been under the control of powerful nations, or corporations, that have the food or means of food production. The issue of controlling people through controlling food sources has been used as a strong argument, which genetically modified (GM) food technology critics use against GM food production. They believe that the large multinational corporations that have complete control of the GM technology that is used to produce GM foods have monopoly and control over food worldwide. They have their own terms that they dictate to the technology users. This has also been regarded as an ethical issue in the sense humans are taken advantage of when they are most in need.

Continuous efforts to improve food production both quantitatively and qualitatively have been ongoing throughout history. Different strategies and plans have been developed, tried, and evaluated. Examples of such efforts include the Green Revolution, various initiatives to address specific or general food and agriculture problems, work of the various UN agencies dealing with agriculture, food, nutrition, and health; FAO and WHO are examples of private and public research efforts carried out by individuals or institutions.

1.1 AGRICULTURAL PRACTICES

Agriculture, which includes plant and animal production, is regarded as the main source of food for the increasing world population estimated to reach 8 billion by the year 2020, with a projection of 9 billion by 2050 (Bruinsma 2009). This exerts enormous challenges on world agriculture for the coming decades. These challenges are multifaceted and encompass increase in agricultural and food production, development and improvement of diet quality, ensuring of food safety, and environment sustainability. The roles agriculture can play in feeding the growing world population and achieving food and nutrition security have always been recognized and under focus, and new developments have been achieved in these directions.

Today, agriculture development occupies about 40% of the earth surface, uses 70% of the water resources, and is responsible for 30% of the CO_2 production (VIB 2013). During the next few years, production of food and feed is expected to be faced with a number of challenges. Multiple factors, including world population growth, climate change, water scarcity, and the further ban on the use of chemicals for crop protection, create these challenges. The scientists in the Life Sciences Research Institute in Belgium (VIB 2013) believe that these challenges can be addressed by adopting an integrated agricultural model that encompasses the best features of conventional and organic farming coupled with the use of modern biotechnologies.

Although the rate of population growth is steadily decreasing, the increase in absolute numbers of people may be such that the carrying capacity of agricultural lands could soon be reached, given current technology. World agriculture faces enormous challenges in the coming decades.

INTRODUCTION

Feeding the world population adequately in 2050 will require a 70% increase in global food output and a near doubling of resources of developing economies (Bruinsma 2009; FAO 2009; Raney and Matuschke 2011).

One of the successful agricultural developments is the "Green Revolution." The Green Revolution refers to the developments in some agricultural practices that took place during the period 1965–1980 and that resulted in large worldwide increases in the yield of many crops such as rice and wheat. The use of improved fertilizers, pesticides, and irrigation techniques and scientific plant breeding helped in achieving the crop yield increase (Freedeman 2009). The Green Revolution achieved its goal, but that was not for long. By the mid-1980s, it was observed that crop yield had reached its maximum, with no possibility of further improvements. Addition of large amounts of chemical fertilizers and pesticides started to raise some concern about their possible effects on the environment, especially on soil and water. In spite of the success that accompanied the Green Revolution, it was felt that the goal of availing and securing adequate food to address world hunger has not been fully achieved.

As part of the continued efforts of scientists to improve human life, it was logical to search for alternatives. The developments in modern biotechnology and the use of GM (also referred to as genetic engineering—GE) to increase food production and to improve its quality were the steps that followed the Green Revolution in a time period referred to as the "GE era."

1.2 THE WORLD FOOD PRIZE

Developments in biotechnology, in general, and specifically in the area of food biotechnology, have been continuous with the aim of achieving and sustaining food and nutrition security at different levels. These developments have been recognized and encouraged. Individuals who participate in these developments are usually recognized by the international community. One such recognition is the awarding of the World Food Prize. This prize is a prestigious honor, often considered as the Nobel Prize for food. It is named for Iowan Norman Borlaug who is considered the father of the Green Revolution and the founder of the prize after winning the Nobel Peace Prize in 1970 for his efforts to combat famine in developing countries. The World Food Prize recognizes achievements of individuals who contributed to human development through working to improve food availability, quality, or quantity for the betterment of humanity. It is presented by the World Food Prize Foundation and sponsored by several foundations, governments, nongovernmental organizations, private donations, and companies, including those active in the field of agribiotechnology, such as Monsanto and Syngenta (The World Food Prize 2014). However, the sponsors have no vote or say in the selection of the laureates. Norman Borlaug believes that the controversy surrounding GM technology should not be a reason to bar active biotechnologists, who have positive efforts in improving the world food situation, from consideration as World Food Prize Laureates. He thinks that they should be evaluated by their contribution to improving food security and alleviating hunger and poverty worldwide.

The 2013 annual symposium for offering the World Food Prize held in October 2013 in Des Moines, IA, USA received much attention and criticism due to the recipients' ties with industrial agriculture and GM crops. The wide international criticism and the rallies that accompanied the occasion came after the announcement of the three laureates for the prize due to their direct involvement in GM technology. The nominees were three prominent and distinguished scientists in the field of agrobiotechnology: Robert Farley, an executive at Monsanto, USA, Mary-Bell Chilton, a scientist at Syngenta, USA, and Marc Van Montagu, the cofounder of Plant Genetic System in Belgium. Their research is making it possible for farmers to grow crops with improved yields, resistance to insects and disease, and the ability to tolerate extreme variations in climate. The prize is meant to recognize their independent, individual breakthrough achievements in founding, developing, and applying modern agricultural biotechnology.

The 2013 awardees represent three international seed biotechnology industries, namely, Monsanto, Syngenta, and the European Federation of Biotechnology. This has drawn wide international criticism, particularly from groups opposing genetic modification (GM). The three Prize Laureates agreed that one of the top priorities needed to portray the real picture of GM technology is the sharing and communication of knowledge of biotech crops with the public. During the prize awarding ceremony, the importance of encouraging professional and objective debate and of increasing public awareness about the potential contribution of biotechnology to improve food security and to help feed the world were highlighted. The three laureates called on a hungry world to "embrace the seeds they helped develop, despite controversy that threatens to limit the reach of biotech crops." After receiving her award, Mary-Dell Chilton said, "My hope is this will put to rest the misguided opposition to the crops." She described the genetically modified organisms (GMOs) as the "wonderful tool" in the fight against hunger. In contrast, Farley of Monsanto believes that GM technology has the potential to feed a growing world population, but at the same time, he blames Monsanto and the agricultural industry in failing to fully explain the importance, benefits, and safety of GM crops. He assured the audience that his company will "do what it takes" to collaborate and share.

The prize annual symposium followed a week of global protests against Monsanto and the use of GM crops. Part of the protests came in the form of a campaign called "Occupy the World Food Prize." During the symposium, the laureates defended their position and research against claims that GM crops and foods are unsafe. While the ceremony was in session inside the Iowa Capitol building, protests by activists and anti-GM groups were held around the building. Press releases and advertisements were distributed by the Union of Concerned Scientists. In addition, and according to the Center for Food Safety (Center for Food Safety 2013) and Iowa Citizens for Community Improvement, a petition signed by more than 345,000 individuals opposing GMOs was delivered to the World Food Prize organizers.

1.3 ADVANCES

Throughout history, plant and animal food resources have gone through many changes and variations in their genetic material as well as in their physical characteristics, that is, they experienced both genotypic and phenotypic changes. These changes can happen naturally or through the intervention of man, ultimately leading to the survival of plants and animals with the most desirable characteristics to regenerate the species. Changes in the genetic makeup of plants and animals are considered as a kind of genetic alteration or GM. In this sense, what is known as "genetic modification" today cannot be regarded as a new phenomenon. Consequently, GM in this sense occurs naturally and forms the fundamental basis of evolution and breeding (Pandey et al. 2010). GM foods are usually developed and produced through modern biotechnological techniques that involve transfer of genes from one living organism to another, a technique referred to as "transgenesis."

Advances in modern biotechnology applied to crop, animal, and food production, also referred to as "agrobiotechnology," are considered as some of the most important technological developments impacting modern agriculture. Agricultural biotechnology, genetically modifying crops for desirable characteristics such as herbicide, pest, and disease resistance, for improved nutritional value, and shelf-life, is being sold as a solution to world hunger. Also promised for the future are drought and frost resistance, nitrogen fixation, and increased yield.

Agrobiotechnolgy can be viewed as an important component in the series of efforts aimed at improving the food and nutrition situation worldwide. It has attracted worldwide attention over the past few decades due to the potential it has and the possible risks it might create.

The application of modern biotechnology to crop and food production is one of the most significant technological advances to impact modern agriculture and food availability (Carter et al. 2011). This technology has resulted in the very wide GM crop production worldwide. It also resulted in changing (for the better or worse, depending on whether you are pro- or anti-GM) many agronomic practices, crop characteristics, and yield quantity and quality. In addition to its use in food production, GM technology has been employed in pharmaceutical and industry crop production, generally referred to as "pharma crop" production. This is viewed as a positive achievement with regard to wider availability and cost reduction of drugs.

1.4 GM TECHNOLOGY

There is a need to view the issue of GM technology and its application to produce foods from both sides. It would be more helpful and beneficial to consumers to get a clear picture of the actual and potential benefits of GM product innovations, as well as knowledge of possible undesirable consequences. In addition, full comprehension of the functions and roles of public and private research institutions and bodies, the local and international regulations and intellectual property rights that govern the technology are needed to help in easing and possibly ending the dispute and controversy around GM technology and GM foods.

With regard to the chronological sequence of GM crops development and introduction into the market, they are categorized as first-, second-, and third-generation GM crops. Some of the specific characteristics of the first-generation GM crops include improved pest and disease control, abiotic stress tolerance, herbicide tolerance, and increased crop yield. These characteristics are also referred to as "input characteristics" (Glover et al. 2014). Characteristics of second-generation GM crops, also known as "output characteristics," define some desirable quality attributes such as improved nutritional profiles leading to functional foods, increased postharvest life, and enhanced processing qualities. The third-generation GM crops, also called "plants as factories" (Glover et al. 2014), are characterized by their suitability for novel uses such as plant-based pharmaceuticals and plant-made industrial products. It is believed that the world stands on the verge of what is called "biotech century," in which GE will cause global changes in quite profound ways from GM foods to "bio-pharming."

The terms "genetically modified organisms" and "genetic modification" are often used to describe transgenic technologies. At the same time, conventional breeding techniques, such as cross-breeding, hybridization, and selection, can also be considered as GM techniques. This might cause some confusion between conventional methods of crop modification and modern techniques that involve transgenesis. For clarity of the meaning and to avoid any confusion, it is suggested to use the general term "genetic modification" to encompass both traditional and modern methods, while using the term "bioengineered" or "genetically engineered" (GE) to denote transgenically produced plants.

The use of GE to develop GMOs is believed to have promising prospects for increased agricultural productivity or improved nutritional value that can contribute directly to enhancing human health—attributes that might not be easily achievable through conventional breeding due to species boundaries (WHO 2005; Marx 2010). At the same time, these potential benefits need to be considered together with the concerns of possible risks associated with human health and the environment. The potential benefits associated with GM crops might not be observed to the same extent in different parts of the world, mainly due to the differences in regional climatic and agricultural conditions and practices. Another issue that needs to be considered is the extent of sustainability of this new technology. The adoption of GM needs to be monitored and evaluated for sustainable benefits, coupled with efforts to minimize potential risks.

GM foods developed through GE have the potential to address the prevailing world hunger and malnutrition problems. At the same time, production of GM foods presents a number of challenges with regard to safety issues, regulations, labeling, and international policy and trade (Whitman 2000). The positive aspects and the promising potential of GE to address some of the global agricultural and food and nutrition problems are accompanied with possible risks and unintended harm to human and animal health as well as to the environment. Whitman (2000) cautions that our enthusiasm about this powerful technology should not make us forget about the possibility of causing unintended harm to humans and the environment.

Figure 1.1 GM corn in the field.

The mid-1990s witnessed the development and introduction of the first GM food, namely, the delayed-ripening tomato known as "FlavrSavr," on the U.S. market. Following that, a number of GM crops including corn, soybean, rape, cotton, rice, papaya, potato, squash, and sugar beet have been developed and entered the international markets. Figure 1.1 shows the GM corn crop in the field. The number of GM crops introduced as well as the area planted with GM crops continued to increase. By the year 2005, the WHO (2005) estimated that GM crops would cover almost 4% of the total global arable land. GM crops reached 9% of global primary crop production in 2007 (World Watch Institute 2015).

GM foods have an interesting history and their development has experienced rapid growth over the last decade (Murnaghan 2014). It is believed that the controversy and debate surrounding the potential benefits and the possible risks of GM foods will continue in the future.

1.5 GM CONTROVERSY

The advent of GM technology has been faced with many obstacles and challenges that have a potential to affect its present and future innovations and developments. This has been mainly, and clearly, manifested in the regulatory and labeling aspects, particularly between the United States and the EU countries. Deciding between mandatory and voluntary labeling has been an important regulatory choice with respect to GM foods. The EU was a little bit cautious in dealing with GM technology and GM foods. It has implemented some restrictive policies that may constrain present and future biotechnology potential. The EU has a long history of conflicts with the United States when it comes to GM products. That collision started in the year 1996 when the United States exported the first GM food (tomato puree) to Europe, which was voluntarily labeled as GE, and which was met great success in Britain because it was cheaper than the traditional tomato puree. That success did not continue for long. When GM soybeans were imported into Europe in the same year, the EU, under the pressure of environmental groups, started to implement mandatory labeling for GM foods. That step was viewed by the U.S. government and the food industry as a barrier that would affect international trade.

The rift between the United States and the EU in issues related to GM crops and GM foods grew wider. The United States continued to favor the adoption of more GM crops and encouraged the use of this technology among U.S. farmers as well as among farmers in other countries including poor and developing countries. In contrast, the EU has been slow in adopting GM technology. This difference of views was clear during the various venues where representatives of these two blocks met, for example, during the World Trade Organization meetings. International trade relations among different countries have been negatively affected, and this was not only between the United States and the EU but included other countries as well.

With regard to GM food production, regulations, and labeling, it has been realized that science and politics cannot be separated and that they are very much interrelated and work together to shape the positions of different countries around the world. Toke (2008) studied, compared, and explained how the different and controversial outcomes have occurred over the GM food and crops issue in regions such as the United States, the United Kingdom, and the EU. The overlap between science and politics and how it relates to globalization were highlighted in the study. The study tries to present logical explanation to the wide controversy about GM foods, particularly among the United States, Britain, and the EU. It compares the GM food politics in these regions and stresses the interrelationship between scientific expertise and citizens' politics.

Many questions are asked about GM foods and GM technology in general. This reflects the confusion that exists among consumers and the lack of adequate and credible information about GM foods. Some of the pertinent questions that need honest answers are as follows: Is it safe to consume GM foods? What happens when people eat food containing GM ingredients? Can GM ingredients consumed by animals be transferred to humans who consume these animals? Is it possible that consumers' health may be affected in ways that scientists will not understand? Can GM foods cause allergies? Satisfactory and credible answers to these questions are not easy to obtain. This is because it depends upon who is giving the answers and who is getting the answers. Which of the two debating and opposing sides (proponents and opponents) does he/she belong to? If the individual giving the answer is a pro-GM food person, it is very unlikely that an anti-GM food person would agree or accept that answer. By the same understanding, GM food proponents would not agree with GM food critics.

In spite of the fact that WHO and many scientific institutions in Europe and elsewhere have given their verdict that GM foods are as safe as other non-GM foods and that they pose no health hazard for humans, politicians in the EU continue to raise the issue of environmental risks caused by GM technology applications. This shows the powerful effect of the environmental lobby groups in Europe and around the globe. One of the eminent scientists in the field of agriculture and Nobel Prize winner, Norman Borlaug, is one of the proponents of GE. Even before the present development of GM, he forecasted that this technology has the potential to produce food needed to feed the world population in the future. Additionally, the former director general of the FAO, Dr Jacques Diouf, also supported the idea that GMOs can help to increase the supply, diversity, and quality of food products and reduce costs of production and environmental degradation (Kariyawasam 2010). He also believed that GM technology is potentially capable of improving world agriculture, hence reducing malnutrition and food insecurity and increasing income of rural farmers.

Concerns over the safety of GM crops, as well as their environmental impact, have resulted in the implementation of regulations that may affect international trade. Global climate change is expected to have negative impacts on various aspects of human life at present and in the future. We need to anticipate and plan for the appropriate interventions to mitigate these effects. GMOs are considered one of the factors that have the potential to minimize the negative effects and at the same time have positive contributions. Developments in and applications of GMOs proved that they can be used to improve the ability to help in feeding the growing the world population, improve crop production, and conserve the environment and the ecosystem.

People have the right to think seriously about GM foods with regard to purchase and consumption. It is also a healthy sign to debate the various issues surrounding them, taking into consideration the potential benefits and the possible risks. Johns (1999) believes that the debate needs to be done without any hysteria, to help to define clearly the issues of concern, and to tackle them rationally and on an informed basis.

The issue of GM foods is an emotionally charged issue, with many concerns such as health, environment, politics, ethics, trade, and economics. It has attracted global interest and controversy because of its potential benefits and possible risks to humans, animals, and the environment. It has also become a battleground involving scientists, commerce, politicians, journalists, lobby groups, and the public (Robinson 1999). With respect to accepting or rejecting GM foods, people are split into three groups: those in favor, those against, and the undecided. The last group is believed to be the majority. Robinson (1999) believes that the arguments are about values, which are neither absolute nor universal. In most cases, the arguments are emotive, using terminologies that are offensive and unbalanced such as "Frankenstein foods" and "Terminator technology." In contrast, trust in scientists has been shaken and has reached an unsteady level and their views are being looked at with a lot of suspicion. This is because some people believe that scientists are paid by multinational biotechnology companies to present research results in line with their aims of marketing their biotech products. Public interest science needs to be practiced by scientists and reflected in their interpretation of the results of their research. Confidence in scientists' work needs to be restored so that consumers benefit from the findings of research related to GM technology.

Socioeconomic and ethical issues sometimes have strong influence on the extent to which different countries accept or reject GM foods. Awareness of benefits and risks of GM foods by consumers in those countries were found to have less effect on accepting or rejecting GM foods. The example of the famine situation in 2002 in Southern Africa, and the reluctance of several recipient countries to accept GM food donations due to socioeconomic, ownership, and ethical issues, is cited by WHO (2005) in support of this view.

Developing countries are far more seriously affected due to the possibility that the GM controversy might seriously reduce the potential role of biotechnology in reducing hunger and helping more people to be fed, as well as reducing food prices. At the same time, these countries stand at the crossroads between the United States and the EU. On the one hand, they want to tap the possibility of applying the GM technology, with the hope that it improves and develops their agricultural systems and food availability, something favored and encouraged by the United States. On the other hand, they are afraid of adopting this technology to avoid any possibility of jeopardizing trade relations with the EU.

GM food is everywhere in the United States and in most of other countries around the world. The shelves of the grocery stores are packed with food that contains GM ingredients or is based on GM food crops (Figure 1.2). From the corn flakes people eat for breakfast to the soy protein energy bars they consume as snacks to other similar foods, all contain GM components. Almost all processed foods contain small amounts of ingredients derived from GM crops. More than 80 transgenic crops have undergone regulatory clearance in the United States; however, only about a dozen are currently marketed for human consumption (Council on Science and Public Health 2012). The most common transgenic crops in the United States are soybeans, corn, sugar beets, and cotton (for cottonseed oil). Byrne et al. (2010) state that an estimate of 70% of

Figure 1.2 GM foods in grocery stores.

the processed foods sold in the U.S. grocery stores contains ingredients derived from transgenic crops. American consumers are thus eating GM foods every day, in most cases even without realizing it. Similar scenarios are encountered in other countries around the globe, but to different extents with regard to the types and amounts of GM foods, the regulations and labeling requirements, as well as the consumers' knowledge and perceptions about GM foods.

Popular rallies and protests against applications of GM technology continue to take place in different parts of the world. Krumboltz (2014) reports that protesters from the Organic Consumers Association and Occupy Monsanto held a rally in Washington, DC on October 2013 against Monsanto and dropped a bag of money (about $500) off the balcony inside the Hart Senate Building. Protesters were heard to shout "Monsanto money." Monsanto opponents claim that the company spends millions to buy the votes it needs to make it easier to bring GM foods to the market. While Monsanto describes itself as a "a sustainable agriculture company" focused on "empowering farmers—large and small—to produce more from their land while conserving more of our world's natural resources such as water and energy," protest leaders are seeking "the permanent boycott of GMOs and other harmful agro-chemicals."

Imports and exports of GM foods form part of international trade among different countries around the world. For the efficient and satisfactory movement of such goods, there needs to be some degree of harmonization among countries with regard to various regulations governing GM foods. This would facilitate trade movement among countries and spread the benefits of new technologies. Intersectoral collaboration is important and helpful in achieving the needed international harmonization.

In an effort to provide international consistency in the assessment of GM foods, and to harmonize the general understanding of the issues surrounding them, the Codex Alimentarius Commission (CAC) developed some guideline principles that cover food safety. In addition to that, the Cartagena Protocol on Biosafety includes components that cover environmental safety of GMOs. CAC recognizes the need for risk assessment of GM foods, as well as the need for continuous evaluation and improvement. Different countries use the CAC guidelines to develop their own regulatory systems for GM foods and their risk assessment. GM foods currently traded on the international market have passed risk assessment in several countries and are not likely, nor have been shown, to present risks for human health (WHO 2005). Risk assessment needs to be coupled with adequate risk communication; otherwise, the benefits and the meaning of risk assessment will not reach consumers and will not be of use to them.

There is growing fear of the use (or abuse) of intellectual property rights by biotech companies. This fear emanates from the possible monopolization of genetic resources, which might lead to market monopolization and creation of dependency on these few companies.

The media carries a huge responsibility in informing and educating the public about various issues affecting their lives and welfare. Currently, the media is very busy covering various issues related to GM foods. Some reporters show a certain degree of bias toward one of the two debating sides, that is, proponents or opponents. They try to portray the views and the arguments of the group they support. In contrast, more reasonable reporters show no specific inclination toward any of the two opposing groups and rather show facts and give due weight to arguments of both sides. The latter group offers the readers a better chance to understand more about the pros and cons of GM foods and be able to make informed decisions. Despite this wide media coverage, it can be difficult for anyone not directly involved to know how to obtain facts (Johns 1999). This is attributed to the multitude of issues raised by the production of GM foods, which include scientific, technological, environmental, social, ethical, economic, and political concerns.

A number of books dealing with various issues of GM foods have been published. Most of them cover a few aspects of GM technology and GM foods. Some of them focus on the potential benefits or the possible risks and hazards of GM foods. A few publications try to present a balanced picture of the controversy of GM foods. One such publication is the book by Fedoroff and Brown (2004). In their book entitled *"Mendel in the Kitchen: A Scientist's View of Genetically Modified Foods,"* they attempt to present a clear and balanced view of the complex GM foods issue. They discuss many misinterpretations of results of scientific data and, at the same time, defend GM foods as a means of increasing global food availability.

1.6 OVERVIEW

This book aims to provide a comprehensive view of GM technology and GM foods. It encompasses the basics of biotechnology, its various applications in the laboratory and in the field, a balanced presentation of the pros and cons of GM foods, giving arguments of both proponents and opponents, regulations governing GM food labeling, and discussion of various consumer issues related to GM foods.

Chapter 2 discusses the basic aspects of biotechnology. It provides a number of definitions of biotechnology considered from different perspectives and links the structure of DNA to GM and its various applications. The development of GMOs and GM foods as related to biotechnology is highlighted. The chapter also discusses the historical developments in biotechnology and gives a detailed timeline of the technology as well as its future potential. The different types of biotechnology based on the way it is used are discussed, and the designation of color to each type is indicated. Views and position statements of a number of international organizations and institutions are reflected at the end of the chapter.

Chapter 3 deals with GM applications at the laboratory and the greenhouse levels. The different techniques of both non-GE and GE, which are in use today, are listed and discussed. Steps followed in each technique, advantages, and shortcomings of each method are discussed. The chapter also covers information about the unintended effects of both traditional and modern GM techniques. Details of the methods used to detect, identify, and quantify GMO components in food and feed are included as well.

The applications of GM technology are continued in Chapter 4. In this chapter, the applications at the field and commercial levels are discussed. It covers up-to-date information on plant, animal, and food biotechnology types and processes, as well as the use of biotechnology to produce biofuels. Details of GM and non-GM foods are included, in addition to information on the use of biotechnology to produce nutraceuticals and functional foods.

Chapter 5 deals with laws and regulations related to GM foods and how they affect mandatory and voluntary labeling. Various issues and arguments from both pro- and anti-GM food labeling groups are discussed. GM food labeling from a global perspective, in addition to positions of selected countries, is included. Consumer views on GM food labeling are also highlighted. Patenting in relation to GM foods and the monopoly that may arise due to having intellectual property rights on them are also discussed.

Chapter 6 presents the issues of concern and controversy around GMOs and GM foods. It presents the arguments of both the supporters of biotechnology and GM foods and those of its opponents and critics. It also gives details of who the supporters and who the opponents are. The chapter concludes by stating that this fueled controversy and heated debate will continue in the future until both sides of the debate view the benefits and risks of GM foods with the same eye and until the consumer welfare remains at the heart of their discussions.

Chapter 7 discusses various consumer issues related to GM foods. Consumer rights to knowledge and information on benefits and risks of GM technology and GM foods, as well as their right to choose, are emphasized. The chapter highlights a number of studies on consumers' perceptions, attitudes, and preferences for GM foods. These studies represent a small fraction of the large number of studies, surveys, and polls conducted around the world. The actual and expected roles of mass media with respect to informing and educating the public and influencing consumers' decisions and views are also discussed.

A glossary of common terms used in biotechnology appears at the end of the book.

REFERENCES

Bruinsma, J. 2009. The Resource Outlook to 2050: By How Much Do Land, Water and Crop Yields Need to Increase by 2050? Paper presented at the FAO Expert Meeting, June 24–26, 2009, Rome, "How to Feed the World in 2050."

Byrne, P., Pendell, D., and Graf, G. 2010. Labeling of genetically engineered foods. http://www.ext.colostate.edu/pubs/foodnut/09371.html (accessed May 3, 2014).

Carter, C.A., Moschini, G., and Sheldon, I. (Eds). 2011. *Frontiers of Economics and Globalization*, Volume 10. Genetically Modified Food and Global Welfare. Emerald Group Publishing Limited, Bingley, UK, pp. 55–82.

Center for Food Safety. 2013. Hundreds of thousands reject World Food Prize sham. Press Releases. http://www.centerforfoodsafety.org/press-releases/2649/# (accessed October 16, 2014).

Council on Science and Public Health. 2012. Labeling of bioengineered foods (Resolutions 508 and 509-A-11) Report 2-A-12. http://factsaboutgmos.org/sites/default/files/AMA%20Report.pdf (accessed April 22, 2014).

FAO. 1996. World Food Summit. Rome declaration on world food security. http://www.fao.org/docrep/003/w3613e/w3613e00.htm (accessed December 10, 2013).

FAO. 2009. Global agriculture towards 2050. How to feed the world 2050. High Level Expert Forum, Rome, October 12–13, 2009.

Fedoroff, N.V. and N.M. Brown. 2004. *Mendel in the Kitchen: A Scientist's View of Genetically Modified Foods.* Joseph Henry Press, Washington, DC.

Freedeman, J. 2009. *Genetically Modified Food: How Biotechnology is Changing What we Eat*. First Edition. The Rosen Publishing Group Inc., New York.

Glover, J., Mewett, O., Cunningham, D., and Ritman, K. 2014. Genetically modified crops in Australia: The next generation. Department of Agriculture, Fisheries and Forestry, Australian Government, Bureau of Rural Sciences. http://data.

daff.gov.au/brs/brsShop/data/biotech_5__2_.pdf (accessed September 28, 2014).

Johns, L. 1999. Science, medicine, and the future. Genetically modified foods. *British Medical Journal*, 318(7183):581–4.

Kariyawasam, K. 2010. Legal liability, intellectual property and genetically modified crops: Their impact on world agriculture. *Pacific Rim Law and Policy Journal Association*, 19(3):459–85.

Krumboltz, M. 2014. Monsanto protesters toss money off balcony in Senate office. *Yahoo News*. http://news.yahoo.com/author/mike-krumboltz-20110327/ (accessed April 25, 2014).

Lovel, H. and Mager, S. 2008. *GM Food Genetically Manipulated Genetically Engineered, 1st Edition*. Aracariaguides.com. http://www.amazon.com/dp/0977577171/ref=asc_df_09775771712660714?smid=ATVPDKIKX0DER&tag=pg-1583-86-20&linkCode=asn&creative=395097&creativeASIN=0977577171 (accessed September 28, 2014).

Marx, G.M. 2010. Monitoring of genetically modified food products in South Africa. PhD Dissertation, Faculty of Health Sciences, Department of Haematology, University of the Free State, Bloemfontein, South Africa.

Murnaghan, I. 2014. Development and history of GM foods. Genetically modified foods. http://www.geneticallymodifiedfoods.co.uk/development-history-gm-foods.html (accessed January 15, 2014).

Pandey, A., Kamle, M., Yadava, L.P. et al. 2010. Genetically modified food: Its uses, future prospects and safety assessment. *Biotechnology*, 9(4):444–58.

Raney, T. and Matuschke, I. 2011. Current and potential farm-level impacts of genetically modified crops in developing countries, Chapter 3, in Carter, C., Moschini, G., and Sheldon, I. (Eds.), *Frontiers of Economics and Globalization*, Volume 10. Genetically Modified Food and Global Welfare. Emerald Group Publishing Limited, Bingley, UK, pp. 55–82.

Robinson, J. 1999. Ethics and transgenic crops: A review. *Electronic Journal of Biotechnology*, 2(2). http://www.ejbiotechnology.info/index.php/ejbiotechnology/article/view/v2n2-3/821 (accessed April 25, 2014).

The World Food Prize. 2014. www.worldfoodprize.org (accessed August 10, 2014).

Toke, D. 2008. *Politics of GM Food. A Comparative Study of the UK, USA and EU*. Routledge; Taylor & Francis Group, Florence, KY.

VIB. 2013. The World Food Prize 2013 recognizes the contribution of agrobiotechnology to world food security. http://www.vib.be/en/news/Pages/The-World-Food-Prize-2013-recognizes-the-contribution-of-agrobiotechnology-to-world-food-security,-a-leading-example-to-the.aspx (accessed September 28, 2014).

Whitman, D.B. 2000. Genetically modified foods: Harmful or helpful? CSA Discovery Guides. http://www.csa.com/discoveryguides/discoveryguides-main.php (accessed February 20, 2014).

WHO. 2005. Modern Food Biotechnology, Human Health and Development: An Evidence-Based Study. Food Safety Department, World Health Organization. WHO Press, Geneva, Switzerland.

World Watch Institute. 2015. Genetically modified crops reach 9 percent of global primary crop production. http://www.worldwatch.org/node/5951 (accessed April 27, 2015).

SUGGESTED REFERENCES

Alexandratos, N. and Bruinsma, J. 2012. World agriculture towards 2030/2050: The 2012 revision. Food and Agriculture Organization of the United Nations. http://www.fao.org/docrep/016/ap106e/ap106e.pdf (accessed November 10, 2014).
Andrews Henningfeld, D. 2009. *Genetically Modified Food*. Greenhaven Press, Detroit, MI.
Byerlee, D. and Fischer, K. 2002. Accessing modern science: Policy and institutional options for agricultural biotechnology in developing countries. *World Development*, 30(6):931–48.
Dyson, T. 1999. World food trends and prospects to 2025. *Proceedings of the National Academy of Sciences USA*, 96(11):5929–36.
Food and Drink Federation. 2000. *Food for Our Future. Genetic Modification and Food*. Food and Drink Federation, London.
Forman, L.E. 2010. *Genetically Modified Foods*. ABDO Publishing, Edina, MN.
Fox, M.W. 1999. *Beyond Evolution, the Genetically Altered Future of Plants, Animals, the Earth Humans*. Lyons Press, Westlake Village, CA.
Gutierrez-Lopez, G.F. 2003. *Food Science and Food Biotechnology*. CRC Press, Boca Raton, FL.
Henninfed, D.A. (Ed.). 2009. *Genetically Modified Food*. Greenhaven Press, Detroit.
Johnson-Green, P. 2002. *Introduction to Food Biotechnology*. CRC Series in Contemporary Food Science. CRC Press, Boca Raton, FL.
Kneen, B. 1999. *Farmageddon: Food and the Culture of Biotechnology*. New Society Publishers, British Columbia, Canada.
Madden, D. 1995. *Food Biotechnology: An Introduction*. ILSI Press, Brussels.
Marshall, E. 1999. *High-Tech Harvest: A Look at Genetically Engineered Foods*. Franklin Watts, London, UK.
Martineau, B. 2002. *First Fruit*. McGraw-Hill Education, London.

WEBSITES

30 Bananas a Day: http://www.30bananasaday.com/profiles/blogs/monsanto-list-of-gmo-foods-to.
ActionBioscience: http://www.actionbioscience.org/biotech/pusztai.html.
AfricaBio: http://www.africabio.com/index.php/biotechnology/biofuels.
AGBIOSAFETY, University of Nebraska, Lincoln: http://agbiosafety.unl.edu.

AgBioWorld: http://agbioworld.org/.
Agricultural Biotechnology Council of Australia: http://www.abca.com.au/.
Agriculture Network Information Center (AGNIC). A guide to quality agricultural information as selected by the National Agricultural Library, land-grant universities, and other institutions: http://www.agnic.org.
Alliance to Feed the Future: www.alliancetofeedthefuture.org.
American Seed Trade Association: http://www.amseed.org/pdfs/ASTA CoexistenceProductionPractices.pdf.
Biofuels: http://dglassassociates.wordpress.com/2010/01/09/biotechnology-and-biofuel-production/.
Biotech Articles—Animal Feed: http://www.biotecharticles.com/Agriculture-Article/Biotechnology-in-Animal-Feed-and-Feeding-162.html.
Biotech Articles: http://www.biotecharticles.com/Biotechnology-products-Article/How-Biotechnology-Helps-Create-Biofuels-152.html.
Biotechnology: http://dglassassociates.wordpress.com/2013/11/.
Biotechnology Industry Organization: https://www.bio.org/articles/2014-isaaa-fact-sheet.
CapitalistShrugged: http://capitalist-shrugged.blogspot.com/2010/02/genetically-modified-foods-pros-and.html.
Center for Environmental Risk Assessment: http://cera-gmc.org/GMCrop Database.
Center for Food Integrity (CFI): http://www.foodintegrity.org/.
Center for Food Safety: http://www.centerforfoodsafety.org/take-action#.
Center for Food Safety—Food Labeling: http://www.centerforfoodsafety.org/issues/976/ge-food-labeling/international-labeling-laws.
Companies Who Use GMO Ingredients: http://www.celestialhealing.net/monsanto/GMO_FoodCo.htm.
Council for Agricultural Science and Technology (CAST): http://www.cast-science.org.
Council for Biotechnology Information (CBI): http://www.whybiotech.com. Information on how biotechnology increases food security: http://www.whybiotech.com/resources/factsheets_food.asp.
EarthOpenSource: http://www.earthopensource.org/index.php/3-health-hazards-of-gm-foods/3-6-myth-gm-bt-insecticidal-crops-only-harm-insects-and-are-harmless-to-animals-and-people.
FAO: http://www.fao.org/ag/magazine/9901sp1.htm.
FAO—Agric Biotechnologies: http://www.fao.org/biotech/en/.
FDA—Background, Questions and Answers, Past Reports and More: http://www.fda.gov/Food/FoodScienceResearch/Biotechnology/default.htm.
Food Standards—Australia New Zealand: http://www.foodstandards.gov.au/consumer/gmfood/gmoverview/pages/default.aspx.
Freaking News—GM Foods Pictures: http://www.freakingnews.com/Genetically-Modified-Foods-Pictures-574-0.asp.
Friends of the Earth: http://www.foe.org/projects/food-and-technology/genetic-engineering.

Future for All: http://www.futureforall.org/bioengineering/biotechnology.htm.
GAIAMTV: http://www.gaiamtv.com/video/how-recognize-and-avoid-gmos#play/34111.
GM Foods: http://www.geneticallymodifiedfoods.co.uk/.
GM Food and Related issues: http://gene-modified.blogspot.com/2011/06/labelling-of-gm-food.html.
GM Food Controversy: http://en.wikipedia.org/wiki/Genetically_modified_food_controversies.
GMO Pundit: http://gmopundit.blogspot.com/2013/02/global-status-of-commercialized.html.
GMO Talk: http://gmotalk.com/.
GMWATCH: http://www.gmwatch.eu/.
Greeniacs: http://www.greeniacs.com/GreeniacsArticles/Food-and-Beverage/Genetically-Modified-Organisms.html?gclid=CPPQzuSoz7wCFXBk7AodajUAcg.
Green Super Rice: http://thegsr.org/.
Industry Experts: http://industry-experts.com/verticals/energy-and-utilities/biofuels-a-global-market-overview.
Information Systems for Biotechnology (ISB): http://gophisb.biochem.vt.edu.
International Food Biotechnology Committee (IFBiC): http://www.ilsi.org/FoodBioTech/Pages/HomePage.aspx.
International Food Information Council Foundation—Food Insight: http://www.foodinsight.org/.
International Rice Research Institute and International Maize and Wheat Improvement Center: http://www/knowledgebank.irri.org/ckb/extras-maize/teosinte-maizes-wild-ancestor.html.
ISAAA: http://www.isaaa.org/resources/publications/pocketk/24/default.asp.
iStock Photos: http://www.istockphoto.com/photos/Genetic+Modification?facets={%2234%22:[%221%22,%227%22,%228%22],%2235%22:[%22Genetic%20Modification%22]}#f0c3163.
Miami Water: http://miami-water.com/blog/2217/product-list-of-gmo-genetically-modified-foods/.
Monsanto: http://www.monsanto.com/pages/default.aspx.
National Agricultural Statistics Service—USDA: http://www.nass.usda.gov/.
National Centre for Biotechnology Education: http://www.ncbe.reading.ac.uk/.
National Centre for Biotechnology Education—GM Food: http://www.ncbe.reading.ac.uk/NCBE/GMFOOD/menu.html.
National Corn Growers Association (NCGA): http://www.ncga.com/. Biotechnology information: http://www.ncga.com/topics/biotechnology.
National Geographic—Biofuels: http://environment.nationalgeographic.com/environment/global-warming/biofuel-profile/.
North American Agricultural Biotechnology Council: http://nabc.cals.cornell.edu/.
North Carolina Biotechnology Center: http://www.ncbiotech.org.

Organic Consumers Association: http://www.organicconsumers.org.
Organisation for Economic Co-operation and Development (OECD): http://www.oecd.org/innovation/inno/keybiotechnologyindicators.htm.
Position Statements on Biotechnology: http://www.isaaa.org/kc/Publications/htm/articles/Position/uneca.htm.
PrecisionNutrition: http://www.precisionnutrition.com/all-about-gm-foods.
Psychology Press-Biofuels: http://www.psypress.com/books/details/9781420051247/.
Scribd—Biotech in Animal Feed: http://www.scribd.com/doc/16539104/BIOTECHNOLOGY-IN-ANIMAL-FEED-FEEDING.
Scribd—Companies Oppose GMO Labeling: http://www.scribd.com/doc/113747190/Companies-Oppose-GMO-Food-Labeling.
Shutterstock: http://www.shutterstock.com/.
Shutterstock GE: http://www.shutterstock.com/s/genetic+engineering/search.html?page=10.
Shutterstock GM Foods: http://www.shutterstock.com/s/genetically+modified+food/search.html?page=7.
Sightings: http://rense.com/politics6/gms.htm.
SOS—Scientific Explorer: http://gmosafety-engl.webnode.com/contact-us/.
Sweet Remedy: http://sweetremedyfilm.blogspot.com/2013/01/new-row-over-gm-food.html.
The GMO Crop Misinformation Page: http://parrottlab.uga.edu/parrottlab/forum2.htm.
The Knowledge Center Inc.: http://theknowledgecenter.org/.
The Pew Charitable Trusts: http://www.pewtrusts.org/en/archived-projects/pew-initiative-on-food-and-biotechnology.
The Royal Society: https://royalsociety.org/.
The Western Producer: http://www.producer.com/2013/07/gm-crops-help-produce-more-with-less/.
The World Food Prize: http://www.worldfoodprize.org/en/about_the_prize/.
Third World Network: http://www.twnside.org.sg/title/genmo-cn.htm.
Turnpoint.Org: http://www.turnpoint.org/?&query=Genetically%20Modified%20Food&afdToken=Cq0BChMI17n0-faCwgIVTj4wCh21awCrGAEgAlD41KABUI7PwwFQj9H5AVCkjoMHULGclQhQwaTNCVC3vNAJUI36pQ1QkIP3D1DqleAWUKewuR1Q6p6-I1Dj35glUKLi_iVQqLmMJlD2oK0pUO_K4GZQ-oeec1Dpgu6XAVCovprGAlD6g9idA2j41KABcUejD-gO88dajQHGPM8ukQECa8ptgc5CtpEBFN3kLtDqZNMSGQCchQJKuBA2t8J1Ph5yczHTaBJ-kvSuCYQ.
Union of Concerned Scientists: http://www.ucsusa.org/.
University of California, Davis Biotechnology Program: http://www.biotech.ucdavis.edu/.
University of California, Davis: Center for Consumer Research (CCR): http://ccr.ucdavis.edu/.
University of Michigan: http://sitemaker.umich.edu/sec006group5/gm_food.

University of Wisconsin Biotechnology Center: http://www.biotech.wisc.edu.
USDA Advisory Committee on Biotechnology and 21st Century Agriculture (AC21): www.usda.gov/documents/ac21_report-enhancing-coexistence.pdf.
USDA: http://www.usda.gov//wps/portal/usda/usdahome?navid=BIOTECH.
U.S. Environmental Protection Agency (EPA): http://www.epa.gov. Biotechnology information: http://www.epa.gov/oppt/biotech/index.htm.
U.S. Farmers and Ranchers (USFRA): http://www.fooddialogues.com/. Biotechnology Information: http://www.fooddialogues.com/foodsource/topics/biotech-seeds.
WHO: http://www.who.int/foodsafety/en/

2

The Basics

2.1 BIOTECHNOLOGY

Any scientific approach or a scientific mindset would start by posing questions starting with one of the following words: *what, how, why, who, when,* and *where*. In the following sections, we will present information about biotechnology intended to answer these questions.

2.1.1 What Is Biotechnology?

The term biotechnology (also referred to as biotech) has been in the spotlight for quite some time, probably due to its close relationship with "genetically modified (GM) foods." Many people think that biotechnology is a relatively new discipline that is recently getting a lot of attention and that it only involves new sophisticated processes. For many, biotechnology is synonymous with the production of genetically modified organisms (GMOs, also known as transgenics). Although GMOs have captured the public attention and led to substantial private investments, a large amount of research has been conducted and various applications of the technology have been made, including agricultural biotechnologies. In fact, various biotechnological processes are commonly used in everyday life, for example, in baked products, and in fermented foods.

The word biotechnology originated from a cross between two Greek words, namely, "bios" (which refers to everything to do with life) and "technikos" (meaning involving human knowledge and skills) (EuropaBio 2013). In that sense, the term biotechnology can be broken down into two parts: "bio" referring to life or living things and "technology" which involves producing substances or developing processes that lead to production of useful materials. Thus, biotechnology could be viewed as the use of living things to make or change products, such as the food we eat. Biotechnology is regarded as the third wave in biological science which represents such an interface of basic and applied sciences, where gradual and subtle transformation of science into technology can be witnessed (Shrestha 2010). Contrary to its name, biotechnology is not a single technology. In fact, it is a group of technologies that share two characteristics: working with living cells and their molecules and having a wide range of practice uses that can improve our lives (Keener et al. 2014).

As a concept, biotechnology includes a large number of various procedures and techniques intended to alter or modify living organisms to achieve one or more desired goals. Some of these procedures go back in history to the days when people started to domesticate animals and cultivate plants for food uses. With time, new developments and improvements in old techniques took place (Section 2.2.1). Biotechnology could be viewed as a multidisciplinary area of study as it is based on, and linked to, other basic biological sciences, for example, genetics, cell biology, microbiology, biochemistry, molecular biology, and tissue culture. In contrast, it benefits from advances made in other areas of knowledge, for example, chemical engineering, bioprocess engineering, bioinformatics, and biorobotics. In some cases, there is an overlap with some of the related fields such as bioengineering and biomedical engineering. Biotechnology is an inclusive term and can be considered as an umbrella that covers a wide spectrum of scientific and technological tools and techniques, including genetic modification (GM) and genetic engineering (GE).

The term itself is largely believed to have been introduced in 1919 by the Hungarian agricultural engineer Károly Ereky (Section 2.2.1).

Some of the commonly used definitions of biotechnology as outlined by Khan (2011) include the following:

- The use of living organisms (particularly microorganisms) in industrial, agricultural, medical, and other technological applications
- The application of the principles and practices of engineering and technology to the life sciences
- The use of biological processes to make products

- The production of GMOs or the manufacture of products from GMOs
- The use of living organisms or their products to make or modify a substance, and biotech includes recombinant deoxyribonucleic acid (DNA) technique, GE and hybridoma technology
- A set of biological techniques developed through basic research and applied to research and product development
- The use of cellular or biomolecular processes to solve problems or make useful products
- An industrial process that involves the use of biological systems to make monoclonal antibodies and GE recombinant proteins

Other accepted definitions include the following:

- Biotechnology is the application of biological organisms, systems, or processes to manufacturing and service industries (Marshall 1994).
- Biotechnology is the use of living systems and organisms to develop or make useful products, or "any technological application that uses biological systems, living organisms, or derivatives thereof, to make or modify products or processes for specific use" (UN Convention on Biological Diversity Art. 2 1992).
- Biotechnology can be described as any technology that uses living organisms to make or modify a product for a practical purpose (GreenFacts 2014).
- The application of biological organisms, systems, or processes by various industries to learning about the science of life and the improvement of the value of materials and organisms such as pharmaceuticals, crops, and livestock (The American Chemical Society 2014).
- "Any technological application that uses biological systems, living organisms, or derivatives thereof, to make or modify products for specific use" (Secretariat of the Convention on Biological Diversity 1992).
- Modern biotechnology encompasses the application of: (a) *in vitro* nucleic acid techniques, including recombinant DNA and direct injection of nucleic acid into cells or organelles, or (b) fusion of cells beyond the taxonomic family, which overcome natural physiological reproductive or recombination barriers and this technique is not used in traditional breeding and selection (Secretariat of the Convention on Biological Diversity 1992, 2000).

- A range of different molecular technologies such as gene manipulation and gene transfer, DNA typing, and cloning of plants and animals (The FAO *Glossary of Biotechnology* [FAO 2001]).
- Biotechnology is defined as a set of tools that uses living organisms (or parts of organisms) to make or modify a product, improve plants, trees, or animals, or develop microorganisms for specific uses (ISAAA 2010).
- Application of technical advances in life science to develop commercial products.

All these definitions are true. They differ in the emphasis and details embodied in each one of them. The bottom line is that they all agree on the fact that biotechnology uses living organisms (bio) to make or modify (technology) a product to achieve a certain goal or an applied purpose.

Broadly, biotechnology can be divided into two major branches (Shrestha 2010).

- Nongene biotechnology: processes and procedures that deal with whole cells, tissues, or even individual organisms.
- Gene biotechnology: techniques that involve gene manipulation, cloning, and similar processes.

Nongene biotechnology is a more popular practice that encompasses widespread practices such as plant tissue culture, hybrid seed production, microbial fermentation, production of hybridoma antibodies, or immunochemicals.

Biotechnology has received a lot of attention in recent years due to its roles in many aspects of life. Following that, a large number of articles have been, and continue to be, published in various areas of biotechnology. Examples of the recently published books and articles include Dubey (2006), Friedman (2008), Gerstein (1999), Hanson (1983), Herdta and Nelson (2011), Kumar (2003), Lemaux (2008, 2009), Levidow and Carr (2009), Pometto et al. (2005), Purohit (2004, 2005), Roy (2011), The Advisory Board for the Research Councils and the Royal Society (1980), Thieman and Palladino (2008), and Zaid et al. (2001).

2.1.2 Genetic Material (DNA)

One type of modern biotechnology is referred to as "gene technology." This technology involves different types of techniques which are employed to control or modify genes or transfer them between related and unrelated species: a technology known as "recombinant DNA technology." This technology can take different shapes, for example, transferring genes from one plant to another plant, from a plant to an animal, from microorganisms to plants or animals, or vice versa. The resulting living organism created through this process is known as "transgenic," signifying the controlled transfer of genes among living organisms. Gene technology is intended to control and manipulate the cell's ability to produce proteins and consequently produce new desired proteins that perform new targeted functions or desired goals. Gene technology is widely applied in areas such as agriculture and medicine. In the medical field, for example, the technology is used in "gene therapy" to manage or treat some diseases that are more related to a person's genetic make-up. It can also be used to manufacture new drugs and vaccines and to improve testing and diagnosis techniques. An important advancement in gene technology is its application to map the "human genome" (the complete set of DNAs in humans), which is expected to help disclose the secrets of many human chronic diseases and consequently help in their treatment or management.

In order to understand how gene technology works and its relationship with GMOs, more specifically, GM foods, we need to look at some details of DNA and genes.

The building units or bricks of all living things on earth including humans, animals, plants, and microorganisms are the cells. Plants and animals are made up of millions of cells. Within the structure of each living cell, we find the genetic material needed and programmed for replication and survival of the living being. Except for some kinds of viruses, DNA represents the genetic material in cells of living organisms. The structure of DNA has been discovered by James Watson and Francis Crick in 1953. Their discovery, which showed that DNA exists in long strands that are twisted in pairs to form a shape called the "double helix," won them the Nobel Prize for Medicine in 1962. The long strands of DNA form the chromosomes which are coiled units wrapped up inside the cells of all living things. DNA has the same structure in all living things. The scientists explained that each DNA strand contains four substances known as "nitrogen bases," which fall under two categories of bases: the purines (adenine [A] and guanine [G]) and the pyrimidines (thymine [T] and cytosine [C]). The nitrogen bases are not the only components of DNA. They rather combine with two other ingredients, namely, the sugar deoxyribose and a phosphate molecule to form a more complex chemical structure known as nucleotide. Thousands of nucleotides combine to make one DNA molecule. Nucleotides are considered the true building units of

GENETICALLY MODIFIED FOODS

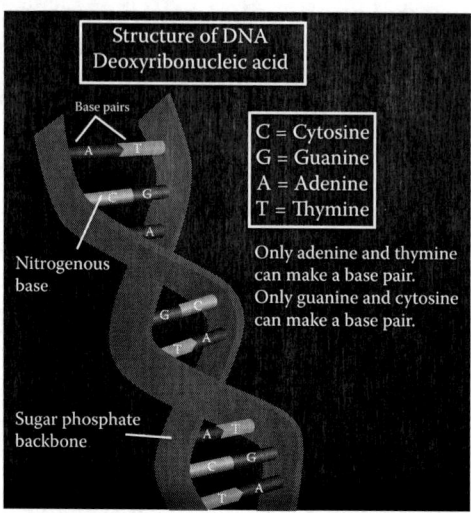

Figure 2.1 Structure of DNA. (From Shutterstock Inc, New York, NY, USA. http://www.shutterstock.com/. With permission.)

DNA. The nitrogen bases are repeated along every length of DNA in linear sequences of A's, G's, T's, and C's (Figure 2.1). DNA could be regarded as the "blueprint of life" that is passed from one generation to the next.

A living organism's DNA contains a large number of genes which are particular parts of the DNA, and only a minor fraction of the DNA, in fact, makes up genes. The remaining major fraction of the DNA represents noncoding sequences, the role of which is not yet clearly explained and is currently being researched. Each gene provides sets of instructions encoded in their internal chemical structure, with the genetic information to produce a protein. There are genes in everything that lives or has lived. There are genes in humans, insects, animals, plants, and microorganisms. The number of genes in these organisms is very high. For example, a 200 g steak contains as much as 750,000,000,000,000 genes (Future Food 2014).

The nature of the protein produced depends on the genetic code or the specific linear sequence of the four nucleotides that make up a gene. Genes are normally passed from one generation to the next, determining and giving all living things their particular physical characteristics and traits. The genes carry all the chemical instructions needed to control the living organism's features, characteristics, and behaviors expressed in a certain way. Because genes are passed on from one generation to the next, the offspring inherits these traits from its parents. This happens in all living organisms, for example, in humans—color of the eyes and type of hair; in plants—size or color of fruits and nutrient content. Naturally, genes do not work on the basis of one gene controlling one trait. They are rather involved in complex reactions and work in a collaborative manner to perform certain biological functions inside the living cell. Many characteristics exhibited by living organisms are the result of a complex collection of different genes working together.

The relevant question of "how does DNA store all the coded information that makes living things what they are?" has been answered in the following simple words stated by Australia Broadcasting Corporation (ABC. net 1999): "It's rather like a library. Molecular information can be compared to our language. Think of cells as the bookshelves. The chromosomes tightly coiled in living cells are the books. The DNA that makes up the chromosomes are the sentences, which in turn are made up of strings of words called genes. Akin to the letters in a words, nucleotides, A, G, T and C, are the fundamental units of genes, and the basis of all biological instructions."

One of the new technological developments related to DNA is "DNA fingerprinting" (Betsch 2014), which is also known as "genetic fingerprinting" and "DNA profiling." It has been introduced by Sir Alec Jeffreys in 1985. It helps to identify a specific individual, rather than simply identifying a species or some particular trait (wiseGEEK 2014). At present, it is widely used to identify paternity or maternity, as well as in the justice system to identify criminals or victims. In the human healthcare system, DNA fingerprinting is applied to diagnose some inherited disorders, such as cystic fibrosis, hemophilia, and sickle cell anemia, in both prenatal and newborn babies. It is also used by genetic counselors to discuss with prospective parents the possible risk of having a baby affected by one of those or similar diseases. In contrast, DNA fingerprinting can be used to develop treatments for inherited disorders (Shrestha 2010). The genetics and genomics revolution have at their core information and techniques that can be used to change humanness itself as well as the concepts of what it means to be human (Kilner and Jones 2004).

2.1.3 GM and GE

The characteristics of a living organism are determined by the coded genetic information contained in its chromosomal DNA. This information is carried and transmitted through individual units of the DNA, namely, the genes, from parents to the offspring. The characteristics expressed on a plant or an animal will be controlled by the type of genes it has inherited from its parents.

Since the dawn of history, man has known and practiced agriculture including plant and animal production. He has been involved in continuous efforts to develop and improve techniques and practices that achieve better yields and higher quality for both crops and animals. What is referred to as "old," "conventional," or "traditional" biotechnology has been used to describe the techniques used by farmers and scientists in old days. That included agricultural techniques such as selection and crossbreeding used for plants and animals. Those conventional techniques, which involved selection and further breeding to achieve certain purposes and to produce desired characteristics, helped to improve some crops and animals. For instance, the controlled selection of plant offspring with improved traits was a positive step in the ongoing process of producing new plant varieties to address different goals of agriculture, for example, high yielding, quick maturing, or more nutritious plants.

Old biotechnology techniques used in agriculture presented some problems that led researchers to look for alternative techniques which were more efficient, less time-consuming, and more precise. One of the major issues that resulted from traditional techniques, for example, in breeding or selection, is the possibility that an undesirable gene may also be introduced. The field of biotechnology witnessed fast developments during the last few decades by the science of molecular biology. Modern techniques have been developed for the modification of the genetic material of living organisms. Further efforts led modern biotechnologists to carry out what is known as GM to change crop characteristics for similar reasons. The developed modern technique of GM is meant to achieve the same goals as those of the process of conventional breeding. In addition to that, the technique has extended the scope of what can be achieved with regard to improving the characteristics of the modified organism.

GM, sometimes known as "transgenic technology," and GE are types of modern biotechnology which comprises a set of techniques used to manipulate, change, or engineer the genetic material, DNA, of a living organism. These techniques enable scientists to cut and join, mutate, copy and multiply genes, and, subsequently, isolate and transfer them from one organism to another, resulting in what is known as a GMO. The removal of individual genes from one species and their insertion into another, which is carried out in GM, do not require sexual compatibility between the two species. The offspring resulting from GM will contain copies of the new gene and can be produced in the traditional manner. GM has also been used to get rid, or suppress the effects, of an undesirable gene already present in a particular variety. It can also be applied to change or control the metabolism of an organism, for example, a plant, to improve some of its quality parameters. (Details of how the process of GM is carried out is given in Chapter 3.)

Although the terms GM and GE are sometimes used interchangeably and synonymously, a clear distinction can be drawn between them. GM is a general term that encompasses both old and new biotechnologies. It should be used to include any technique that changes or modifies the natural genetic make-up of a living organism. This includes simple selection, crossbreeding, which are regarded as "old technologies," as well as new technologies which transfer selected genes from one living organism to another. It is more appropriate to name the latter techniques as GE. In that sense, GE can be regarded as one type of GM. A living organism is considered GM if its genetic material has been changed by any technique, including traditional biotechnology such as breeding. The resulting organism is referred to as GMO. On the contrary, an organism is considered genetically engineered (GE) if it was GM through direct insertion of gene or genes from another living organism. GM can be used to refer to "changing of a specific gene in an organism," whereas GE can be used to denote "transfer of genes from one living organism to another living organism."

Sometimes, a distinction between GM and GE is based on geographical aspects, for example, GE is the standard U.S. term for the process in which foreign genes are spliced into a nonrelated species, creating an entirely new organism; GM, the same as GE, is more widely used in Europe because it translates more easily among different languages. In today's world, GE touches the life of every human being in one way or the other. There has been lots of literature and research conducted to examine and discuss how GE affects everybody's life. Bains (1987) attempted to explain the potential GE has, what it can do and what it will do.

GM leads to a number of changes in the genes of the host organism. This would subsequently cause changes in the characteristics of the subject to which genes have been transferred. For example, vegetables can be GM to keep them fresh for longer periods, and cereals can be GM to increase their nutrient content.

Because all genes of living organisms, be they human, plant, animal, or bacteria, are created from the same material, geneticists are able to transfer genes from one species to another. This gives the scientists large numbers of genetic characteristics to choose from and make their work more precise and more efficient. Scientists can now (literally) pinpoint which genes control which characteristics in a living organism, engineer, and transfer these genes, and subsequently the characteristics they control into another organism to produce a new modified organism. GE can be applied for all living organisms. It is widely used in agriculture. The process of plant GE is described in the Bitterroot Restoration (2011) website. More details of this process are given in Chapter 3 which describes the various applications of GE.

Some authors (Rees 2006) believe that GM is "imprecise because it is impossible to guide the insertion of the new gene into the intended host." This group states that "this process can lead to unpredictable effects since genes do not work in isolation but in complex relationships, which are not understood." Additionally, they believe that "any change to the DNA at any point will affect it throughout its length in ways scientists cannot predict." However, these claims remain debatable and more research is needed to confirm or nullify them. The debate on GM and GM foods is discussed in depth in Chapter 6.

2.1.4 GMOs, GM Crops, and GM Foods

2.1.4.1 Genetically Modified Organisms

GMOs are plants, animals, or microorganisms that have their genetic structure modified. According to the WHO (2014), GMOs can be defined as organisms in which the genetic material (DNA) has been altered in a way that does not occur naturally; they are produced because they can add some advantages such as lower price, longer durability or higher nutritional value, or both. Any organism in which the genetic material has been changed in a manner that does not occur naturally by mating and/or natural recombination would be regarded as a GMO. Generally speaking, any living thing which has been GM would be referred to as GMO.

2.1.4.2 Genetically Modified Crops

This is also referred to as GMCs, GM crops, and biotech crops. When GM is carried out in plants used in agriculture, the resulting organism is known as GM crop (Figure 2.2). Generally, the aim is to introduce new traits into the plant, which do not occur naturally in the species. The technique is now being used to insert herbicide tolerance, virus resistance, slow ripening, and other traits into plants for food use. "Biopharm" is the term used to describe GE crops designed to produce pharmaceutical drugs and industrial chemicals.

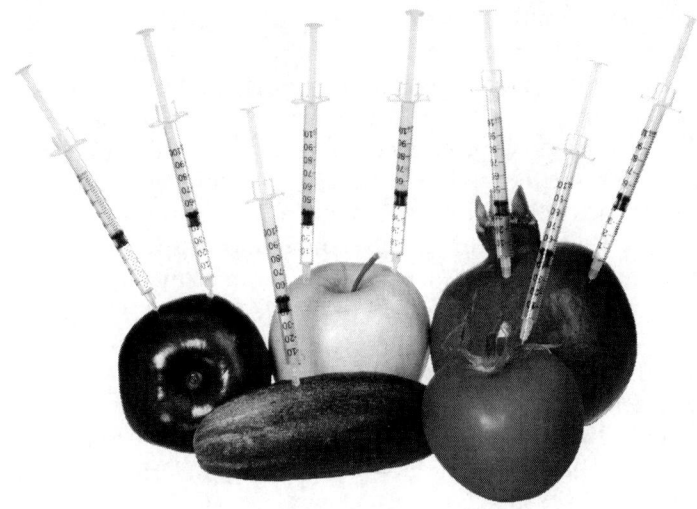

Figure 2.2 Illustration of GM fruit and vegetables. (From Shutterstock Inc, New York, NY, USA. http://www.shutterstock.com/. With permission.)

2.1.4.3 GM Foods
This is also referred to as GM foods, biotech foods, gene foods, bioengineered foods, gene-altered foods, transgenic foods, and foods that have been created through GE; Frankenfoods: another term for GM foods referring to the story of Frankenstein and science gone bad. GM foods are foods produced from organisms (plants or animals) which have had specific changes introduced into their DNA using one of the methods of GM. GM has been used in a variety of ways to assist food manufacturing and to improve factors such as storage or nutritional value of food. Many processed foods contain GM ingredients. GM foods differ from non-GM (conventional) foods, in that they contain or are produced from a GMO or they contain GM ingredients.

2.2 BIOTECHNOLOGY—PAST, PRESENT, AND FUTURE

2.2.1 Historical Developments in Biotechnology

Since the dawn of history and for more than 10,000 years back, man has known and practiced agriculture and biotechnology to produce crops and foods, domesticate animals, and prepare medicine. During that era, man has noticed that living organisms can adapt to various and changing environmental conditions, thus improving their characteristics by themselves. At the same time, he observed that this process takes very long periods before any positive change could be observed. This led man to think about controlled methods to manage the desired positive changes. Biotechnology has developed over the years to address emerging human needs. As stated earlier in this chapter, sometimes, the terms "old, ancient, traditional, and classical" are used to describe biotechnological techniques practiced in the past and which have developed (and still developing) through time to lead to what is now described as "new or modern" biotechnology. The cultivation of plants practiced by early civilizations could probably be considered as the earliest biotechnological technique. Ancient biotechnological techniques, for example, the use of natural yeast in the process of fermentation to produce bread, alcohols, fermented milk, and cheese, have been practiced by ancient civilizations, for example, Greeks, Romans, Chinese, Sumerians, Egyptians, and Indians.

Traditional biotechnologies mainly included agricultural practices used by ancient farmers to produce better crops in addition to the use of microorganisms in food processing. During those times, farmers have tried to improve the characteristics of their crops through selection and breeding. They used to select the best looking plants and seeds on the basis of their physical features such as faster growth, higher yields, pest and disease resistance, and larger seeds, and saved them to plant for the next year. With time, man discovered that plants could be artificially mated or cross pollinated and that could lead to improvements in the characters of the plant. That was how and when the technology of plant breeding developed. Continuous selection and breeding with improved properties each time led to production of plants with characteristics that are substantially genetically different from their ancestors. These plants were also changed in a way that they can be grown in new and different environments. Further developments in the science and technology of plant breeding were observed early in the twentieth century. Plant breeders started to better understand the mechanism of selecting plants with superior qualities in order to plant and breed them. Their aim was to create new and improved varieties of different crops. These new developments led to improvements in both quantitative and qualitative attributes of agricultural products.

Classical crossbreeding involves crossing of two plants of the same or of two closely related species. Because each half of the genetic material comes from one of the two parents and is transmitted to the offspring, it is possible that undesirable characteristics are introduced along with the desired ones. However, this conventional method of breeding proved to be time-consuming, imprecise, and demanding because many generations of the plant offspring might be needed to remain with, or strengthen, the good characteristics and to eliminate the bad ones. Modern agricultural biotechnologies, further described in Chapters 3 and 4, helped to make the process more efficient.

Some of the processes introduced and used during ancient times, which apply microorganisms in food fermentation and production, continue to be used today. As expected, there are many advances in these technologies but the principles remain the same. The fermentation process could be regarded as the first use of biotechnology to convert one food source into a different one. The process was not fully understood until Louis Pasteur's work in 1857.

Efforts to find more precise, reliable, and sustainable techniques in plant breeding continued. The results of those efforts led to production of some new crop varieties such as open-pollinated (OP for corn), inbred (for rice), or an F1 (first filial generation) hybrid variety. If OP and inbred varieties are properly managed, they will retain the same inherent traits when propagated. Hybrid seeds are produced by the process of hybridization or by crossing of diversely related parent plants. Pure lines are the offspring produced through several cycles of repeated self-pollination that "breed true" or produce sexual offspring that closely resemble their parents (ISAAA 2010). The impact, created by such developments in breeding, has been significant, particularly for major food crops, for example, wheat, corn, rice, sorghum, and some vegetables, thus participating in the improvement of the global food situation (ISAAA 2005).

Developments in knowledge, generally and specifically in science and technology, have no specific or clear cut boundaries that separate one era from the other. These developments build on each other and represent a continuum of information. This could be said for the field of biotechnology. We do not find a dividing line between old and new biotechnologies. Rather, there are continuous efforts, new discoveries, and improvements in existing technologies. Similarly, modern biotechnology could be viewed in that context. New developments and discoveries in plant breeding and more focused research on gene technology led to the birth of new biotechnology, which is considered to have taken place in the year 1971 when Paul Berg successfully experimented on gene splicing. This new technology was further developed through the collaborative work of Herbert Boyer and Stanley Cohen. In 1972, they were able to transfer the genetic material into a bacterium in such a manner that the transported material would be reproduced. The late twentieth and early twenty-first century witnessed wide expansion of biotechnology in the form of development and introduction of new and diverse technologies such as recombinant gene technology and genomics. Greater understanding of the nature of the genetic code of DNA helped the fast developments in biotechnology. GM (GE) is one of the modern biotechnology tools that utilize recombinant DNA technology. This new technology could be regarded as an extension of traditional plant breeding. It involves direct modification of DNA and is more precise. Additionally, it is used to direct, monitor, and predict changes with no possibility of introducing undesirable traits. This new technique has the distinct feature of breaking the natural barriers among species through transferring genes across unrelated species, for example, from a plant to an animal or from a microorganism to a plant, something that cannot be achieved in selective breeding. Modern biotechnology is currently used industrially to make useful products such as vaccines, antibiotics, enzymes, and hormones. Discussion and details of this field are outside the scope of this book, which focusses on GM foods.

2.2.2 Biotechnology Timeline*

BC

About 10,000 BC: Civilizations harvest from natural biological diversity, domesticate crops and animals, begin to select plant materials for propagation and animals for breeding

8500–5500 BC: People begin to settle in one place and raise plants and animals; the best of their crop is saved to use as seed the next year

4000 BC: Development of classical biotechnology: dairy farming develops in the Middle East; Egyptians use yeasts to bake leaven bread and to make wine. Fermentation and malting are the mainstays in Mesopotamia

2000 BC: Egyptians, Sumerians, and Chinese develop techniques of fermentation, brewing, and cheese-making

---

* Compiled from FAO 2004; Institute of Medicine and National Research Council of the National Academies 2004; Andrews 2013; Citizens Concerned About GM 2013; International Food Information Council Foundation 2013.

1800 BC: The Babylonians improve the quality of date palms by pollinating female trees with pollen from male trees with desirable characteristics

1500s

1500: Application of acidic cooking techniques to produce sauerkraut and yogurt—two examples of using beneficial bacteria to flavor and preserve food. Aztecs make cakes from *Spirulina* algae

1600s

1694: Discovery of sexual reproduction in plants

1700s

1719: Development of the first record of plant hybrid (intraspecific hybridization)
1799: First report of cereal hybrid

1800s

1859: Charles Darwin publishes the first edition of *The Origin of the Species* (Theory of Natural Selection), which, among other things, gives extensive information on the knowledge of breeding at that time
1861: French chemist Louis Pasteur develops pasteurization—preserving food by heating it to destroy harmful microbes
1863: From observing pea plants in a garden, renowned scientist Mendel concludes that certain "unseen particles" (later described as genes) pass traits from parents to offspring in a predictable way—the laws of heredity begin to be understood
1865: The Austrian monk, botanist, and plant scientist Gregor Mendel publicizes his discoveries on the breeding of garden peas. His experiments provide the basis for theories of inheritance and becomes the foundation of modern genetics
1869: The Swiss physician and biologist Friedrich Miescher isolates various phosphate-rich chemicals, which he called *nuclein* (now nucleic acid), from the nuclei of white blood cells. His discovery leads to the identification of DNA as the carrier of inheritance
1876: Development of interspecific and intergeneric crossings (e.g., leading to the hybrid triticale, which is a cross between wheat and rye)

1879: The American botanist William James Beal develops the first experimental hybrid corn

1900s

1900: Beginning of hybrid maize breeding in the United States

1900: Rediscovery of Mendel's experiments—Mendelian's Era of Genetics. The Dutch botanist Hugo Marie de Vries suggests the concept of genes, rediscovering the laws of heredity while unaware of Mendel's work. He introduces the term "mutation" and develops the "mutation theory" of evolution which forms the basis of mutagenesis

1902: Walter Sutton and Theodor Boveri propose the "Boveri–Sutton Chromosome Theory" (also known as the chromosome theory of inheritance or the Sutton–Boveri theory). The theory identifies chromosomes as the carriers of the genetic material and are the basis of genetic inheritance

1909: Protoplast fusion reported

1910: The American biologist Thomas Hunt Morgan presents chromosomal theory of heredity, advancing the Boveri–Sutton chromosome theory and demonstrating that the chromosomes are the concrete entities which carry the genes

1913: The American geneticist Alfred Henry Sturtevant constructs the first genetic map of a chromosome. He works on the organism *Drosophila melanogaster*

1914: The German physician and chemist Robert Feulgen develops a method for staining DNA (currently known as the Feulgen stain) and demonstrates that DNA exists in all living cells

1917: The Hungarian agricultural engineer Károly Ereky coins the word *biotechnology*. His work which is published in 1919 in Berlin lays the groundwork for this new discipline. He is regarded by some as the "father of biotechnology"

1927: The American geneticist Hermann Joseph Muller conducts work on the physiological and genetic effects of x-rays (x-ray mutagenesis). He works with *Drosophila* and demonstrates that genetic mutation could be induced by x-rays

1928: The American geneticists George Wells Beadle and Edward Lawrie Tatum discover the role of genes in regulating biochemical events within cells, show that genes control individual steps in metabolism. They hypothesize that one gene directs the production of one protein

1931: The American geneticists Harriet Baldwin and Barbara McClintock work in the field of maize cytogenetics. They describe chromosomal crossover for the first time and demonstrate the direct physical recombination (the linking of DNA from different chromosomes) by examining maize chromosomes microscopically

1936: Modern development of artificial insemination in livestock (dairy cattle)

1937: Introduction of the "polyploidization" process

1940: The development of "single seed descent" technique (SSD)

1941: George Beadle and E.L. Tatum pin a gene defect down to a single step in a biochemical pathway that would normally be carried out by an enzyme. They restore normal growth to a mutant microorganism by adding the missing enzyme

1946: Scientists first discover that DNA can transfer between organisms

1953: The molecular biologists James Watson (American) and Francis Crick (English) codiscover the structure of the DNA molecule. They propose and describe the double helix structure of DNA. Their work is founded on research reported by the British scientist Rosalind Franklin

1957: Francis Crick proposes the "Central Dogma" theory (which states that "genetic information is passed from DNA to RNA and then to proteins, but it cannot be passed from proteins to DNA")

1958: The American geneticists Matthew Meselson and Franklin Stahl demonstrate the semiconservative replication of DNA. This is when the DNA forms a copy of itself: one strand remains the same, and the other contains newly synthesized DNA

1960: Embryo rescue technique refined

1960: Discovery of messenger RNA (mRNA)

1961: U.S. Department of Agriculture (USDA) registers *Bacillus thuringiensis* as the first biopesticide

1961: The American geneticist Marshall Warren Nirenberg succeeds in "breaking the genetic code" and describing how it operates in protein synthesis

1966: Marshall Nirenberg and Har Gobind Khorana finish unraveling the "genetic code"

1970: The "double haploid" techniques are introduced

1970: The American scientists Howard Temin and David Baltimore discover the enzyme "reverse transcriptase" in RNA viruses, which has implications for GE

1970: Introduction of recombinant DNA technology (start of modern biotechnology)

1972: The American biochemist Paul Berg produces recombinant DNA (rDNA) molecules

1973: First somatic hybridization (tomato and potato)

1973: The American scientists Stanley Cohen and Herbert Boyer invent the technique of DNA cloning, which allow genes to be transplanted between different biological species

1974: Ti plasmid found to be the tumor-inducing factor in *Agrobacterium tumefaciens*

1974: Stanley Cohen, Annie Chang, and Herbert Boyer create the first GM DNA organism

1975: Development of electrophoresis enables separation of DNA fragments

1975: Conference held in the United States where scientists met behind closed doors to reach a consensus on self-regulation and how the newly discovered recombinant DNA technology (GE) should proceed

1976: The National Institutes of Health in the United States produce guidelines for GM research

1977: The British biochemist Frederick Sanger develops "chain termination DNA sequencing," allowing scientists to read the nucleotide sequence of a DNA molecule

1980s to early 1990s: China put GM crops on sale for the first time, namely, a virus-resistant tobacco and a tomato

1980: The American scientists Jon Gordon and Frank Ruddle develop the first transgenic (GM) mouse, using pronuclear injection

1980: Patenting of GE microbes is permitted in the United States

1981: Introduction of the term "somaclonal variation"

1982: First GE product—human insulin produced by Eli Lilly and Company using *Escherichia coli* bacteria—is approved for use by diabetics

1982: Giant mouse produced by transferring growth hormone genes from a rat

1983: First transgenic plant: *Agrobacterium* is used to transfer a gene from one plant species to another (used in tobacco plant)

1983: Polymerase chain reaction (PCR) technology enables faster analysis of DNA and RNA

1983: The American biochemist Kary Mullis invents the PCR, which is a technique enabling scientists to reproduce bits of DNA faster than ever before

1983: Four separate groups of scientists create GM plants: three groups insert bacterial genes into plants and one inserts a bean gene into a sunflower plant

1983: Richard Palmiter and Ralph Brinster place the gene for human growth hormone in an early mouse embryo. The resulting adult is double the normal size

1984: Development of genetic fingerprinting, a technique greatly helps the police force in finding and identifying criminals

1985: Development of first GE farm animals (e.g., transgenic "Beltsville Pigs")

1985: Development of "oncomouse" (a transgenic animal)

1986: Environmental Protection Agency (EPA) approves commercial growing of the first GE crop—tobacco plants resistant to tobacco mosaic virus

1987: John Sanford and Theodore Klein develop the "Gene Gun" for microprojectile bombardment

1987: A series of transgenic mice produces carrying human genes. A transgenic plant produces resistant to a particular kind of herbicide

1988: First transgenic plant producing a pharmaceutical. Transgenic maize (corn) produced

1988: Oncomouse becomes first patented transgenic animal

1989: Publication (*Science* 254: 1281–8) of data about the "Beltsville pig"; the transgenic pig which suffered a range of pathological conditions because it had a gene for human growth hormone

1990: Pfizer Inc. produces and introduces "Chymax chymosin," an enzyme used in cheese-making—first commercially approved GE food product of recombinant DNA technology in the U.S. food supply

1990: The first successful field trial of GM herbicide-tolerant cotton is conducted in the United States

1990: The first GM cereals

1991: The first gene therapy trials on humans

1992: FDA issues a policy stating that foods from biotech plants would be regulated in the same manner as other foods. Premarket consultation with FDA is encouraged, consistent with industry practice

1993: U.S. Food and Drug Administration (FDA) approves "Bovine somatotropin" (bST), a naturally occurring metabolic protein hormone used to increase milk production in dairy cows for commercial use

1994: The American company Calgene, Inc. starts marketing Flavr Savr™ tomato (after approval by the U.S. FDA) as the first commercially approved GE whole food crop in the U.S. food supply. Virus-resistant squash is also planted

1994: The use of plant *in vitro* fertilization in maize (corn)

1995: A transgenic tobacco variety developed producing hemoglobin

1995: (*B. thuringiensis*) Potato is approved safe by the EPA, making it the first pesticide-producing crop to be approved in the United States

1995: In the United States, the following transgenic crops receive marketing approval: canola with modified oil composition (Calgene), *B. thuringiensis* corn/maize (Ciba-Geigy), cotton resistant to the herbicide bromoxynil (Calgene), *B. thuringiensis* cotton (Monsanto), *B. thuringiensis* potatoes (Monsanto), soybeans resistant to the herbicide glyphosate (Monsanto), virus-resistant squash (Monsanto-Asgrow), and additional delayed ripening tomatoes (DNAP, Zeneca/Peto, and Monsanto)

1995: First bacterial genome sequenced

1996: Biotech varieties of soybean, cotton, corn, canola, tomato, and potato seed are planted on 4.5 million acres in Argentina, Australia, Canada, China, Mexico, and the United States

1996: J. Sainsbury and Safeway grocery stores in the United Kingdom introduce Europe's first genetically modified food product. A variant of Flavr Savr is used by Zeneca to produce tomato paste

1996: Brazil nut genes are spliced into soybeans by Pioneer Hi-Bred Company

1996: The birth of the first cloned animal, Dolly, the sheep, is announced

1996: GM tomato paste approved in the United Kingdom, first GM herbicide-tolerant soya beans (Roundup Ready Soybeans), and insect-protected maize approved in the European Union

1996: Council Directive 90/220/EEC of April 23, 1990 on the deliberate release into the environment of GMOs

1997: Sequencing of *E. coli* genome

1997–1999: GM ingredients appear in two-thirds of all U.S. processed foods

1997: The cloning of a transgenic lamb (Polly) cloned from cells engineered with a marker gene and a human

1997: EC Novel Foods Regulation (258/97) [Regulation (EC) No. 258/97 of the European Parliament and of the Council of January 27, 1997 concerning novel foods and novel food ingredients] comes into effect, requiring a safety assessment for novel and GM foods before they go on sale

1998: "Terminator technology" moves a step closer to the fields: U.S. Patent No. 5,723,765, grants to Delta & Pine Land Co. (an American cotton seed company) and the USDA

April, 1998: A U.K. supermarket chain bans use of GMOs in its products; a move that is adopted by other U.K. supermarket chains

1998: First GM labeling rules introduced to provide consumers with information regarding the use of GM ingredients in food

1998: Virus-resistant papaya, developed through biotechnology to save the crop from devastation, is planted in Hawaii. Insect-protected sweet corn is also planted

1999: Sequencing of *Drosophila* genome

1999: Enviropig™ is GE in Canada to produce an enzyme in its saliva that would allow it to get more phosphorus from its feed. This would reduce phosphorus runoff into waterways

1995–2005: The total surface area of land cultivated with GMOs has increased by a factor of 50, from 17,000 km² (4.2 million acres) to 900,000 km² (222 million acres), of which 55% are in Brazil

2000s

2000: Bioinformatics, genomics, proteomics, and metabolomics era (*known as the "omics" era*)

2000: Development of the GM golden rice (rice with increased nutrient value)

2000: Sequencing of *Arabidopsis* genome

2000: Draft sequence of human genome

22 January 2001: U.K. Parliament passes a regulation believed to allow the cloning of human embryos for the purposes of research into serious disease. Embryos may be experimented on only up to their 14th day of life

2001: Directive 2001/18/EC on the deliberate release into the environment of GMO and repealing Council Directive 90/220/EEC. This contains a so-called "safeguard clause" (Art. 23). According to this clause, member states may provisionally restrict or prohibit the use and/or sale of the GM product on its territory. The member state must have justifiable reasons to consider that the GMO in question poses a risk to human health or the environment. Six member states currently apply safeguard clauses on GMO events: Austria, France, Greece, Hungary, Germany, and Luxembourg

2002: Patent law proposes for biotechnology industries to protect their "intellectual property"

2002: Draft sequence of the mouse genome

2002: Draft sequence of the rice genome

2003: Human genome is sequenced

2003: European GMO-free regions' network is established. Ten European regions sign a joint declaration at the European Parliament to safeguard their agriculture policies (mainly based on support to high quality, traditional, and low impact production systems), which can be disrupted by the introduction of GMOs. The network is based on a political agreement with no binding juridical status

2003: Japanese researchers develop a biotech decaffeinated coffee bean

2003: Countries that grew 99% of the global transgenic crops are the United States (63%), Argentina (21%), Canada (6%), Brazil (4%), China (4%), and South Africa (1%)

2004: EC Regulation on GM Food and Feed (EC 1829/2003) and EC Regulation on Traceability and Labeling of GMOs (EC 1830/2003). The EC Regulations become legally binding on 18 April 2004. Regulation 1830/2003 requires labeling of all GM food and feed, which contain or consist of GMOs or are produced from or contain ingredients produced from GMOs, regardless of the presence or absence of GM material in the final food or feed product. This is an extension to the previous labeling rules which are only triggered by the demonstrable presence of GM material in the final product

2005: Principles for the European GMO-free regions are formally laid down in February in Florence during the Network's 3rd Conference with the subscription of a joint document called "Charter of Florence"

2006: A pig is engineered to produce omega-3 fatty acids through the insertion of a roundworm gene

2006: In the United States, 89% of the planted area of soybeans, 83% of cotton, and 61% corn are GM varieties

2006: GM rice is approved for human consumption in the United States

2007: The USDA approved the planting of 11 new pharmaceutical or industrial GM crops

2008: Sugar beets produced with biotechnology are commercialized

2008: FDA releases its risk assessment on animal clones, concluding that food from clones is as safe as other foods

2008: The European Commission authorized the GM maize GA21 for feed and food use and for import and processing. GA21 is not approved for cultivation in the European Union

2010: Amflora is approved for industrial applications in the European Union by the European Commission. Amflora is a GM potato: the result of two decades of research efforts. The Amflora potato is selected for its special starch properties, which is used in paper making and adhesives

2011: "High-oleic" soybean varieties higher in heart-healthy monounsaturated fats are available in the United States

2011: Additional whole foods enhanced by biotechnology are submitted for government review, including nonbrowning apples and low-acrylamide potatoes

2011: The United States leads a list of multiple countries in the production of GM crops, and 25 GM crops has received regulatory approval to be grown commercially

2012: Researchers report that the first "hypoallergenic" cow, Daisy, has been GE to remove a protein that can trigger whey allergy in humans

2012: Biotech crops are planted on 420.8 million acres by 17.3 million farmers in 28 countries. More than 90% of the farmers planting biotech seed are small, resource-poor farmers in developing countries

2013: About 85% of corn, 91% of soybeans, and 88% of cotton produced in the United States are GM

2.2.3 The Future of Biotechnology

What can biotechnology achieve? And what does the future hold for biotechnology? These are two of the pertinent questions that are always asked and that need to be answered and explained in some detail, with clear evidence to support any views presented.

Depending on the perspective, for example, technological, ethical, religious, humane, or the focus, for example, agricultural, medical, and environmental, we are considering, the answers to these questions would be very much different. Many people believe that based on the rate at which developments occurred in the field of biotechnology during the past 50 years or so, it would be reasonable and logical to say that its future is bright and promising. Another group of people believes that further developments in biotechnology would be catastrophic to humanity since we are acting as if we are "God" and interfering in the norms of "nature." Nevertheless, it will be worthwhile to review some of the expected occurrences and developments in biotechnology, some predicted claimed benefits, and some views of both sides of the argument.

Biotechnology is a multidisciplinary area of knowledge because a multitude of disciplines, mainly those dealing with living organisms, are involved in its study and it requires good understanding of these disciplines. These disciplines include, but not limited to, biochemistry, molecular biology, genetics, food technology, molecular engineering, medicine, and pharmacy. It is thus logical to say that any expected future development requires combined and collaborative advancements in some or all of these areas. Many manufacturing industries are integrating these sciences to advance biotechnology from different angles.

There are many future promises made by biotechnologists, enthusiasts, and those supporting the discipline of biotechnology. These promises mainly revolve around improvement in the quality and safety of human life, addressing global food and hunger problems, managing and treating chronic and deadly diseases, and effectively managing the environment. Research on the human genome project is considered a major project in modern biotechnology. It aims at designing a detailed map of human DNA. Chromosome maps, which are used as genetic markers for different diseases, are being developed in different research stations throughout the world. Further work is expected to be conducted to unravel the nature of various diseases. This would result in prompt and more efficient diagnosis and treatment of diseases such as cystic fibrosis, cancer, sickle cell anemia, and diabetes. In contrast, organisms produced through recombinant gene technology could be used to produce new vaccines needed to manage some diseases. These future endeavors are expected to help improve and maintain human health and, consequently, improve the quality of life.

In the medical field, biotechnology could be used in what is known as gene therapy. This technology aims at correcting any defects in the genetic material. The process involves the insertion of a normal healthy gene to replace a malfunctioning one. Additional technologies with future potential in the medical field include: identifying the genetic profiles of most diseases; studying and understanding the biochemical mechanisms and environmental interactions of diseases; understanding how human behavior is affected by genetic and physiological factors; development of liposomes that would help delivery of cytotoxic drugs to tumor positions with minimal effect on adjacent healthy cells; and the isolation of monoclonal antibodies to be used in cancer treatments. Developing protein-based "biochips" that would replace the silicon chips currently used as implants in the human body to deliver drugs is a new technology that has a lot of future potential. It is expected to be faster and more precise in drug delivery to manage heart rate or hormone secretion.

In the field of agriculture, current biotechnology applications are expected to continue in both plants and animals. Developing high yielding, disease- and herbicide-resistant crops, and disease-resistant animals are some of these applications. Efforts to enable the effective management of the organic environment will help to further the understanding of the genetic make-up of plants and animals. This will lead to wider genetic intervention activities and programs aiming at genetic enhancement ultimately producing new forms of GMOs with the desired traits. Research on nutraceuticals and functional foods to produce foods with health-promoting characteristics would reduce reliance on drugs in the management and treatment of some diseases. This field can also benefit from information generated through agribiotechnology.

Although traditional plant breeding has contributed to the improvement of crops in the past, transgenic technology has immense potential to achieve this objective as an additional/supplementary method. GM crops represent one of the most rapidly adopted technological innovations that have been commercialized in the history of agriculture despite significant and regulatory barriers (Dunwell 2000; Fernandez-Cornejo 2005).

Different views on biotechnology and what it is achieving today in addition to its future prospects have been widely expressed by both supporters and opponents. An article in the "Future For All" website (FutureForAll.org 2014) highlights some of the current and predicted capabilities of biotechnology. It focusses on the issues of ending the world hunger, production of foods with improved qualities, disease- and pest-resistant crops, medical and pharmaceutical benefits, biological development of clothing, plastics and building materials, and the creation of environmentally friendly manufacturing processes that minimize waste.

The American Chemical Society (2014) expressed its views on the future of biotechnology in an article published in the website acs.org.

The Society supported biotechnology and stated that it is the science of the future. The focus of the article was on the positive role of chemistry in biotechnology and on the wide opportunities for chemists in the dynamically evolving field of biotechnology.

The article "Future of biotechnology" by Treohan (1993) highlighted the main areas in which biotechnology is playing, and will continue to play, critical roles for human welfare. It expressed optimistic views on the potential roles of biotechnology and molecular biology as important sources of knowledge in the coming century. The present influence of biotechnology on medicine, agriculture, the industry, and the environment indicates a positive direction of its roles in the future. The human genome project, which aims at detailed mapping of the human DNA, is cited as a major biotechnological endeavor. To date, the genetic markers of more than 4000 diseases caused by single mutant genes have been mapped. The development of protein-based biochips, using modern biotechnological techniques, is expected to replace silicon chips. Biochip implants in the human body are believed to be faster, precise, and more efficient in delivering drugs that affect heart rate and hormone secretion or to control artificial limbs. Another area of promising future for biotechnology is gene therapy, in which new delivery systems known as "liposomes" are being developed to supply cytotoxic drugs to tumor areas with minimal damage to nearby healthy tissues. Biotechnology techniques can also be used to isolate new monoclonal antibodies that can be used in cancer treatment, diagnostic laboratory work, bone marrow transplant, and other medical applications.

Another optimistic view on the future roles of biotechnology has been expressed by Oldham (2014). He indicated that biotechnology has improved the quality of life, managed and treated diseases, and terminated hunger. The anticipation is that the next chapter in the information age is possibly the age of biotechnology.

The progress and development of biotechnology could be evaluated and monitored through studying certain indicators. The directorate for science, technology, and industry of the Organization for Economic Cooperation and Development (OECD) publishes and updates information and statistics on the key biotechnology indicators. Information and data on the following indicators have been published by OECD (2013): biotechnology firms, biotechnology R&D, public-sector biotechnology R&D, biotechnology applications, and biotechnology patents.

In today's world, biotechnology represents an undeniably important venture, particularly in the scientific and industrial sectors. More and more applications of it are advanced everyday, which led to the existence of diverse uses in different areas of professions. This diversity necessitated the design of a system to classify the different biotechnology uses, a system that relies on common features or end results. A convenient system that identifies the different uses and which identify them by colors has been designed and used. Five colors have been used to define the main types of biotechnology. These are described in the next sections.

2.3 TYPES OF BIOTECHNOLOGY

The diversity in the use of biotechnology as a tool in the scientific and industrial arenas has made it necessary to have a system to classify biotechnology applications on the basis of common features or final purpose. As a result, nowadays, there exist five main groups in biotechnological applications, which have been identified by a color system (Martínez 2011). The following sections give details of the different types of biotechnology based on the color code assigned to them. They are designated as red, white, green, gray, and blue biotechnologies.

2.3.1 Red Biotechnology

The red color is used to define the various biotechnology uses in, or related to, medicine. It includes areas such as designing of organisms to produce antibiotics and vaccines, pharmaceutical drug discovery, development and production, molecular diagnostic techniques and procedures, genetic manipulation to treat diseases, and various regenerative therapies. Red biotechnology is also used to describe pharmacogenomics, which analyzes how genetic make-up affects an individual's response to drugs (Ermak 2013) by correlating gene expression to the drug's efficacy or toxicity (Wang 2010). Pharmacogenomics aims at introducing rational means to match drug therapy with the patient's genotype, thus achieving what is referred to as "personalized medicine," where a certain drug is optimized for each individual's genetic make-up. This is similar to the "individualized nutrition" approach used in the field of human nutrition.

2.3.2 White Biotechnology

This is also referred to as "industrial biotechnology" as it encompasses all the biotechnology applications related to industrial processes. One of the main aims of white biotechnology is to design industrial processes and products that tend to consume less in resources than traditional ones (Shrestha 2010), making them more energy efficient and less environment-polluting. Some of the uses of white biotechnology include: production of different chemicals utilizing microorganisms (industrial fermentation); design and production of detergents; plastics, paper and pulp, and textiles; development of new sustainable energy sources such as biofuels; and the use of enzymes as industrial catalysts to either produce beneficial chemicals or destroy hazardous or polluting chemicals.

2.3.3 Green Biotechnology

As the green color might imply, this type focusses on agriculture. It includes all approaches, activities, and procedures aimed at developing new plant or animal types with desired and improved characteristics, producing biofertilizers and biopesticides, and improving quality of plant and animal products. Both traditional (e.g., selection and crossbreeding) and modern agricultural technologies (e.g., transgenesis) are included in this group. Green biotechnology has been employed to achieve a multitude of objectives. One objective is to develop crop varieties resistant to pests and diseases and that can tolerate extreme or adverse environmental conditions. Green biotechnology is expected to produce more environmentally friendly conditions than traditional agricultural practices. Another objective is to develop agricultural products with improved nutritional attributes and with higher yields. A third objective is to develop plants which can be used for the production of biomedical and industrial substances.

2.3.4 Gray Biotechnology

This term is used to describe all applications of biotechnology which are directly related to the environment. The two main issues related to the environment that can be addressed by biotechnology are biodiversity maintenance and sustenance and contaminants control and removal. Biodiversity maintenance could be achieved through the application of molecular biology to genetic study and analysis of populations and species, which are part of the ecosystems. With regard to pollutants' removal or bioremediation, microorganisms and plants can be used to isolate, degrade, and dispose harmful or hazardous substances and pollutants, for example, heavy metals and hydrocarbons. Gray biotechnology helps to remain with a cleaner and healthier environment.

2.3.5 Blue Biotechnology

This term is used to describe technologies applied for marine and aquatic resources. It aims at exploiting sea and freshwater resources to develop products of industrial interests and values. Although the use of blue technology is relatively rare, a number of applications have already been successfully and effectively tried. Some of these applications include the use of marine resources as suppliers of hydrocolloids and gellings which are widely included in foods and health items. Other useful substances extracted from marine sources include enzymatically active molecules, which can be used in diagnostics research, and pharmacologically or regeneratively active agents. Further potential uses of blue technology relate to the agricultural and cosmetic sectors.

2.3.6 Bioeconomy

This new term has been used to show the link between "economics" and biotechnology. It is also referred to as "biobased economy" and "biotechonomy." The term refers to all economic activities derived from biotechnology. It deals with the investment and the economic output of the different types of biotechnologies described earlier. It is commonly used by biotechnology companies, international organizations, and regional and global development agencies. The flourishing biotechnology industry and its wide applications in the fields of agriculture, medicine, pharmaceuticals, chemical, and energy are a good indicator of the success of the bioeconomic activity.

2.4 POSITION STATEMENTS ON BIOTECHNOLOGY

A number of international organizations around the globe have published statements expressing their positions with regard to biotechnology and its various applications. Some of these organizations fall under the United Nations umbrella, for example, FAO (2000), WHO (2001), UNDP (2001), and UNEP (2014). Other international organizations include OECD (1999), the International Service for the Acquisition of Agri-biotech Applications (ISAAA 2014), Third World Academy of Sciences (TWAS 2000), which includes the Mexican Academy of Sciences, the Brazilian Academy of Sciences, the Chinese Academy of Sciences, and the Indian National Academy of Sciences, International Council for Science Union (ICSU 2014), and International Life Science Institute (ILSI 2014). In the African continent, some societies such as the International Society of African Scientists (ISAS 2001) and the United Nations Commission for Africa (UNECA 2014) have also stated their positions on biotechnology. The Asian Development Bank (ADB 2001) is one of the Asian institutions which expressed their views on biotechnology. In Europe, the Royal Society of London (RSL 2002), the European Commission (EC 2014), the Food Safety Authority of Ireland (FSAI 2013), and the French Academy of Sciences (FAS 2014) have also stated their positions on biotechnology. Some of the societies and associations in the United States that published their positions on biotechnology include the American Medical Associations (AMA 2001), the American Society for Microbiology (ASM 2014), the National Academy of Sciences (NAS 2014), the National Research Council (NRC 2014), the American Society of Plant Biologists (ASPB 2014), and the Federation of Animal Science Societies (FASS 2014). The Canadian Biotechnology Advisory Committee (CBAC 2014) has also stated its position on biotechnology. Other societies expressing their views on biotechnology include the Australia New Zealand Food Authority (ANZFA 2014) and New Zealand Royal Commission (2014).

All these organizations, societies, institutions, and professional bodies support the application of biotechnology in one way or the other. They present their views and positions on biotechnology in a balanced manner in the sense that they acknowledge both the benefits and the risks associated with the application of biotechnology in different areas. The general issues covered and the main views that transpired from these statements are summarized in the following.

- Biotechnology (including GE) has powerful tools that can contribute to sustainable development in different areas at present and in the future.
- Both benefits and risks associated with the application of biotechnological techniques are acknowledged. It is important to weigh both risks and benefits and to determine which ones outweigh the others and base our final decision on the outcome of this comparison. Positions of the above professional bodies indicate that the benefits of GM technologies applied in agriculture would outweigh their risks particularly in developing countries, provided that appropriate controls are in place.
- There is a legitimate concern about the potential risks associated with the use of certain aspects of biotechnology. Risks that affect

human and animal health as well as risks that may lead to environmental degradation or pollution need to be considered with caution and addressed appropriately.
- The issue of labeling of GM foods needs to be addressed carefully. Different views have been expressed in position statements about labeling. Some support mandatory labeling and others believe that there is no need for special labeling of GM foods because there is no scientific justification that makes it necessary.
- There is need for general awareness and education, particularly among policy-makers in developing countries, regarding the opportunities as well as the potential risks of biotechnology.
- Biotechnology research should be encouraged and increased.
- There is need for developing regulatory protocols for GM products and for providing the relevant training for individuals who would be in charge of implementing these regulations.
- Designing and promoting appropriate international standards for biotechnology products would harmonize and streamline international trade.

Detailed position statements of each of the above professional bodies are included in their relevant websites listed in the references section.

REFERENCES

ABC.net. 1999. What is gene technology? http://www.abc.net.au/science/slab/consconf/genes.htm (accessed March 5, 2014).

American Medical Associations (AMA). 2001. AMA report on genetically modified crops and foods. http://www.ama-assn.org (accessed April 12, 2014).

American Society for Microbiology (ASM). 2014. Statement of the American Society for Microbiology on genetically modified organisms. http://www.asmusa.org (accessed April 10, 2014).

American Society of Plant Biologists (ASPB). 2014. Statement on genetic modification of plants using biotechnology. http://www.aspb.org (accessed April 10, 2014).

Andrews, R. 2013. "All about" series. All about genetically modified foods. http://www.precisionnutrition.com/all-about-gm-foods (accessed December 11, 2013).

Asian Development Bank (ADB). 2001. http://www.adb.org (accessed April 12, 2014).

Australia New Zealand Food Authority (ANZFA). 2014. http://www.foodstandards.gov.au/_srcfiles/gm_and_consumer_pub02_00.pdf (accessed April 8, 2014).

Bains, W. 1987. *Genetic Engineering for Almost Everybody: What does it do? What will it do?* Penguin, London, UK.

Betsch, D.F. 2014. DNA fingerprinting in human health and society. http://www.accessexcellence.org/RC/AB/BA/DNA_Fingerprinting_Basics.html (accessed April 20, 2014).

Bitterroot Restoration. 2011. The process of plant genetic engineering. http://www.bitterrootrestoration.com/genetic-engineering/the-process-of-plant-genetic-engineering.html (accessed January 7, 2014).

Canadian Biotechnology Advisory Committee (CBAC). 2014. http://www.cbac-cccb.ca/documents/en/cbac.report.pdf (accessed April 6, 2014).

Citizens Concerned About GM. 2013. gmeducation.org—making sense of science and evidence. A brief history of genetic modification. http://www.gmeducation.org/faqs/p149248-a%20brief%20history%20of%20genetic%20modification.html (accessed December 11, 2013).

Dubey, R.C. 2006. *A Textbook of Biotechnology*. S. Chand, New Delhi, India.

Dunwell, J.M. 2000. Transgenic approaches to crop improvement. *Journal of Experimental Botany*, 51:487–96.

Ermak, G. 2013. *Modern Science and Future Medicine.* Second Edition. CreateSpace Independent Publishing Platform, an Amazon company.

EuropaBio. 2013. What is biotechnology? http://www.europabio.org/what-biotechnology (accessed December 11, 2013).

European Commission (EC). 2014. EC publishes review on bio-safety research of GMOs. http://europa.eu.int/ (accessed April 12, 2014).

FAO. 2000. Agricultural biotechnologies in crops, forestry, livestock, fisheries and agro-industry: FAO Statement on Biotechnology. http://www.fao.org/biotech/fao-statement-on-biotechnology/en/ (accessed January 11, 2014).

FAO. 2001. Glossary of biotechnology for food and agriculture: A revised and augmented edition of the glossary of biotechnology and genetic engineering. FAO Research and Technology Paper 9, Rome. http://www.fao.org/biotech/index_glossary.asp?lang=en (accessed March 10, 2014).

FAO. 2004. *The State of Food and Agriculture.* FAO Agriculture Series No. 35. The Food and Agriculture Organization of the United Nations.

Federation of Animal Science Societies (FASS). 2014. http://www.fass.org (accessed April 6, 2014).

Fernandez-Cornejo, J. 2005. Adoption of GE crops in US: The economics of food, farming, natural resource and rural America. Economic Research Service, USDA, Washington, DC.

Food Safety Authority of Ireland (FSAI). 2013. Genetically modified food GM leaflet 2013 FINAL.pdf. www.fsai.ie-2013 (accessed March 5, 2014).

French Academy of Sciences (FAS). 2014. French Academy of Sciences announces support for genetically modified crops. http://www.academie-sciences.fr/index_gb.htm (accessed April 13, 2014).

Friedman, Y. 2008. *Building Biotechnology: Starting, Managing, and Understanding Biotechnology Companies.* Logos Press, Washington, DC.

Future Food. 2014. http://www.bionetonline.org/english/content/ff_tool.htm (accessed January 7, 2014).

Future For All (FutureForAll.org). 2014. http://www.futureforall.org/bioengineering/biotechnology.htm (accessed January 10, 2014).

Gerstein, M. 1999. *Bioinformatics Introduction*. Yale University. http://www.primate.or.kr/bioinformatics/Course/Yale/intro.pdf (accessed December 16, 2013).

GreenFacts. 2014. Facts on health and the environment. http://www.greenfacts.org/en/gmo/3-genetically-engineered-food/1-agricultural-biotechnology.htm (accessed March 15, 2014).

Hanson, E.D. 1983. *Recombinant DNA Research and the Human Prospect*. American Chemical Society, Washington, DC.

Herdta, R.W. and Nelson, R. 2011. Biotechnology and agriculture: Current and emerging applications, Chapter 1, in Colin, A., Carter, C.A., Moschini, G., and Sheldon, I. (Eds.), *Frontiers of Economics and Globalization*, Volume 10. Genetically Modified Food and Global Welfare. Emerald, Bingley, UK, pp. 1–27.

Institute of Medicine and National Research Council of the National Academies. 2004. *Safety of Genetically Engineered Foods: Approaches to Assessing Unintended Health Effects.* p. 19. The National Academies Press, Washington, DC.

International Council for Science Union (ICSU). 2014. http://www.icsu.org (accessed April 10, 2014).

International Food Information Council Foundation. 2013. Presentation Handout. Food Biotechnology Timeline. http://www.foodinsight.org/foodbioguide.aspx (accessed December 6, 2013).

International Life Science Institute (ILSI). 2014. http://www.ilsi.org (accessed April 12, 2014).

International Service for the Acquisition of Agri-biotech Applications (ISAAA). 2005. Agricultural biotechnology (a lot more than just GM crops). Global Knowledge Center on Crop Biotechnology. ISAAA SE*Asia* Center. Second Printing. http://www.isaaa.org/kc/ (accessed March 10, 2014).

International Service for the Acquisition of Agri-biotech Applications (ISAAA). 2010. Agricultural biotechnology (a lot more than just GM crops). Global Knowledge Center on Crop Biotechnology. http://www.isaaa.org/kc/ (accessed March 10, 2014).

International Service for the Acquisition of Agri-biotech Applications (ISAAA). 2014. Position statements on biotechnology. http://www.isaaa.org/kc/Publications/htm/articles/Position/ama.htm (accessed January 27, 2014).

International Society of African Scientists (ISAS). 2001. Position statement on agricultural biotechnology applications in Africa and the Caribbean. http://www.dca.net/isas (accessed April 12, 2014).

Keener, K., Hoban, T., and Balasubramanian, R. 2014. Biotechnology and its applications. FSR0031. http://fbns.ncsu.edu/extension_program/documents/biotech_applications.pdf (accessed October 6, 2014).

Khan, F.A. 2011. *Biotechnology Fundamentals*. CRC Press, Boca Raton, FL.

Kilner, J.F. and Jones, N.L. 2004. Genetics, Biotechnology and the Future. The Center for Bioethnics and Human Dignity, Exploring the nexus of biomedicine, biotechnology and our common humanity. Trinity International University. http://cbhd.org/content/genetics-biotechnology-and-future (accessed January 27, 2014).

Kumar, H.D. 2003. *Modern Concepts of Biotechnology*. Vikash Publishing House, New Delhi, India.

Lemaux, P.G. 2008. Genetically engineered plants and foods: A scientist's analysis of the issues (part I). *Annual Review of Plant Biology*, 59:771–812. http://arjournals.annualreviews.org/eprint/9Ntsbp8nBKFATMuPqVje/full/10.1146/annurev.arplant.58.032806.103840?cookieSet=1 (accessed December 10, 2013).

Lemaux, P.G. 2009. Genetically engineered plants and foods: A scientist's analysis of the issues (part II). *Annual Review of Plant Biology*, 60:511–59. http://arjournals.annualreviews.org/doi/abs/10.1146/annurev.arplant.043008.092013 (accessed December 10, 2013).

Levidow, L. and Carr, S. 2009. *The GM Food on Trial: Testing European Democracy (Genetics and Society)*. Routledge, Oxford, UK.

Marshall, S. 1994. Genetically modified organisms and food. *Nutrition and Food Science*, 94(1):4–7.

Martínez, V.D. 2011. The colors of biotechnology. BiotechSpain. http://biotechspain.com/en/article.cfm?iid=colores_biotecnologia (accessed December 16, 2013).

National Academy of Sciences (NAS). 2014. http://www.nasonline.org/ (accessed April 6, 2014).

National Research Council (NRC). 2014. http://www.nationalacademies.org/nrc/ (accessed April 10, 2014).

New Zealand Royal Commission. 2014. http://www.royalcommission.govt.nz/ (accessed April 15, 2014).

OECD. 1999. Policy brief. Modern biotechnology and the OECD. http://www.oecd.org/science/biotech/1890904.pdf (accessed January 12, 2014).

OECD. 2013. Key biotechnology indicators. www.oecd.org/sti/biotechnology/indicators (accessed April 20, 2014).

Oldham, M. 2014. The future of biotechnology. University of South Florida. http://www.the-elite.net/ClarkPage/Biotech/Future.html (accessed January 15, 2014).

Pometto, A., Shetty, K., Paliyath, G., and Levin, R.E. 2005. *Food Biotechnology*. Second Edition. CRC Press, Boca Raton, FL.

Purohit, S. 2004. *Biotechnology: Fundamentals and Applications*. Agrobios, India.

Purohit, S. 2005. *Agricultural Biotechnology*. Agrobios, India.

Rees, A. 2006. *Genetically Modified Food: A Short Guide for the Confused*. Organic Consumer Association. Pluto Press, London, UK.

Roy, M.J. 2011. *Biotechnology Operations: Principles and Practices*. CRC Press, Boca Raton, FL.

Royal Society of London (RSL). 2002. http://www.royalsoc.ac.uk (accessed April 12, 2014).

Secretariat of the Convention on Biological Diversity. 1992. Convention on Biological Diversity. http://www.biodiv.org/convention/articles.asp (accessed March 10, 2014).

Secretariat of the Convention on Biological Diversity. 2000. Cartagena Protocol on Biosafety to the Convention on Biological Diversity: Text and annexes. Montreal, Canada. http://www.biodiv.org/biosafety/protocol.asp (accessed March 12, 2014).

Shrestha, R. 2010. The importance of biotechnology in today's time. Aakhayan. A Chapter Scripting Life. http://raunakms.wordpress.com/2010/06/04/the-importance-of-biotechnology-in-todays-time/ (accessed December 28, 2013).

The Advisory Board for the Research Councils and the Royal Society. 1980. Biotechnology, Report of a Joint Working Party of the Advisory Council on Applied Research Development, HMSO, London.

The American Chemical Society (ACS). 2014. Biotechnology is the science of the future. http://www.acs.org/content/acs/en/careers/whatchemistsdo/careers/biotechnology.html (accessed January 10, 2014).

The American Chemical Society. 2014. Biotechnology. http://www.acs.org (accessed March 15, 2014).

Thieman, W.J. and Palladino, M.A. 2008. *Introduction to Biotechnology*. Second Edition. Pearson/Benjamin Cummings, London, UK.

Third World Academy of Sciences (TWAS). 2000. Seven science academies urge expanded use of crop biotechnology. http://www.twas.org (accessed April 10, 2014).

Treohan, A. 1993. Future of biotechnology. Woodrow Wilson Biology Institute. http://www.accessexcellence.org/AE/AEPC/WWC/1993/future.php (accessed January 10, 2014).

UN Convention on Biological Diversity, Art. 2. 1992. http://www.cbd.int/convention/text/ (accessed March 15, 2014).

UNDP. 2001. Report supports biotechnology. http://www.undp.org (accessed January 15, 2014).

UNEP. 2014. http://www.unep.org (accessed January 20, 2014).

United Nations Commission for Africa (UNECA). 2014. http://www.uneca.org/harnessing (accessed April 14, 2014).

Wang, L. 2010. Pharmacogenomics: A systems approach. *Wiley Interdisciplinary Reviews: Systems Biology and Medicine*, 2(1):3–22.

WHO. 2001. Safety assessment of foods derived from genetically modified microorganisms. http://www.who.int/foodsafety/publications/biotech/ec_sept 2001/en/ (accessed January 11, 2014).

WHO. 2014. Food safety. Frequently asked questions on genetically modified foods. http://www.who.int/foodsafety/areas_work/food-technology/faq-genetically-modified-food/en/ (accessed October 16, 2014).

wiseGEEK. 2014. http://www.wisegeek.org/what-is-dna-fingerprinting.htm (accessed April 10, 2014).

Zaid, A., Hughes, H.G., Porceddu, E., and Nicholas, F. 2001. *Glossary of Biotechnology for Food and Agriculture—A Revised and Augmented Edition of the Glossary of Biotechnology and Genetic Engineering*. FAO, Rome.

SUGGESTED REFERENCES

Ashok Ganguli, A. 2009. *Biotechnology Fundamentals and Applications*. Oxford Book Co., Oxford, UK.

British Medical Association. 1999. *Biotechnology, Weapons and Humanity*. London BMJ Bookshop.

Bud, R. 1994. *The Uses of Life. A History of Biotechnology*. Cambridge University Press, Cambridge.

Carpenter, J.E. and Gianessi, L.P. 2001. *Agricultural Biotechnology: Updated Benefits Estimates*. National Center for Food and Agricultural Policy, Washington, DC.

Heldman, D.R., Hoover, D.G., and Wheeler, M.W. 2010. *Encyclopedia of Biotechnology in Agriculture and Food*. CRC Press, Boca Raton, FL.

Lee, B.H. 1996. *Fundamentals of Food Biotechnology*. VCH, New York.

Messina, L. (Ed.). 2000. *Biotechnology*. H.W. Wilson, New York.

Wieczorek, A.M. and Wright, M.G. 2012. History of agricultural biotechnology: How crop development has evolved. *Nature Education Knowledge*, 3(10):9.

3

Applications of Genetic Modification at the Laboratory and Greenhouse Levels

3.1 GENETIC MODIFICATION OF PLANTS

Although the term "genetic modification" (GM) is widely used today to denote modern biotechnology, or more specifically "genetic engineering" (GE), it can be argued that any human intervention to change the natural genetic make-up of a living organism is regarded as GM. As the records show, thousands of years back (as reflected in Section 2.2.1), farmers and scientists have been actively engaged in continuous efforts to improve farm plants and yard animals both quantitatively and qualitatively. Before the advent of new biotechnologies during the 1970s, a number of techniques have been discovered and used for GM. The following sections highlight and discuss some of these techniques, conveniently referred to by the Institute of Medicine and National Research Council of the National Academies (2004) as "techniques other than GE" or "techniques not involving GE."

3.2 NONGENETIC ENGINEERING TECHNIQUES (CONVENTIONAL BIOTECHNOLOGY)

3.2.1 Simple Selection

As the term "simple selection" implies, this method does not involve complex or many steps, and that what is basically done is choosing from among a population of plants or animals based on exhibiting desirable or superior characteristics. Since the beginning of agriculture, 8000–10,000 years ago, farmers have been altering the genetic make-up of the crops they grow (ISAAA 2005). Early farmers used the simple selection method and it is still in use today. Together with domestication of wild animals, this method is regarded as the oldest and easiest method of GM. Farmers would inspect and identify which of their plants (or animals) look better than the others with respect to the desired characteristics, for example, higher yields, faster growth, superior sensory properties, pest and disease resistance, larger seeds, or sweeter fruits. The identified plants are then selected to be used for planting during the next growing season, and their seeds were stored for further propagation of the species. Those plants which were not chosen for propagation and which do not possess the desired characteristics would be used for food or thrown away. The process of sowing the seeds of the selected plants with the desired traits is continued and is expected to produce plants having all or most of the desired traits. Further selection could be carried out in order to produce a higher number of plants with the desired traits or plants with a higher number of the desired traits. Continued sowing of the seeds from the selected superior plants is expected to create a new set of plants which will manifest the desired traits. Over the years, this process would lead to an increase in the numbers of plants with the desired characteristics and subsequently results in dominance of the superior genotype. Modern biotechnologies, for example, marker-assisted selection (discussed in Section 3.2.2), have been used and helped in many advances of this old technique. This advanced technique uses molecular analysis to identify plants with the potential to show the desired features. It proved to be more precise and faster in identifying plants with the desired features.

3.2.2 Marker-Assisted Selection

Simple selection, described in the previous section, involves selection of plants by breeders or farmers based on the phenotype of the plant, that is, its visible physical and measurable characteristics resulting from the interaction between the genetic make-up of the plant (genotype) and the environment. Due to the observed shortcomings of the simple selection technique, for example, its slow development, the effect of the environment, and its high cost, plant breeders started using a more advanced technique known as "molecular marker-assisted (or marker-aided) selection," abbreviated as MAS. It is an indirect process in which a desired trait is selected on the basis of a molecular marker linked to it, rather than the trait itself. Molecular markers are short strings or sequence of nucleic acid which makes up a segment of the DNA. The markers are positioned adjacent to the DNA of the gene of interest. It is expected that the marker and the gene stay close together as new generations of plants are produced. This type of association, which helps scientists to predict whether the plant will have the desired gene or not, is known as "genetic linkage." The presence of the marker would indicate the presence of the desired gene. A map of the genes and the markers on specific chromosomes can be designed when breeders are able to locate the position of each marker on the chromosomes. This map helps to show where genes and markers are located and how far they are from other genes. Currently, MAS is widely used in breeding of most crops, provided that the gene and the marker for a specific trait are identified.

3.2.3 Crossing

"Crossing" is a more advanced technique of plant modification than simple selection. It is carried out under controlled conditions and is conducted by plant breeders who have the required knowledge, skill, and experience of the process. The plant breeder removes pollen from the flowers of one plant and rubs it onto the pistil of a flower of a sexually compatible plant. The resulting progeny known as a "hybrid" is expected to have genetic material and to exhibit characteristics from both parents. At maturity, the hybrid plant may be used as a parent for further propagation of the species. Hybrid seeds are developed by the hybridization or crossing of parent lines that are "pure lines" produced through inbreeding. Pure lines are plants that "breed true" or produce sexual offspring that closely resembles their parents (ISAAA 2005). When pure lines are crossed, a uniform population of seeds (referred to as F1 hybrid seeds) can be developed with predictable and desired characteristics.

The technique of crossing usually aims at combining desired features of two plants. Naturally, it is not possible to find a plant that has all the desired features, and at the same time, we do not find a plant with features that are all undesirable. In other words, it has been observed that all plants possess both desirable, useful features, for example, high yield, disease or insect resistance, and quick maturing, and undesirable ones, for example, low yield and susceptibility to insects or diseases. Plant breeders usually work to combine the useful features of two compatible plants. The technique of crossing has been successfully used to achieve this purpose. As expected, the process of recombining genes and characteristics through crossing is random in nature. This means that breeders need to carry out large numbers of hybrid progeny to produce and be able to find out the ones with more desirable features and a minimum number of undesirable ones. Despite its shortcomings, crossing is regarded as an effective method of plant breeding that could further be improved and become more efficient.

3.2.4 Interspecies Crossing

Interspecies crossing is the type of crossing that can take place among closely related plant species. It may occur naturally or through human intervention. It can also happen naturally, though rarely, among more distant relatives. The crossing leads to development of hybrid progeny carrying characteristics of both parents.

3.2.5 Embryo Rescue

Under natural conditions, interspecies crossing can take place among different plant species. In some cases of natural interspecies crossing, the resulting hybrid embryo is not able to complete its full life cycle to maturity. Scientists observed this natural phenomenon and decided that human intervention is needed in such cases to help save the embryo and to complete the process to allow the plant reach its maturity. This process is referred to as "embryo rescue" or "embryo culture." In embryo rescue/culture, the embryo is removed before seed abortion occurs and is subsequently grown outside the parent plant to produce a new plant. Embryo rescue involves the culture of immature embryos of plants in a special medium, which contains the nutrients required for growth and development, to prevent abortion of the young embryo and to support its germination. This is used routinely in breeding parental lines having different or incompatible genomes, such as in introducing important traits of wild relatives into cultivated crops. The intervention of the plant breeder comes after the natural pollination takes place, but before the embryo seizes to develop. The plant embryo is removed and placed in a tissue culture (TC) environment in which a growth medium and development hormones are supplied, making it affordable to the embryo to complete its development. Embryo rescue enables crosses to be made between species which would not normally be sexually compatible. Such *in vitro* embryo rescue techniques are often used to rescue plant embryos from aborting progeny seeds that result when two distantly related plants, for example, two species, are crossed together. Such "wide crosses" are often desirable to transfer genetic traits from the secondary and tertiary gene pools, that is, crop wild relatives, to the cultivated primary gene pools of crop plants (FAO 2005). This technique has been successfully used in breeding of a number of crop species, a typical example of which is the production of triticale, the hybrid species resulting from a cross between wheat and rye.

3.2.6 Somatic Hybridization

Somatic hybridization, also known as "cell fusion," is a nonsexual technique of genetic combination. It involves the fusion of somatic protoplasts which are isolated from different plants to produce a heterokaryon that later develops into a hybrid variety. The first step of the process is the removal of the plant protective cell walls with the help of enzymes capable of degrading the polysaccharides present in the cell wall, for example, pectinase, cellulase, and hemicellulase. These enzymes digest the cell walls leaving the protoplasts. Protoplasts isolated from different sources are pooled, and in the presence of a fusogen, they fuse together to form a heterokaryon. The fusogen could be a chemical substance such as potassium or sodium chloride solution or an electric current. The somatic hybrid, resulting from the protoplast fusion, contains the nuclei of the fused species and subsequently their genetic material. The hybrid then grows into cells that are capable of regenerating the entire plant if appropriate conditions are maintained.

3.2.7 Somaclonal Variation

Somaclonal variation is the term used to describe the high variability observed in plants that have been regenerated by plant TC. The main reason for this type of variation is the chromosomal rearrangement leading to spontaneous mutations. It is a phenotypical variation of genetic origin, that is, it is a variation in the chromosomes which is inherited by future plant generations or of epigenetic origin, where it is considered a transitory variation caused by the physiological stress suffered during the *in vitro* cultivation in TC. Somaclonal variation is regarded as an important source of genetic diversity in plants.

For many years, plants regenerated from TC sometimes had novel features. It was not until the 1980s that two Australian scientists thought that this phenomenon might provide a new source of genetic variability, and that some of the variant plants might carry attributes of value to plant breeders (Larkin and Scowcroft 1981). Through the 1980s, plant breeders around the world grew plants *in vitro* and scored regenerants for potentially valuable variants in a range of different crops. New varieties of several crops, such as flax, were developed and commercially released (Rowland et al. 2002). Molecular analyses of these new varieties were not required by regulators at that time, nor were they conducted by developers to ascertain the nature of the underlying genetic changes driving the variant features. Somaclonal variation is still used by some breeders, particularly in developing countries, but this nongenetic engineering technique has largely been supplanted by more predictable GE technologies.

Plant regeneration from cell cultures is central to the application of gene transfer techniques such as biolistics and *Agrobacterium*-mediated transformation. Somatic variation can be beneficial in crop improvement, especially on traits for which somaclonal mutants can be enriched during *in vitro* culture, including resistance to disease pathotoxins, herbicides, and tolerance to environmental or chemical stress, as well as for increased production of secondary metabolites (FAO 2005).

3.2.8 Mutation Breeding

Early plant breeders used to search for individual plants that show desirable traits. Some of these traits occasionally arise spontaneously through a process called "mutation." The natural rate of mutation is very slow and unreliable to produce all plants that breeders would like to see (ISAAA 2005). Through further research, plant breeders discovered that they could greatly increase the number of these variations or mutations by exposing plants to x-rays and certain chemical substances. Mutation breeding is the process or technique in which plants or seeds are exposed to ionizing radiation, for example, x-rays and gamma-rays, or treated with chemical mutagens, for example, sodium azide and ethyl methanesulfonate. Both the ionizing radiation and the chemicals are regarded as mutagenic agents because they lead to production of plant mutants. Mutagenesis results in random changes in the plant or seed DNA sequence. The aim of mutation breeding is to produce mutants that have desirable traits to be bred with other cultivars. The mutagenic agent (radiation or chemical) dose can be controlled in a way that would be enough to produce the desired mutation, but at the same time not too high to destroy the plant. It is only through adjusting the mutagenic agent dose that the effect on the plant can be controlled. Similar to what happens in somaclonal variation, a high percentage of mutations taking place due to this technique is both deleterious and undesirable. To overcome this undesirable outcome, a large number of plants or seeds are mutagenized, allowed to develop to reproductive maturity, and progeny are derived. The progeny are then evaluated for the presence of desirable new traits. When breeders observe a potentially useful mutation, they work to minimize the deleterious mutations and other undesirable features of the mutant. Changes in the genetic material of the plant or seeds, beyond the mutations that resulted in the desired superior traits, are still expected to take place. This shows the random nature of this technique, which may lead to the occurrence of mutations other than those resulting in desired useful traits. Due to this reason, it is not an easy task to regulate induced-mutation crops for food or environmental safety, something that is observed in most countries. Regardless, the technique continues to be used, and by the year 2007, the estimated number of mutagenic plant varieties released worldwide was 2540. The lack of records, proper identification, and labeling of these plant varieties make it difficult to know how many of them are currently being used in agriculture worldwide. Examples of plants that were produced via mutation breeding include wheat, barley, rice, potatoes, soybeans, and onions (ISAAA 2005).

3.2.9 Plant TC and Micropropagation

Naturally, plants reproduce through forming seeds by sexual reproduction, in which egg cells present in the flowers are fertilized by pollen from the stamens of the plants. Each of these sexual cells contains genetic material in the form of DNA. The process of sexual reproduction involves combining genetic material from both parents. Usually, this process happens in new and unpredictable ways, creating unique organisms. If this process of sexual reproduction is aimed at producing plants or seeds with desirable characteristics, breeders may need to carry out careful greenhouse work and wait for a number of years to achieve the desired goal. The problems of unpredictability and the long time needed to develop plants with desired characteristics lead plant breeders to search for alternative ways to sexual reproduction. Through dedicated efforts and hard work, scientists were able to come up with plant propagation methods other than sexual reproduction. These efforts led to the introduction of the technique "plant TC." This technique has been in use for more than 40 years.

Plant TC refers to a group of techniques developed to maintain or propagate individual plant cells, tissues, or organs using a nutrient culture substrate of specific known composition, carried out under sterile conditions (Bhojwani and Razdan 1996). The plant TC technique applied in the production of clones of plants is known as micropropagation. Plant TC is also used in plant breeding, as reported by Faraz (2014).

TC technique has been developed by scientists to shorten the time plants usually take to reproduce through sexual means, that is, using flowers and seeds to create the next generation of plants. Plant TC is an *in vitro* practice used to propagate plants under sterile conditions, often to produce clones, that is, identical genetic copies, of the parent plant. In TC, single cells or small pieces of tissue are removed from a plant that has desirable traits. The isolated plant part is placed in a liquid or solid specially formulated nutrient growth medium along with plant hormones (Figure 3.1). Under

Figure 3.1 Plant TC process. Scientist performs an experiment. Forceps are holding a small piece of leaf up to a microfuge tube. (From © iStockphoto. With permission.)

the right conditions, an entire plant can be regenerated from a single cell, which is an exact copy of the parent plant from which tissue was taken. In TC, no GE steps are carried out. The genetic structure of the regenerated plant remains the same. The techniques of TC can be applied to produce plants of superior quality. During the *in vitro* growth, plants can be "prepared" for optimal growth after transfer to the *ex vitro* conditions.

Plant TC has multiple benefits. These benefits include the potentially unlimited multiplication of superior plant lines or elite individuals, avoidance of contamination with pathogens, and production of true-to-type multiplication material of desirable plant lines suited for different storage conditions. It is thus considered as an important and useful technology for developing countries to produce disease-free, high-quality plant material and for shortening the period required for producing uniform plants. To date, many developing countries opted to use this technique and have benefited from it. In contrast, the major limitation of the application of this technology is the need for technically skilled personnel and some essential equipment. It is also labor-intensive and time-consuming and can be expensive. This may limit their commercial application to a few high value-added crops, which deserve investing in them. In spite of these limitations, a wide range of plants in developing countries have been produced using TC technique. Some examples of these plants are oil palm, plantain, pine, banana, date, eggplant, pineapple, cassava, yam, sweet potato, tomato, and jojoba (ISAAA 2005).

There are a number of types of TC, depending on the part of the plant (explant) used. One type, known as "anther culture," is a TC method used to develop enhanced plant varieties in a short period of time. Anther culture involves the aseptic culture of immature anthers to generate fertile haploid plants from microspores. The pollen within an anther of a flower contains half the amount of the genome, that is, haploid, which spontaneously doubles to become diploid during culture. Doubling of the genome will allow the expression of recessive traits, which were suppressed, masked, or undetected in routine plant breeding. Anthers are placed in a special medium, and immature pollen within the anther divides and produces a mass of dividing cells known as "callus." Healthy calli are picked and placed in another medium to regenerate and produce shoots and roots. Stable plantlets are allowed to grow and mature in the greenhouse. The plants with desired characteristics are then selected from the regenerated.

The production of haploid plants through anther culture is widely used for breeding purposes, as an alternative to the numerous cycles of inbreeding or backcrossing usually needed to obtain pure lines in conventional breeding. The success achieved with anther culture has led to the development of microspore culture. This involves the isolation of the microspores from the anthers, culturing them in specialized media, and subsequent regeneration of fertile homozygous plants (FAO 2005).

Currently, *in vitro* anther culture is widely used to improve the characteristics of some vegetable crops such as asparagus, sweet pepper, eggplant, watermelon, and *Brassica* vegetables. It is also used in cereal crop improvement both as a source of haploids and new genetic variation (FAO 2005).

Micropropagation is the use of TC methods to propagate plants. It aims to produce, maintain, and mass propagate disease-free, superior quality planting material and achieve rapid production of many uniform plants. Millions of new (clonal) plants can potentially be derived from a single plant using micropropagation. Basically, tissue from a plant (explant) is isolated to create a sterile culture of that species *in vitro*. When that culture is stabilized and is developing well *in vitro*, multiplication of the tissue or regeneration of entire plants can be carried out. Although in most cases, shoots and leaf pieces are used, cultures can be generated from many different plant tissues. In this technique, actively dividing meristem cells (a zone of cells with intense divisions of about 0.1 mm in diameter, which are situated at the top of buds and extremities of roots) are placed in a special nutrient medium and treated with plant hormones to produce many similar sister plantlets. Usually, the meristem divides faster than disease-causing virus. This process helps the propagation of clean materials, leading to the development of large numbers of uniform plantlets during a short period of time. In the past, micropropagation has been used successfully in the removal of viruses from many plants. Now, it is used to eradicate many viral diseases to produce clean and uniform planting materials in a range of crops, for example, oil palm, banana, date, eggplant, tomato, cassava, and sweet potato.

Some of the successful applications of micropropagation include:

- Production of plant species which proves to be difficult to grow from seed
- Production of disease-free plant material
- Production of genetically uniform plant material, that is, clones
- Production of some horticultural plants, for example, some rose varieties, throughout the year
- Production of plant culture systems to be used for genetic transformation, for example, to introduce disease resistance (FAO 2005)

Micropropagation is among the most widely used plant biotechnologies worldwide. Twenty-one African countries, ten Asian countries, nine East European countries, nine Latin American countries, and eight Near East countries have been reported to benefit from it (FAO 2005).

TC can also be used at the cellular level in a process known as "cell selection." In the cell selection process, plant scientists isolate some cells from chosen plants which have superior agricultural characteristics. These cells are then grown in a suitable culture which contains nutrients and hormones necessary for growth and development. During the first stages of development, genetic homogeneity throughout the whole population is usually observed. This homogeneous nature can change spontaneously or be induced through application of mutagenic agents. Cell selection is carried out and would be based on exhibition of a desired phenotypic variation. The selected cells are then allowed to develop into whole plants. Several commercial crop varieties have been developed using cell selection, including varieties of soybeans (Sebastian and Chaleff 1987), canola (Swanson et al. 1988), and flax (Rowland et al. 1989, 2002).

The main advantages of TC over conventional breeding are low cost and short time needed to complete the process and achieve desired goals. This is due to the fact that TC is carried out in a petri dish and the whole process is completed in a short time when compared with breeding in the field throughout the whole growing season.

3.3 UNINTENDED EFFECTS FROM CONVENTIONAL BIOTECHNOLOGY

Unintended or unexpected effects do not only occur in GE technologies, but they can also be introduced through traditional biotechnologies, for example, plant breeding. However, in such situations, the plant breeder is able to perform a selection process through which the lines that express undesirable traits are discarded and eliminated from further consideration. In contrast, the best plants, which only show desirable characteristics, would be selected and kept for future propagation or for commercial purposes.

Foods with undesirable traits, some of which exhibit negative health effects in humans, can be produced through conventional crop production practices. Some crops produce antinutritional substances, toxic components, or substances that cause allergic reactions in humans. These substances are produced naturally and as part of the physiological processes of these plants. Examples of these substances include: various types of steroidal glycoalkaloids produced by crops belonging to the family Solanacea, such as potato and tomato, and glucosinolates produced in canola. The approach successfully followed by plant breeders is to monitor the level of these potentially hazardous substances with the aim of eliminating any breeding line which exhibits higher than normal levels of the harmful substances.

3.3.1 Mutations

Mutations, defined as any change in the basic sequence of DNA, either can occur naturally or be man-induced. Both modes are known for production of new crop varieties. Most mutations are undesirable because they lead to deleterious effects. In some cases, mutations can result in desirable traits and may be selected for breeding. Plant breeders can induce mutations, by using ionizing radiation or mutagenic chemicals such as ethyl methanesulfonate, to affect random changes in DNA. Foods produced from plants with induced mutations have been accepted and consumed without unexpected negative health effects. Furthermore, there is no documented evidence of removal of foods produced through mutant varieties from the market due to adverse health incidents.

3.4 GE TECHNIQUES

3.4.1 Overview

The general principles of the different techniques of GE are basically the same for any type of plant. The time duration needed to carry out all the steps varies. This is controlled by a number of factors including the plant species, the gene(s) to be transferred, and the availability of other resources required for the process. It is estimated that it may take between 6 and 15 years for the release of a newly developed GE hybrid and its growing on large scale in the field (Bitterroot Restoration 2014).

The two terms GM and GE are sometimes used interchangeably or synonymously, although they do not exactly mean the same thing. GM is a general term that describes a change in the genetic material of a living organism through any method, including conventional breeding. On the contrary, GE involves a change in the genetic material using techniques that allow the direct transfer or removal of genes from one living organism to another.

Different terms are used to describe GE technology. One of these terms is "recombinant gene technology," which simply means recombining or mixing of genes from different sources. This description stems from the fact that genes from different living organisms are brought together in a manner that does not exist naturally.

The physical characteristics and the appearance of a living organism result from different protein combinations that occur in living cells. The types and nature of proteins synthesized in living cells are dictated by the genetic material of the organism. Genes send chemical messages that give orders to the cells to perform their functions by making different types of proteins including enzymes. Thus modifying the genetic make-up of an organism through changing a gene would be expected to change the traits of the living organism.

An initial step in the GE process is identifying the gene that controls the desired trait. This gene may be obtained from another member of the same plant or animal species to be modified or from a different type of living organism, being plants, animals, or bacteria. This gene is then inserted into the plant or animal to be modified to acquire the desired trait. Gene technologists use a "cutting-copying-pasting" approach to transfer genes from one organism to another. Usually, bacterial enzymes are used that recognize, cut, and join DNA at specific locations, acting as molecular "scissors-and-tape." However, the selected gene is copied billion-fold, with the result that the amount of the original genetic material in the modified organism is immeasurably small (Institute of Food Science and Technology [IFST] 2008).

It is now possible to modify one single trait or a number of traits during one operation. "Stacked-trait organism" is the term used to describe an organism that has multiple traits genetically modified. A typical example for crops with more than one trait changed is Roundup Ready/YieldGard corn, developed and produced by Monsanto. This type of corn is resistant both to herbicides and to insects.

GM can be achieved through addition or deletion of genes in living organisms. Genes may be taken from an animal, a plant, or a microorganism. When gene transfer is carried out artificially between living organisms that could be conventionally bred, the process is known as cisgenesis, where cis means same or of similar nature. In contrast, if gene transfer is done among living organisms from different species, the term used to describe the process is transgenesis and the resulting genetically modified organism (GMO) is referred to as "transgenic." The latter type of gene transfer is considered as a form of horizontal gene transfer.

Sometimes, the genetic material does not move easily from one organism to another. In such cases, "vehicles," for example, plasmids (small rings of bacterial DNA), may be used; alternatively, some plant cells may be transformed by "shooting" small particles, incorporating the new DNA into the target cell using a special type of gun, known as the "Gene Gun." The modified cell can then be used to regenerate a new organism (IFST 2008).

Traditional or conventional methods of GM in plants and animals have been practiced and developed for centuries. Those traditional methods involve random mixing of genes and were mainly applied for living organisms of related nature. These methods are also not precise in nature and require long periods of time to make a change or achieve a desired goal. On the contrary, GE techniques are not random, in which specific genes controlling desirable traits are manipulated and positioned to add a desired trait.

3.4.2 From the Laboratory to the Greenhouse

The process of GE starts in the biotechnology laboratory. A number of steps are carried out in the laboratory before the developing plant is transferred to the greenhouse for hardening. Transgenic seedlings are first developed in petri dishes containing the required nutrients for plant growth (Figure 3.2). The development of the seedlings is regularly inspected and monitored (Figure 3.3). The seedlings are then allowed to sprout and develop shoots (Figure 3.4). Further development of the seedlings to produce leaves is also carried out in the laboratory (Figure 3.5). The next step is removal of the GE plants, which have been developed in the laboratory, to the greenhouse (Figure 3.6). The greenhouse allows the GE plants to develop under controlled conditions of temperature, humidity, and light. The greenhouse stage allows close monitoring and inspection before the GE plants are planted in the field.

3.4.3 Techniques of GE

A number of techniques used to genetically modify living organisms, including foods, have been reported in the literature. Murnaghan (2013) describes some of these techniques. The following sections discuss some of the techniques developed to modify living organisms and which are widely used today.

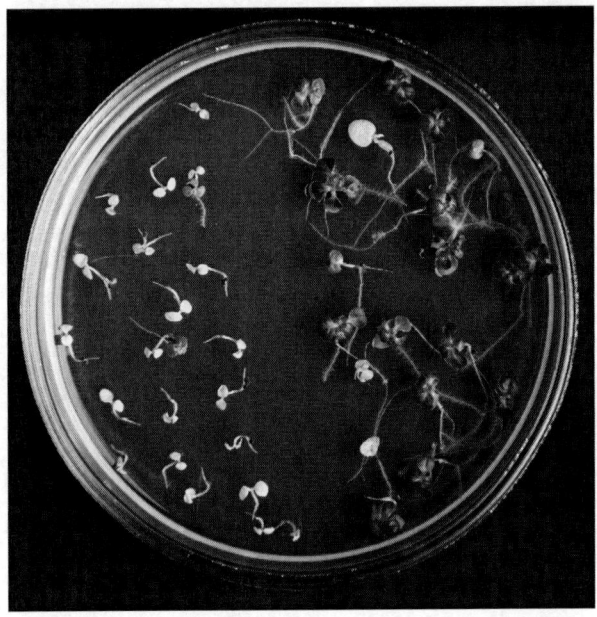

Figure 3.2 Transgenic plants in petri dishes in the biotechnology laboratory. (From Shutterstock Inc, New York, NY, USA. http://www.shutterstock.com/. With permission.)

Figure 3.3 Transgenic inspection of GE plants in the biotechnology laboratory. (From Shutterstock Inc, New York, NY, USA. http://www.shutterstock.com/. With permission.)

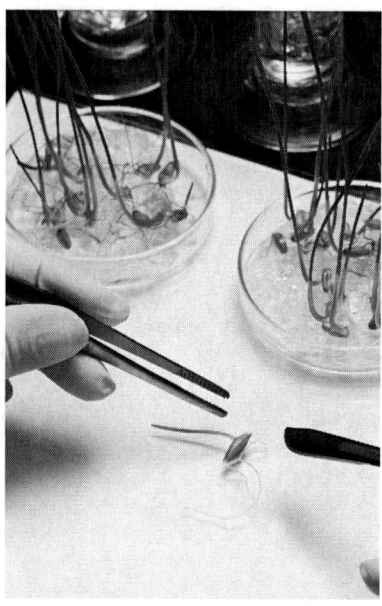

Figure 3.4 GE seedlings allowed to sprout in the biotechnology laboratory. (From Shutterstock Inc, New York, NY, USA. http://www.shutterstock.com/. With permission.)

3.4.3.1 Microbial Vectors

Microorganisms, mainly bacteria and viruses, have been used successfully to carry and transfer genetic material and the desired traits into plants and animals. This GE technique is known as microbial vectors. A vector is a DNA molecule used as a vehicle to artificially carry foreign genetic material into another cell, where it can be replicated and/or expressed. A vector containing foreign DNA is termed recombinant DNA (rDNA). The purpose of a vector employed to transfer genetic information to another cell is typically to isolate, multiply, or express the transgene in the target cell. One type of vectors, known as expression vectors or expression constructs, carries out the function of expressing the transgene in the target cell and generally has a promoter sequence that helps expression of the transgene. Another type of vectors, with the capability of only being transcribed but not translated, is known as transcription vectors. The latter is a simpler type of vectors, which, unlike the expression vectors, can only be replicated but not expressed in the target cell. The two

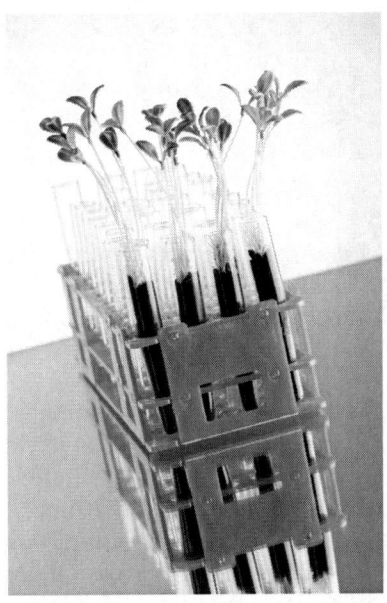

Figure 3.5 GE seedlings allowed to develop in the biotechnology laboratory. (From Shutterstock Inc, New York, NY, USA. http://www.shutterstock.com/. With permission.)

Figure 3.6 GE plants growing in the greenhouse. (From Shutterstock Inc, New York, NY, USA. http://www.shutterstock.com/. With permission.)

main types of microbial vectors are known as bacterial carriers and viral carriers. Insertion of a vector into the target cell is usually referred to as transformation if it is accomplished through bacterial cells, whereas the insertion of a viral vector is referred to as transduction.

3.4.3.1.1 Bacterial Carriers

Bacterial carriers have been found to be very effective for delivering DNA. This technique is sometimes referred to as "*Agrobacterium*-mediated gene transfer" because the type of bacteria used belongs to the genus *Agrobacterium*. The bacterium *Agrobacterium* can infect plants, which makes it a suitable carrier for delivering DNA to a new organism (Figure 3.7).

Agrobacterium tumefacien is a species of bacteria found in the soil and is known for causing crown gall disease on some plant species. When it infects a host plant, the bacteria transfer a portion of its own DNA into the plant cell. The transferred DNA is stably integrated into the plant DNA. This results in marked changes in the plant which would then be able to read and express the transferred genes as if they were its own. The transferred genes help the development of a crown gall. During the early 1980s, scientists were able to develop strains of *Agrobacterium* devoid of the disease-causing genes, but still had the ability to attach to susceptible plant cells and transfer DNA. By substituting the desired DNA for the disease-causing DNA, it was possible to obtain new strains of *Agrobacterium* that

Figure 3.7 Bacterial colony picking for DNA cloning. (From Shutterstock Inc, New York, NY, USA. http://www.shutterstock.com/. With permission.)

deliver and stably integrate specific new genetic material into the cells of target plant species. The process involves placing the bacterium in a special solution which makes its cell wall more porous and permeable. The desired selected gene is inserted into a bacterium extrachromosomal DNA molecule (a plasmid) and dropped into the solution. To facilitate the entrance of the plasmid into the bacterium, the solution is heated. The genetically altered bacterium (recombinant) is allowed to recover (is "rested") and grow and, depending on the plasmid, make extra copies of the new gene. Finally, the bacterium is allowed to infect the target plant cells so that it can deliver the plasmid and the new gene into the cells to be transformed. Through regenerating the transformed cell into a whole plant, all cells in the progeny will also carry and be able to express the inserted genes. *Agrobacterium* is regarded as the naturally occurring GE tool responsible for most of the GE crops produced on commercial scale.

3.4.3.1.2 *Viral Carriers*
This technique is based on the fact that some viruses can invade target cells, but not cause cell damage or death. Viruses have the ability to efficiently transport their genetic material inside the cells they infect. The process of delivering genes by a virus is called transduction, and the infected cells are described as transduced. It has been found that a virus can be used as an effective carrier for modifying living organisms. The virus chosen will be one that does not cause any kind of disease or death. Through the addition of the selected DNA to the carrier virus genome, the virus is allowed to infect the target plant. Once the virus invades the cell and replicates making copies of itself, the chosen DNA would automatically be added to the targeted cell.

3.4.3.2 Electroporation
In principle, electroporation or electropermeabilization is a large increase in the electrical conductivity and permeability of the cell plasma membrane caused by an externally applied electric field. It is usually used in molecular biology as a way of introducing some substance into a cell, such as loading it with a molecular probe, a drug that can change the cell's function, or a piece of coding DNA (Sugar and Neumann 1984).

Purves et al. (2001) describe electroporation as a "mechanical method used to introduce polar molecules into a host cell through the cell membrane. In this procedure, a large electric pulse temporarily disturbs the phospholipid bilayer, allowing molecules like DNA to pass into the cell." The concept of electroporation capitalizes on the relatively weak nature of the phospholipid bilayer's hydrophobic/hydrophilic interactions and its ability to spontaneously reassemble after disturbance (Purves et al. 2001). Thus, a quick voltage shock may disrupt areas of the membrane temporarily, allowing polar molecules to pass; however, the membrane may then reseal quickly and leave the cell intact.

Electroporation allows cellular introduction of large and highly charged molecules such as DNA, which would normally find it difficult to pass through the hydrophobic "phospholipid" bilayer. Electrical impulses of high field strength are used to reversibly permeabilize cell membranes to facilitate uptake of large DNA molecules.

In using electroporation to create GMOs, the target cells are prepared and immersed in a special buffer solution together with the desired foreign DNA. A short intense electric shock is then transmitted through the solution. This electric shock is expected to cause little pores or tears in the cell walls, which consequently allows the new genetic material to enter the nuclei. Afterwards, the damaged cells are placed in a different solution that would help repair their damaged walls. This process helps to lock up the donor DNA inside the cell. The selected DNA is incorporated into the host as a plasmid, which is then incorporated into the plant DNA to give the host this new gene. Electroporation is limited by the poor efficiency of most plant species to regenerate from protoplasts.

The success of the electroporation process depends greatly on the purity of the solution used, particularly on its salt content. High salt concentration of the solution might cause an electrical discharge which often lowers the efficiency of the process.

The equipment used to carry out this process is known as "electroporator" (Figure 3.8). One company that manufactures this equipment is BTX Harvard Apparatus, Holliston, MA, USA (http://btxonline.com/).

3.4.3.3 Microinjection

Microinjection, also known as DNA microinjection and pronuclear microinjection, refers to a simple direct physical method involving the mechanical insertion of the desirable DNA into a living target cell. It is more popular among mammals than in plants and is used to transfer genes between animals, to create transgenic organisms. The technique involves manual injection of the DNA through an extremely slender glass micropipette into the nuclei of living cells. Some proportion of these cells will survive and integrate the injected DNA. It is generally used for precious cells such as mammalian eggs. In such cases, the selected DNA is injected into a fertilized ovum through an extremely slender glass capillary tube. Following

Figure 3.8 Electroporator. (From BTX Harvard Apparatus, Holliston, MA, USA. With permission.)

the DNA injection, the genetically modified egg is transferred into the oviduct of a recipient female where it develops to full term. This technique ensures that nearly all cells in the body of the developing organism contain the induced DNA. Microinjection can also be used in the cloning of organisms, in the study of cell biology and viruses. The equipment used to carry out the process of microinjection is known as "Microinjector."

3.4.3.4 Microprojectile Bombardment (Biolistics)

Other commonly used terms describing this technique are biolistics, bioballistics, particle bombardment, bioblaster, and gene gun (Figure 3.9). It was developed by John Sanford, Ed Wolf, and Nelson Allen at Cornell University, USA (Sanford et al. 1987), together with Ted Klein of DuPont, during the period 1983–1986. In 1987, Klein and coworkers discovered that naked DNA could be delivered to plant cells by "shooting" them with microscopic pellets to which DNA had been adhered. Though crude, this technique was found to be an effective physical method of DNA delivery, specifically in species such as corn, rice, and other cereal grains, which *Agrobacterium* does not naturally transform. Many GE plants in commercial production were initially transformed using microprojectile delivery.

GENETICALLY MODIFIED FOODS

Figure 3.9 Helios PDS 1000/He Biolistic Particle Delivery System (Gene Gun) from BioRad. (From Wikimedia Commons, http://commons.wikimedia.org/wiki/File:Genegun.jpg)

In this technique, the selected DNA to be inserted and integrated into the plant genome is attached to microscopic particles of gold or tungsten. The microparticles—now "carrying" DNA—are shot or fired into the target cells using an intense burst of pressurized gas. The microscopic particles release the DNA into the target plant cells. Genes can thus be "shot" into almost any type of cell including plants. Whole plants are then regenerated from isolated tissues, or, in some cases, biolistic treatment of plant meristems can yield transgenic plants.

3.4.3.5 Gene Splicing

Gene splicing (Figures 3.10 and 3.11) is a term used in biotechnology to refer to the process in which cut pieces of DNA from one or more organisms are combined to form rDNA and are made to function within the cells of a host organism.

The concept of gene splicing involves cutting a gene with desired traits from one organism and inserting it into the DNA of another organism, so that the desired characteristic can be transferred from the donor organism

APPLICATIONS OF GM AT THE LABORATORY AND GREENHOUSE LEVELS

An example of genetic engineering:

1. Scientists take *Bacillus thuringiensis*, a commonly occurring soil bacteria...

2. ...and use enzymes to remove from it the Bt gene, which produces a protein that turns toxic in the digestive tract of caterpillars.

3. The Bt gene is then incorporated into the chromosomes of cotton and corn, killing caterpillars that feed upon these plants.

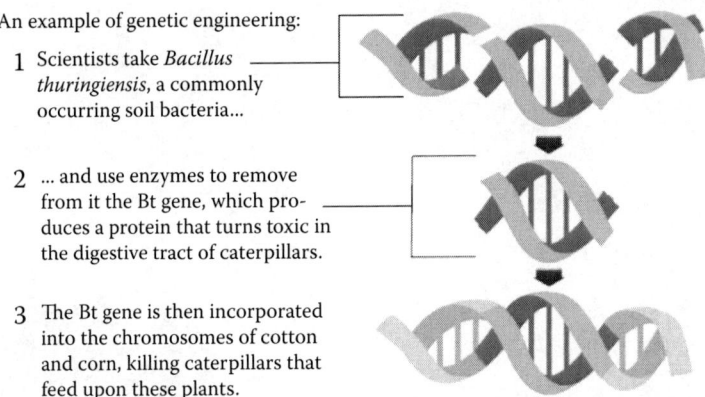

Figure 3.10 Splicing genes together. (From Andrews, R. 2014. Splicing genes together. All about genetically modified foods. All about series. http://www.precisionnutrition.com/all-about-gm-foods (accessed January 23, 2014). With permission from Precisionnutrition.)

into the recipient organism. Cutting is performed using special chemicals called restriction enzymes. Restriction enzymes used in the process of cutting could be considered as chemical "scissors" (BIOL1020 Genetics Blog 2012; Andrews 2014). Bacteria contain restriction enzymes that form part of their natural defense system against invasion by other organisms or bacteriophages (bacterial virus). Usually, the restriction enzymes attack the foreign DNA by cutting it into precise sections and preventing it from being inserted into the bacterium's chromosome.

Different bacteria produce different types of restriction enzymes that have the ability to cut any DNA at different spots. Sometimes, this can make the DNA "sticky," making it easy to be "pasted" directly onto the target organism's prepared DNA.

Restriction enzymes from bacteria are used in the laboratory to "genetically engineer" the DNA for "insertion" into target cells to modify gene traits. Another enzyme is then used to fuse the newly added gene into the chromosome.

3.4.3.6 Gene Silencing

Gene silencing is a general term used in biotechnology to describe the processes employed in gene regulation in living organisms. It is generally used to describe the technique of "switching off" or "turning off" of a gene. This process results in preventing the gene from its normal

GENETICALLY MODIFIED FOODS

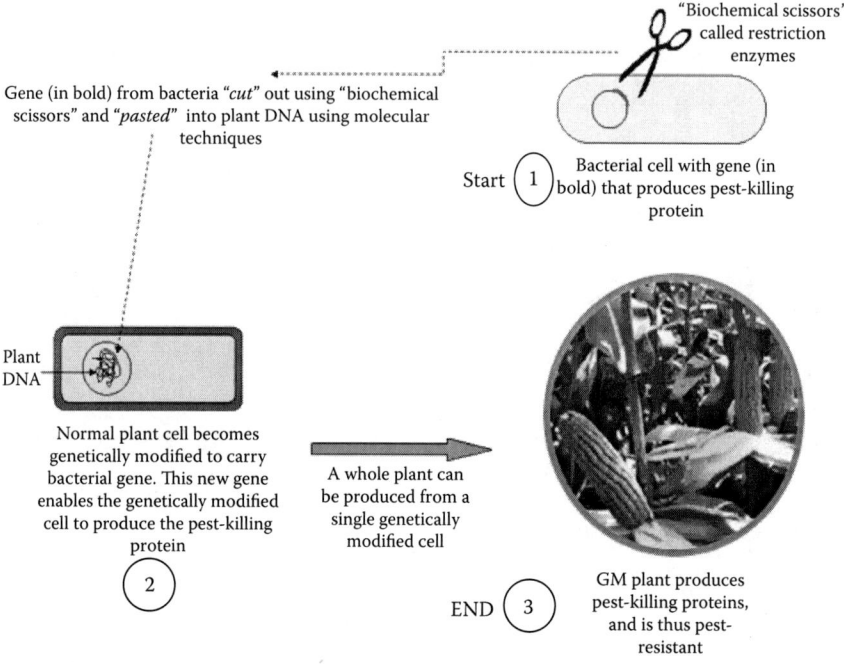

Figure 3.11 Gene transfer process. (From BIOL1020 Genetics Blog. 2012. Genetically modified food. http://biol1020-2012-1.blogspot.com/2012/05/genetically-modified-food.html (accessed January 26, 2014). With permission.)

functions of expressing in the form of protein production or other forms of gene expression. That is, a gene which would be expressed ("turned on") under normal situations is turned off by some mechanism inside the cell. Gene silencing occurs when RNA is unable to produce a protein during the translation process. The process of gene silencing may take place naturally in many instances to achieve the goal of regulating the expression of genes, consequently preventing possible damage from viruses. Gene expression usually occurs when DNA is converted to RNA through a process known as "transcription." The RNA is then converted to proteins through a process called "translation." Gene silencing can interfere with the process of transcription or with some other subsequent processes that result in gene expression. The gene *per se* is not modified or damaged and remains intact, but the different steps that lead up to expression are not allowed to proceed and complete the process.

In the biotechnology laboratory, gene silencing technique may be used to disable some genes with the intention of determining the role(s) of that gene. There are many different ways and mechanisms through which gene silencing, and consequently gene expression, may be accomplished. One method of "silencing" a particular gene is to attach a second copy of the gene the wrong way around. The gene responsible for the organisms' undesirable trait is first identified. Then, another copy of the gene is attached, but in the other direction, which prevents the expression of that trait. This technique is used to prevent crops such as peanuts and wheat from producing proteins (allergens) that trigger allergic reactions in human beings. Another approach is to insert foreign DNA within a gene to inactivate it.

3.4.3.7 Calcium Phosphate Precipitation

In this technique, the selected DNA would be exposed to calcium phosphate. This mixture results in the creation of miniscule granules. The targeted cells react with these granules by essentially "swarming," surrounding, and ingesting them—a process known as endocytosis. This makes it easy for the granules to release the DNA and subsequently deliver it to the host's nuclei and chromosomes.

3.4.3.8 Lipofection

Lipofection (also known as liposome transfection) is a technique in which tiny bubbles of fat, called liposomes, are used as the carriers of, and to inject, the selected DNA. The target cells and the liposomes are placed in a special solution. The liposomes merge with the cell membrane, allowing the DNA into the cells for inclusion in the chromosome. The principle of the lipofection process relies on the use of a positively charged, that is, cationic, lipid to produce an aggregate with the negatively charged, that is, anionic, genetic material.

3.4.4 The Steps Followed in GE

As discussed in Chapter 2, there are many different techniques used in GE. The process of GE requires the successful completion of a series of five steps, as summarized by Bitterroot Restoration (2014) and presented below.

Step 1: Nucleic acid (DNA/RNA) extraction. Nucleic acid extraction (Figure 3.12), either DNA or RNA, is the first step in the GE process. In order to conduct any work with the genetic material DNA, it has to be isolated from the desired living organism. Tested and reliable methods need to be used for isolating these components from the cell. In any isolation

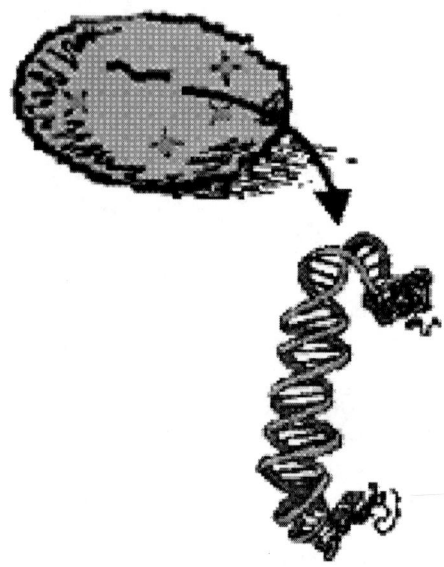

Figure 3.12 Nucleic acid extraction. (From Bitterroot Restoration. 2014. http://www.bitterrootrestoration.com/genetic-engineering/the-process-of-plant-genetic-engineering.html (accessed January 22, 2014). With permission.)

procedure, the initial step is the disruption of the cell in order to break the cell wall of the desired organism, which may be viral, bacterial, or plant cells. This helps in the extraction of the nucleic acid. A number of chemical and biochemical steps are used to accomplish the extraction. The extracted nucleic acid can then be precipitated to form thread-like pellets of DNA/RNA.

Step 2: Gene cloning. The second step of the GE process is gene cloning (Figure 3.13). The cloning process involves four stages, namely, generation of DNA fragments, joining to a vector, propagation in a host cell, and selection of the required sequence. All DNA from the desired organism is extracted at once. This genomic DNA is treated with special enzymes called restriction enzymes which act to cut it into smaller fragments with clear ends to help it to be cloned into bacterial vectors. Copies of the vector will then harbor many different inserts of the genome. These bacterial vectors are transformed into bacterial cells, and thousands of copies are generated. Using information relating to specific molecular marker sequences and the desired phenotype, the vector harboring the desired sequence is detected, selected, and isolated, and clones are produced.

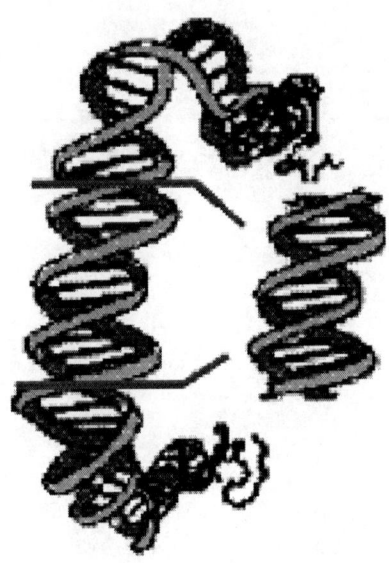

Figure 3.13 Gene cloning. (From Bitterroot Restoration. 2014. http://www.bitterrootrestoration.com/genetic-engineering/the-process-of-plant-genetic-engineering.html (accessed January 22, 2014). With permission.)

Restriction enzymes are used once more to determine whether the cloning process of the desired gene was correctly completed.

Step 3: Gene design and packaging. Once the desired gene has been cloned, it has to be linked to pieces of DNA that will control its expression inside the plant cell. These pieces of DNA will control the gene expression by switching on (through a promoter, which allows differential expression of genes) and off (through a terminator). Gene designing/packaging can be performed by replacing an existing promoter with a new one, incorporating a selectable marker gene and a reporter gene, adding gene enhancer fragments, introns, and organelle-localizing sequences, among others (Figure 3.14).

Selectable marker genes. Selectable marker genes are typically attached to the gene of interest to facilitate its detection once inside the plant tissues. This helps the selection of cells that have been successfully incorporated with the gene of interest, thus making the process more efficient.

Reporter genes. Reporter genes are cloned into the vector very close to the gene of interest, to facilitate the identification of transformed cells as well as to determine the correct expression of the inserted gene.

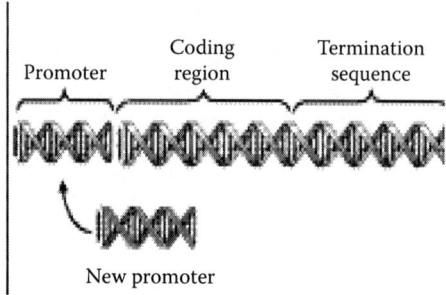

Figure 3.14 Promoter and terminator. (From Bitterroot Restoration. 2014. http://www.bitterrootrestoration.com/genetic-engineering/the-process-of-plant-genetic-engineering.html (accessed January 22, 2014). With permission.)

Enhancers. Several genetic sequences can also be cloned in front of the promoter sequences (enhancers) or within the genetic sequence itself (introns or noncoding sequences) to promote gene expression.

Step 4: Transformation. The next step in the process for the modified gene is transformation or gene insertion (Figure 3.15). As plants are made up of millions of cells, it would not be practical to insert a copy of the transgene into every cell. Therefore, the technique of TC is used to propagate masses of undifferentiated plant cells called callus. The callus will be the cells to which the new transgene is added.

One of the different GM techniques discussed earlier, for example, gene gun, *Agrobacterium*, or electroporation, may be used to insert the new gene into a few of the cells. The main goal of each of these methods is to transport the new gene(s) and deliver them into the nucleus of a cell without killing it (Figure 3.15). Transformed plant cells are then regenerated into transgenic plants. The transgenic plants are grown to maturity in greenhouses, and the seed they produce, which has inherited the transgene, is collected for further use.

The integrity of the transgene into the plant cells could be checked and detected using a reaction known as polymerase chain reaction (PCR). This reaction represents a quick test to verify whether the regenerated cells or plants contain the desired gene.

Step 5: Backcross breeding. The fifth and final step of the process is known as "backcross breeding" (Figure 3.16). This process involves crossing the transgenic plants produced with elite breeding lines or commercial varieties, which already have the desired agronomic traits but lack

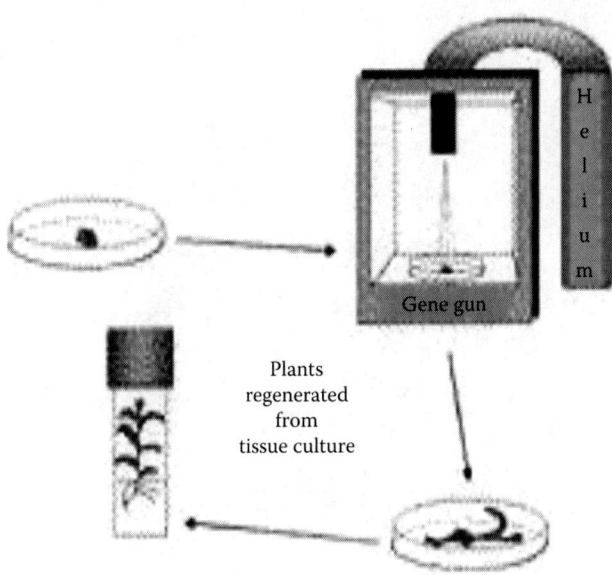

Figure 3.15 Gene insertion (transformation). (From Bitterroot Restoration. 2014. http://www.bitterrootrestoration.com/genetic-engineering/the-process-of-plant-genetic-engineering.html (accessed January 22, 2014). With permission.)

the trait of the transgene. Traditional plant breeding methods are usually used in this crossing. Backcross breeding achieves the combination of the desired traits of the elite parent and the transgenic line in the progeny. The offspring is repeatedly crossed back with the elite line in order to end up with a high-yielding transgenic line. The resulting offspring is expected to be a plant with a yield potential close to current hybrids and which expresses the trait transmitted by the new transgene.

The genetically modified plant will be ready for commercialization if it shows stability in several generations and upon successfully passing and fulfilling varietal registration requirements. The length of time required to develop transgenic plants depends on many factors, for example, the desired gene, the plant species, nature and level of available resources, and regulatory approval. Experience has shown that a period of 6–15 years may be required before a new transgenic plant or hybrid is ready for commercialization.

A brief summary illustrating the steps followed in GE is also shown in Figure 3.17 (SOS 2014).

GENETICALLY MODIFIED FOODS

Figure 3.16 Backcross breeding. (From Bitterroot Restoration. 2014. http://www.bitterrootrestoration.com/genetic-engineering/the-process-of-plant-genetic-engineering.html (accessed January 22, 2014). With permission.)

3.4.5 Methods for Detection, Identification, and Quantification of GMOs in Food and Feed

3.4.5.1 Introduction

A large number of GMOs and transgenic plants have already been approved or are under approval for use. These numbers are increasing every day. Inspection authorities would find it difficult to distinguish GMOs from non-GMOs in the absence of techniques and instruments to detect, identify, and quantify GM ingredients in food or feed. Qualitative detection methods of GM-DNA sequences in foods and feeds have evolved fast during the past few years (Schreiber 1999). Information gathered by detection of GMOs in food and feed would be helpful for setting regulations that govern the use of GM components as well as for labeling purposes. It is also recognized that the implementation of postrelease monitoring and labeling of GMOs, as required by national and international regulations (e.g., the Novel Food Regulation [EC/258/97, EC/1139/98, EC/49/2000, EC/50/2000, and EC/1829/2003] applied in European countries), will require reliable, standardized, and

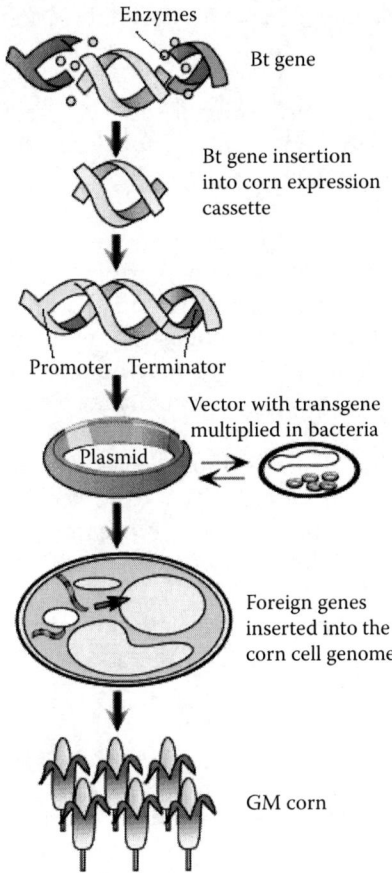

Figure 3.17 Summary of steps followed in GE. (From SOS. 2014. http://gmosafety-engl.webnode.com/contact-us/ (accessed January 18, 2014). With permission.)

accurate analytical detection methodologies for GMOs (Tung Nguyen et al. 2008; Marx 2010). On the contrary, GM components present in food or feed are considered as contaminants by some legislation (Hemmer 1997). This created a high demand for test methods for detecting, identifying, and quantifying GM material in agricultural products handled at different levels, for example, farm gate, processing, and retailing. A number of analytical methods have been set in order to address those issues. Marmiroli et al. (2008) have reviewed the different aspects and the analytical methods related to detection, identification, and quantification of GM material within the food and feed chain. Currently, the existing methods for GMO testing proved to be of great help to inspection authorities, giving them a wide range of choices. The choice of a specific method to be used is dictated by a number of factors, mainly for what purpose the information is needed. Dong et al. (2008) compiled a "GMO Detection Method Database," covering almost all the previously developed and reported GMO detection methods grouped by different strategies as well as a user-friendly search service of the detection methods. Some of the potential future challenges that might face GMO detection, as reported by Schreiber (1999), include the need for the development of more advanced, multidetection systems, the possibility of decreasing relevance of methods which screen for sequences commonly found in GMOs, the inability to detect GM foods for which the modified sequence is unknown, and the need for continuous updates of databases. The website Science.gov (Science.gov 2013) contains sample records for the topic "detecting genetically modified." Information on this website is continuously updated.

In the next sections, an overview of the different existing methods used for sampling and analyzing GMOs in food and feed is presented.

3.4.5.2 Sampling

It is important to realize that in any type of experiment, following the correct steps of sampling is crucial and critical to obtain accurate and credible results. This needs to be taken into consideration when sampling for GMO testing. A specific sampling plan needs to be designed and followed, depending on the nature of the GMO product to be tested, for example, seeds, food, and food ingredient. Both the sample size and the sampling procedure need to be taken into consideration when testing for GMOs in raw material, food, food ingredients, or feed. The sample size varies and depends on the level of detection required and on what the researcher is looking for. It must be taken in sufficient amounts to ensure adequate sensitivity and to achieve reliable statistical significance. The important point to consider is that the sample should be a statistically representative one, so that the result obtained reflects the nature of the whole lot of material under investigation. Samples that are too small and not statistically representative would increase the chance of obtaining false results which do not accurately reflect the actual nature of GMOs in the material at hand.

3.4.5.3 GMO Testing Methods

A number of different analytical methods and strategies have been in use for the analysis of GMOs in food and feed. Some of the common detection methods routinely used in biotechnology laboratories have been reported in the literature (Feriotto et al. 2002; Cardarelli et al. 2004; Nesvold et al. 2005; Dinon et al. 2008; Bahrdt et al. 2010).

Analytical methods used to detect, identify, and/or quantify GMOs in food or feed need to be selected carefully and should have the following important characteristics:

- They should be able to detect all types of GMOs present in the tested material.
- They should be able to provide quantitative data that show how much GMO is available in the sample.
- They should be suitable to be used for a wide range of foods, feeds, and other related products.
- They should be highly sensitive, reproducible, and reliable, so that they might be used in different laboratories and under different conditions, obtaining the same results.

The assessment of GMOs in the test material can be considered at three different levels, namely, detection, identification, and quantification.

Detection: This is a general screening step to be conducted at the initial stages of the assessment. It aims to determine whether the sample contains any GMO. The information collected at this stage is only a positive or negative (yes or no) evidence of the presence of GMOs. Detection is usually carried out using the PCR, further discussed in detail in Chapter 4. A positive (yes) indication of the presence of GMOs obtained in this step will need further investigation, as described in the next step. A negative (no) indication of GMOs does not require further investigation.

Identification: This step is required if GMOs have been detected in the previous step. After the detection of GMOs, the types of GMOs present in the test sample are identified. Identification of the nature of the GMO present is usually carried out using the PCR method, which is the most widely accepted method for this purpose.

Quantification: The aim of quantification is to determine exactly how much of each GMO is present in the test material. This information would be helpful if GMO regulations and labeling are required. Semi-quantitative PCR or real-time PCR is usually used to complete this step.

Researchers need to remember that in addition to the components we are testing for, DNA and protein, foods, and feeds contain other biological materials such as lipids and carbohydrates. The results of the GMO assessment may be affected by the interference of some of those components, for example, the PCR may be negatively affected by the presence of polysaccharides. Some control measures need to be in place to account for such situations and that effect should be taken into consideration, otherwise a negative result might be reported.

Lübeck (2014) classifies GMO analysis and assessment methods as "low technology methods" and "high technology methods." The low technology methods include phenotypic characterization (herbicide bioassays) and immunoassays. These methods can be set up and conducted in most laboratories concerned with GMO testing. In contrast, PCRs and microarrays are considered as high technology methods because they require more special equipment as well as trained personnel to conduct tests and interpret the results.

A different system of classification of GMO detection methods was reported by Marx (2010). Here, the analytical methods used to detect (i.e., qualitative or indicating presence [yes] or absence [no] and quantitative, i.e., percentage content) GMOs were grouped into two categories. The first category is the protein analysis which is used to detect the specific protein expressed by the transgene in the GMO and which includes the use of enzyme-linked immunosorbent analysis (ELISA) and lateral flow strip tests. The second category is the DNA analysis which is used to detect the specific transgene in the GMO or specific elements associated with the transgene. Protein methods can be used on raw and semi-processed samples, as long as the protein is not denatured or destroyed by processing (Anklam et al. 2002). Protein identification requires the use of antibodies raised against the transgenic protein.

3.4.5.3.1 Low Technological Methods
a. *Phenotypic characterization (herbicide bioassays)*

   Phenotypic characterization methods, also known as herbicide bioassays, are used to detect the presence or absence of a very specific characteristic. Currently, the only trait tested using this technique is resistance or tolerance for herbicides. The presence or absence of herbicide-resistant GMOs is tested using this method. Seed germination tests are conducted on solid germination media in the presence of a specific herbicide, in which GMO and non-GMO seeds may exhibit different distinct characteristics. It is important to make sure that all tested seeds germinate. Further confirmation tests for the seeds that show positive signs need to be conducted. Some known features of the herbicide bioassays include low cost, accuracy, and their use as preventive tests, particularly by seed companies, which apply them as part of their quality assurance programs.

b. *Immunoassays (protein methods)*

These methods are employed for new foreign proteins added through GM of plants, both qualitatively (detection) and quantitatively (amount). Immunoassays use antibodies of high level of specificity for the target molecule. A number of immunoassay techniques are in use. These include ELISA and lateral flow sticks.

i. ELISA

In ELISA, the antigen–antibody reaction takes place on a solid phase (microtiter plates). ELISA requires a basic protein extraction followed by antibody detection in a microwell plate (Marx 2010). Antigen and antibody react and produce a stable complex, which can be visualized by addition of a second antibody linked to an enzyme. Addition of a substrate for that enzyme results in a color formation, which can be measured photometrically or recognized by the naked eye (Lübeck 2014). In order to determine that a product is non-GMO, different tests must be used for as many different transgenes as are commercially available (Marx 2010). ELISA proved to be a suitable method for screening new materials and basic ingredients when the presence of modified proteins can be detected and is not degraded by factors such as high temperature, pH, and salts (Ahmed 2004). Some of the positive features of the ELISA assays include their robustness, suitability for routine testing, low costs, and the short period of time required to perform the test. However, their application is limited and depends on the degree and nature of processing that the test material, for example, food, has undergone (Kuiper 1999).

Advances in biotechnology have made it possible to develop diagnostics, which made it easy for farmers around the world to detect and manage different diseases related to their crops. For this purpose, different types of ELISA test kits have been developed and used. The kits are usually easy to use, and part of the tests can be performed in the field. The range of crops in which the ELISA test kits have been successfully used includes cassava, beet, potato, banana, apple, grape, wheat, rice, and some vegetables. ELISA test kits have been used to detect ratoon stunting disease of sugarcane, tomato mosaic virus, papaya ringspot virus, watermelon mosaic virus, and banana bract mosaic virus.

ii. *Lateral flow sticks*

Lateral flow sticks are used to detect GMOs present in leaves, seeds, and grains. In this technique, paper strips or plastic paddles are used as support materials to help in capturing the antibodies. The process involves homogenizing the sample to the appropriate particle size, buffer added for simplified protein extraction, and the strip/paddle is dipped in vials containing different solutions. Each dip is followed by a rinsing step. After several minutes, a positive result is indicated by a discolored test line due to antibody–protein recognition (Marx 2010). Lateral flow sticks technique is regarded as the simplest method to qualitatively detect a GMO.

3.4.5.3.2 High Technological Methods

a. PCR

PCR is the most widely used technique for GMO detection and analysis. Raw and processed products can be tested with the PCR method, as long as DNA can be extracted from the sample (Marx 2010). Fagan (2004) lists the key elements in the PCR process as follows:

1. "Primers," which are small DNA molecules whose sequences correspond to the target sequence
2. A heat-stable DNA polymerase—typically Taq polymerase
3. A thermocycler—an apparatus that can be programmed to carry the contents of the PCR reaction vessel through multiple, precisely controlled temperature cycles

The basic steps of PCR include isolation and purification of the DNA, exponential amplification of the inserted DNA via enzymatic replication, without using a living organism, and confirmation of the amplified PCR product. It makes the detection of specific strands of DNA easy by making millions of copies of a target genetic sequence. The target sequence is essentially photocopied at an exponential rate, and simple visualization techniques can make the millions of copies easy to see. PCR involves pairing the targeted genetic sequence with custom designed complementary bits of DNA known as primers. Matching of the target sequence and the primers will help to trigger a chain reaction. DNA replication enzymes use the primers as docking points and start multiplying the target sequences. This process is repeated a number of times by sequential heating and cooling until several millions of the target sequence are obtained. Purification of the resulting identical fragments follows. The purified fragments are then dyed and can be visualized under UV light.

It needs to be emphasized that the accuracy of sampling procedures affects the PCR results (as it does with other tests). In addition to preparing a representative sample and to be able to obtain reliable and true results, the researcher needs to observe the following points: optimal and high-yield extraction, avoidance of DNA breakdown, and getting rid of any chemical contaminant to avoid inhibition of PCR amplification. Adequate control measures and the use of authentic standards needed for comparison and verification are necessary to obtain reliable results. The most specific method to identify a GMO is event-specific detection, in which the PCR target sequence is a junction between the host DNA and the inserted gene construct (Viljoen 2005).

PCR tests may be used to detect any of the inserted material: promoter, structural gene, stop signal, or marker gene (Lübeck 2014).

"Quantitative PCR" (Q-PCR) is the method used to determine the amount of a PCR product. It is widely used for the quantitative measurement of transgenic DNA in food or feed. It helps to determine whether a certain DNA sequence is present in the sample material and how many copies of it exist. Currently, the most frequently used, reliable, accurate, and preferred PCR method is quantitative real-time PCR. The quantification process is called real-time PCR because amplification of the target DNA sequence is visualized during PCR in "realtime" (Marx 2010). This method uses fluorescent dyes such as SYBR Green or fluorophore-containing DNA probes such as TaqMan to determine the quantity of the amplified product in real-time. For GMO quantification, it is preferable to use specific detection probes to avoid problems of nonspecific amplification (Viljoen 2005).

b. *Microarrays*
Microarray technology, also known as DNA chip technology or biochip, has been developed and used for automated rapid screening of gene expression and sequence variation of large numbers of samples. It measures the expression levels of large numbers of genes at the same time. A DNA microarray is a collection of thousands of microscopic DNA spots attached to a solid surface (usually a glass slide or a silicon thin-film cell). Each spot on a microarray contains multiple identical strands of DNA and each spot represents one gene. The exact position and the sequence of each spot are usually recorded in a computer program.

3.4.5.4 Validation and Standardization of Methods

Validation of methods is the process of showing that the combined procedures of sample extraction, preparation, and analysis will yield acceptably accurate, reliable, and reproducible results for a given analysis in a specified matrix. For validation of an analytical method, the testing objective must be defined and performance characteristics must be demonstrated. Performance characteristics include accuracy, extraction efficiency, precision, reproducibility, sensitivity, specificity, and robustness (Lübeck 2014). There has been international interest in designing strong analytical methods for GMO testing and detection. These methods need to be validated and standardized using official noncommercial guidelines. Some of these guidelines are preparation of certified reference material (CRM), sampling, treatment of samples, and production of stringent analytical protocols to help determining the validity of a specific GMO detection technique. Ahmed (2004) argued that there is need for more efforts to develop standardized reference material for GMO testing assays. He further stated that false-positive results should be avoided because they result in the destruction or unnecessary labeling of foods. Fagan (2004) has also emphasized the urgent need for the establishment of internationally standardized methods for the analysis of GMOs both for qualitative and for quantitative evaluation. It has been noted that the development of standardized methods for GMO testing has lagged behind the introduction of GMOs into the food system, as well as the enactment of labeling laws. Due to the global nature of the food production systems, the process of methods standardization needs to be global as well. This would help in harmonizing the international food trade and gives more confidence to food importers and exporters in the test results.

3.4.5.5 Reference Material for GMO Testing

The relevant and appropriate reference material needs to be used, alongside the test sample under the same conditions. It is usually used for calibrating or for evaluating and validating the performance of a test method. The reference material has to have a reasonable stability and should be homogeneous in order to be suitable for applications. Different GMOs require different reference materials. A CRM is the one that holds a certificate from a recognized authority, showing the value of one or more properties and their uncertainty. CRMs are essential tools in quality assurance of analytical measurements. They help to improve precision and trueness of analytical measurements and to ensure comparability of achieved results. CRMs are produced, certified, and used in accordance with relevant International Organization for Standardization (ISO) and Community Bureau of Reference guidelines (Trapmann et al. 2004). In general, CRMs are designed for specific intended use, and possible fields of application are laid down in the certificate. Standards validated by a laboratory may be used if CRMs are not available. The Institute of Reference Materials and Measurements at the Joint Research Center in Geel, Belgium offers a set limited number of reference materials, through Fluka (Buchs, Switzerland), for modified soya, corn, and maximizer maize (Markoulatos et al. 2004). For international reliability of GMO testing, internationally standardized reference materials are absolutely required.

3.4.5.6 The Future of GMO Testing

Before forecasting the future of GMO testing, it might be worthwhile to evaluate the tests used presently to detect the presence of GMOs. Fagan (2004) believes that those tests do not completely fulfill the needs of industry and regulators. The ever-increasing number of approved GMOs and GM foods would widen the gap between the need for appropriate tests and the technological capabilities. Examples of the required technological capabilities, as listed by Fagan (2004), include the following:

- Ability to simultaneously analyze hundreds of GMOs
- Be highly sensitive and highly specific
- Carry out accurate quantification
- Be rapid
- Be economical
- Flexibility for effective use with diverse food matrices
- Easy to use and portable
- Field operable

The most widely used method for GMO detection, i.e., PCR, does not have some of the requirements stated above, for example, it is expensive if used to screen large number of samples. Additionally, it is also time-consuming. In contrast, immunodetection methods are less expensive and faster than PCR, but they are less sensitive. Generally, all the methods in use today for the detection of GMOs lack some of the required capabilities indicated earlier. New technological developments for the methods would be required (Fagan 2004).

Some of the challenges facing techniques of GMO detection are as follows:

- Improving sampling techniques and statistical analysis
- Designing optimum methods for isolation of DNA from different sources
- Ability to differentiate GM material carrying stacked transgenes (transgenic cultivars derived from crosses between transgenic parent lines and combining the transgenic traits of both parents) from mixtures of single events
- Unavailability of DNA sequence information for some nonauthorized GM material
- Developing suitable portable equipment to facilitate analysis outside the laboratory boundaries

Currently, the majority of laboratories conducting work on GMO qualitative detection use the PCR technique. Quantitative PCR is expected to gain more focus of attention in response to the need for more detailed results. It is also expected that detection of GMOs will become more complicated in the future due to their increased usage. Continuous revision and updating of the methods employed and developing alternative methods would be necessary to parallel this situation. There are some efforts in progress aiming to develop multiplex PCR methods that can simultaneously detect many different transgenic lines. There is an obvious need for international collaboration to ensure that the methods used by different companies and research laboratories worldwide are harmonized, consistent, and have future potential.

3.4.5.7 ISO Standards for Detecting GMOs in Food

The ISO has developed a new ISO standard (ISO 24276, Foodstuffs) that gives general requirements for identifying and quantifying GM materials in foods. This standard is expected to add some degree of transparency to the widely controversial issue of GM, particularly when it comes to food and health.

ISO 24276, *Foodstuffs—Methods of analysis for the detection of genetically modified organisms and derived products—general requirements and definitions*, lists the following steps that need to be taken after sample collection for analysis:

- Nucleic acids or proteins are extracted from the test portion
- Extracted analytes are further purified
- Quantified (if necessary) and diluted (if necessary)
- Analyzed using the PCR or other procedures

The standard main focus is on the PCR-based technique, a precise method for generating unlimited copies of specific fragments of the DNA, but other methods using protein as the analytical target are also covered.

It comprises the validation of methods and a checklist of the type of information the test reports need to cover and supplies some guidelines for laboratory setup.

- Procedural requirements
- Laboratory design
- Personal regulations
- Apparatus and equipment maintenance
- Materials and reagents to be used for the analysis

ISO 24276 gives details of how to use the standards for protein analysis (ISO 21572), nucleic acid extraction (ISO 21571), qualitative nucleic acid analysis (ISO 21569) and quantitative nucleic acid analysis (ISO 21570), and their relationship in the analysis of GMOs in foodstuffs.

Although this ISO standard had been designed for foodstuffs, it could also be used for GMO analysis in other areas such as seeds, feeds, or various plant samples.

ISO 24276 was prepared by the European Committee for Standardization (CEN) Technical Committee CEN/TC 275, *Food analysis—Horizontal methods*, in collaboration with ISO/TC 34, *Food products*. The standard is available from ISO national member institutes and from ISO Central Secretariat, as shown below:

ISO Store: to order ISO24276:2006, Foodstuffs—Methods of analysis for the detection of genetically modified organisms and derived products–General requirements and definitions.

3.5 UNINTENDED EFFECTS OF GE

It has been observed that in different types of GM, either additional traits could, theoretically, be acquired or existing traits lost. Such occurrences have been referred to as "unintended effects." These effects take place in addition to the insertion of the specific trait to the target organism. The latter is usually referred to as "intended effects." The possible occurrence of unintended effects is not limited to the use of rDNA techniques. It can occur in conventional breeding as well.

Unintended effects may be due to factors such as random insertion events, which might result in the disruption of existing genes, modification of protein expression, or formation of new metabolites (FAO/WHO 2000). The concept of unintended effects from GM crops and products, that is, effects that go beyond those of the original desired modification and which might have impact on human and animal health as well as on the environment, has been of much concern and discussion recently. Many pertinent questions arise in connection with the possible unintended effects of GE on human and animal health. Bodnar (2011) raises the following question: "are there unanswered questions about the health effects of GE foods?" She is soliciting views which are based on reliable sources that would help clarify this issue. Freedman (2009) discusses the extent to which biotechnology is affecting and changing the food that we eat today.

It is an undeniable fact that a range of those unintended effects occur. It only depends on which side of the coin you are on: are you a "pro- or anti-GM technology?" It seems that people who are "pro" would pay little or no attention to these effects or try to justify or explain them. In contrast, opponents of GM technology would try to give all details and discuss all the studies that point out the negative consequences of the unintended effects of GE.

Unexpected and unintended changes in composition can occur and be observed with all methods of GM, whether old (e.g., selection and crossing) or new (GE). Traditionally, breeders observe such off-types regularly; they disregard and discard these individuals with no further interest or consideration.

It has been observed and reported that all living organisms go through natural spontaneous changes in their genomes, resulting in the production of novel traits, which may carry some hazard. Such incidences occur relatively rarely, and the nature of hazard related to them is generally not predictable. Alternatively, mutations induced by man, such as through rDNA, which allow a gene or genes from any species of living organism to be inserted into and expressed by another living organism, may lead to such changes. It would be logical to say that such a product would have the potential to be hazardous if the introduced gene leads to the production of a hazardous material. The process of GM in itself might not be inherently hazardous, but the end product resulting from it might be.

In all types of biotechnological methods, whether old or new, some degree of genetic manipulation takes place in the form of adding, removing, or altering of genes. It is possible that this genetic manipulation results in hazards or benefits to the health or to the environment. Research results on various aspects of GE are widely reported and disseminated in the current literature. Some of these reports focus on the issue of unintended effects of GE, alternatively referred to as "nontarget effects" by Holdrege (2014). Among other terms used to describe such effects are unexpected effects, unintended consequences, and pleiotropic effect of the gene. Pleiotropy has been described by Omar (2009) as "unintended consequences—what is intended may not be the same as what actually happens with the gene insertion." This differentiates them from the targeted results of any experiment. Unintended consequences may result from the following conditions:

- Genes are inserted at random into the genome of the target organism, for example, some are inserted within the organism gene and some are inserted near the organism gene.
- The placement of the inserted gene can affect the working of the organism's genome (Omar 2009).

Pleiotropic effects are indirect effects that might be caused by the insertion of the genes into the GM plant. If a gene attached to a strong promoter is inserted into the plant's chromosome, it might cause a problem if it was inserted adjacent to a toxin gene that had previously been present but not expressed in the plant. This insertion can lead to the expression of the toxin, along with the desired novel gene product.

Pleiotropic effects do occur in plants developed through modern methods of GM. Plant breeders, using traditional technologies, sometimes use such unexpected effects that might result in the production of useful varieties. In contrast, if undesirable or harmful effects are produced from traditional breeding, the resulting line would be discarded as it would not be agronomically useful. Pleiotropic effects arising from

GE would be easier to predict because information about the inserted gene is available.

Experiments are usually conducted with a specific objective in mind, and certain intended results are expected. Along with, or instead of, those expected results, other unexpected or unintended results/effects are often encountered. GE experiments are not an exception. It would be helpful in the ongoing debate about GMOs (specifically GM foods), if all the unintended effects of GE and their possible consequences are clearly stated and explained. This will allow the consumer to understand what they are dealing with, what they are consuming, and if there are any potential health or environmental risks due to these effects.

In GE processes which aim at creating a transgenic organism, the results are unpredictable, and both targeted and untargeted results would be expected. Some degree of transparency is needed when reporting or interpreting those results. It will be useful and helpful to clearly and accurately report all the results, whether positive or negative, expected or unexpected. It has been observed that in reporting results of GE experiments (and possibly all types of experiments), some researchers put more emphasis, or even only report, on the intended or expected results. As Holdrege (2014) explains, researchers handle unintended results in a variety of ways, as follows:

- Simply not reporting them, although some effects may have been noted
- No explicit reporting of unintended effects
- Unintended effects are reported, but the main focus is on the intended effects
- Unintended effects are the primary subject of the research

As can be seen, some researchers would be biased toward what they want to emphasize or prove and will not show the whole truth but part of it. It will be misleading and not honest to state that there are no unintended effects in reporting results of any experiment, if there are. Researchers need to be specific and cite statements such as "no unintended effects were noticed under the present study conditions." The following statement by Holdrege (2014) "in scientific terms, the absence of evidence is not evidence of absence" should always be taken into consideration when dealing with or reporting unintended effects.

Unintended effects may further be differentiated into "predictable" and "unpredictable" (Cellini et al. 2004). Predictable unintended effects go beyond the target effects, but may be explicable based on our current knowledge of the subject. Unpredictable unintended effects usually fall beyond the boundaries of the researchers' present knowledge and understanding, which makes difficult for them to interpret the results related to these effects.

The Nature Institute (2014) has a project, directed by Craig Holdrege, called "Unintended effects of genetic manipulation." This project aims to highlight the main facts about the unintended effects of GE: what do they mean, how they are detected, and their different categories. A clear and complete picture needs to be portrayed about the unintended effects. A better understanding of these facts would help people to evaluate the GE technology from different perspectives.

In the future, GM of plants is likely to be more complex, perhaps involving multiple between-species transfers and this may lead to an increased chance of unintended effects. Where differences are observed using profiling techniques, the possible implications of the differences with respect to health need to be considered (FAO/WHO 2000).

REFERENCES

Ahmed, F.E. 2004. Protein-based methods: Elucidation of the principles, Chapter 5, in F.E. Ahmed (Ed.), *Testing of Genetically Modified Organisms in Foods*. Food Products Press. An Imprint of the Haworth Press, Inc., New York, London, Oxford, pp. 117–46.

Andrews, R. 2014. Splicing genes together. All about genetically modified foods. All about series. http://www.precisionnutrition.com/all-about-gm-foods (accessed January 23, 2014).

Anklam, E., Gadani, F., Heinze, P., Pijnenburg, H., and Ven Den Eede, G. 2002. Analytical methods for detection and determination of genetically modified organisms in agricultural crops and plant-derived food products. *European Food Research and Technology*, 214:3–26.

Bahrdt, C., Krech, A.B., Wurz, A., and Wuff, D. 2010. Validation of a newly developed hexaplex real-time PCR assay for screening for presence of GMOs in food, feed and seed. *Analytical and Bioanalytical Chemistry*, 396:2103–12.

Bhojwani, S.S. and M.K. Razdan. 1996. *Plant Tissue Culture: Theory and Practice*. Elsevier, Amsterdam, Netherlands.

BIOL1020 Genetics Blog. 2012. Genetically modified food. http://biol1020-2012-1.blogspot.com/2012/05/genetically-modified-food.html (accessed January 26, 2014).

Bitterroot Restoration. 2014. The process of plant genetic engineering. http://www.bitterrootrestoration.com/genetic-engineering/the-process-of-plant-genetic-engineering.html (accessed January 22, 2014).

Bodnar, A. 2011. Are there unintended health effects of genetic engineering? The Biofortified Blog. http://www.biofortified.org/2011/01/are-there-unintended-health-effects-of-genetic-engineering/ (accessed January 24, 2014).

Cardarelli, P., Branquinho, M.R., Ferreita, R.T.B., Da Cruz, F.P., and Gemal, A.L. 2004. Detection of GMO in food products in Brazil: The INCQS experience. *Food Control*, 16:859–66.

Cellini, F., Chesson, A., Colquhoun, I. et al. 2004. Unintended effects and their detection in genetically modified crops. *Food and Chemical Toxicology*, 42:1089–125.

Dinon, A.Z., De Melo, J.E., and Arisi, A.C.M. 2008. Monitoring of MON810 genetically modified maize in foods in Brazil from 2005 to 2007. *Journal of Food Composition and Analysis*, 21:515–8.

Dong, W., Yang, L., Shen, K. et al. 2008. Database GMDD: A database of GMO detection methods. *BMC Bioinformatics*, 9:260. doi:10.1186/1471-2105-9-260 (pdf). http://www.biomedcentral.com/1471-2105/9/260 (accessed January 20, 2014).

Fagan, J. 2004. DNA-based methods for detection and quantification of GMOs: Principles and standards, Chapter 7, in F.E. Ahmed (Ed.), *Testing of Genetically Modified Organisms in Foods*. Food Products Press. An Imprint of the Haworth Press, Inc., New York, London, Oxford, pp. 163–220.

FAO. 2005. Status of research and application of crop biotechnologies in developing countries: Preliminary assessment. Food and Agriculture Organization of the United Nations, Rome.

FAO/WHO. 2000. Safety aspects of genetically modified foods of plant origin. Report of a Joint FAO/WHO Expert Consultation on Foods Derived from Biotechnology, World Health Organization.

Faraz, N.H. 2014. Application of plant tissue culture in plant breeding. Plant Tissue Culture. http://www.scribd.com/doc/18340374/Plant-Tissue-Culture- (accessed January 22, 2014).

Feriotto, G., Borgatti, M., Mischiati, C., Bianchi, N., and Gambari, R. 2002. Biosensor technology and surface plasmon resonance for real-time detection of genetically modified Roundup Ready Soybean gene sequences. *Journal of Agricultural and Food Chemistry*, 50(5):955–62.

Freedman, J. 2009. *Genetically Modified Food: How Biotechnology is Changing What We Eat*. First Edition. The Rosen Publishing Group Inc., New York.

Hemmer, W. 1997. Foods derived from genetically modified organisms and detection methods. BATS Report 02/1997, Agency for Biosafety Research and Assessment of Technology Impacts of the Swiss Priority Program, Biotechnology of the Swiss National Science Foundation, Basel, Switzerland. http://www.bats.ch/bats/publikationen/1997-2_gmo/gmo_food.pdf.

Holdrege, C. 2014. Understanding the unintended effects of genetic manipulation: An introduction. http://natureinstitute.org/txt/ch/nontarget.php (accessed March 3, 2014).

Institute of Food Science and Technology (IFST). 2008. Information Statement. Genetic modification and food. www.ifst.org (accessed April 20, 2014).

Institute of Medicine and National Research Council of the National Academies. 2004. *Safety of Genetically Engineered Foods: Approaches to Assessing Unintended Health Effects*. The National Academies Press, Washington, DC.

International Service for the Acquisition of Agri-biotech Applications (ISAAA). 2005. Agricultural biotechnology: A lot more than just GM crops. Global Knowledge Center on Crop Biotechnology, ISAAA SE*Asia* Center. Second Printing. http://www.isaaa.org/kc/ (accessed March 10, 2014).

ISO Standards. ISO 24276:2006. 2009. Foodstuffs—Methods of analysis for the detection of genetically modified organisms and derived products—general requirements and definitions. http://www.iso.org/iso/home.html (accessed January 20, 2014).

ISO Store: to order ISO24276:2006. 2009. Foodstuffs—Methods of analysis for the detection of genetically modified organisms and derived products—general requirements and definitions.

Kuiper, H.A. 1999. Summary report of the ILSI Europe workshop on detection methods for novel foods derived from genetically modified organisms. *Food Control*, 19:339–49.

Larkin, P.J. and W.R. Scowcroft. 1981. Somaclonal variation: A novel source of variability from cell cultures for plant improvement. *Theoretical and Applied Genetics*, 60:197–214.

Lübeck, M. 2014. Detection of genetically modified plants. Methods to sample and analyse GMO content in plants and plant products. http://www2.sns.dk/erhvogadm/biotek/detection.htm (accessed January 26, 2014).

Markoulatos, P., Siafakas, N., Papathoma, A. et al. 2004. Qualitative and quantitative detection of protein and genetic traits in genetically modified food. *Food Reviews International*, 20(3):275–96.

Marmiroli, N., Maestri, E., Gullì, M. et al. 2008. Methods for detection of GMOs in food and feed. *Analytical and Bioanalytical Chemistry*, 392:369–84.

Marx, G.M. 2010. Monitoring of genetically modified food products in South Africa. Dissertation submitted in fulfilment of requirements for the degree Doctor of Philosophy in the Faculty of Health Sciences, Department of Haematology, University of the Free State, Bloemfontein, South Africa.

Murnaghan, I. 2013. Types of techniques used to genetically modify food. http://www.geneticallymodifiedfoods.co.uk/types-techniques-used-genetically-modify-food.html (accessed January 10, 2014).

Nesvold, H., Kristoffersen, A.B., Holst-Jensen, A., and Berdal. K.G. 2005. Design of a DNA chip for detection of unknown genetically modified organisms (GMOs). *Bioinformatics*, 21(9):1917–26.

Omar, R. 2009. The genetic engineering of rice and other crops. The Genetic Engineering industry: Impacts and regulation. http://www.fomca.org.my/v2/images/stories/pdf/071009_GE_Seminar_ppt6.pdf (accessed January 20, 2014).

Purves, W.K., Sadava, D., Orians, G.H., and Heller, H.C. 2001. *Life: The Science of Biology*. Sixth Edition. pp. 316–7. Sinauer Associates, Inc., New York, NY.

Rowland, G.G., McHughen, A.G., and Bhatty, R.S. 1989. Andro flax. *Canadian Journal of Plant Science*, 69:911–3.

Rowland, G.G., McHughen, A.G., Hormis, Y.A., and Rashid, K.Y. 2002. CDC Normandy flax. *Canadian Journal of Plant Science*, 82:425–6.

Sanford, J.C., Klein, T.M., Wolf, E.D., and Allen, N. 1987. Delivery of substances into cells and tissues using a particle bombardment process. *Journal of Particulate Science and Technology*, 5:27–37.

Schreiber, G.A. 1999. Challenges for methods to detect genetically modified DNA in foods. *Food Control*, 10(6):351–2.

Science.gov. 2013. Your gateway to U.S. Federal Science. Detect genetically modified food. http://www.science.gov/topicpages/d/detecting+genetically+modified.html# (accessed January 26, 2014).

Sebastian, S.A. and Chaleff, R.S. 1987. Soybean mutants with increased tolerance for sulfonylurea herbicides. *Crop Science*, 27:948–52.

SOS. 2014. Genetically modified organisms and their influence on the people! http://gmosafety-engl.webnode.com/contact-us/ (accessed January 18, 2014).

Sugar, I.P. and Neumann, E. 1984. Stochastic model for electric field-induced membrane pores electroporation. *Biophysical Chemistry*, 19(3):211–25. http://www.sciencedirect.com/science/article/pii/0301462284870039 (accessed January 10, 2014).

Swanson, E.B., Couman, M.P., Brown, G.L., Patel, J.D., and Beversdorf, W.D. 1988. The characterization of herbicide tolerant plants in *Brassica napus* L. after *in vitro* selection of microspores and protoplasts. *Plant Cell Reports*, 2:83–7.

The Nature Institute. 2014. Nontarget effects of genetic manipulation. A project of the Nature Institute. http://www.natureinstitute.org/nontarget/ (accessed January 20, 2014).

Trapmann, S., Corbisier, P., and Schimmel, H. 2004. Reference materials and standards, Chapter 4, in F.E. Ahmed (Ed.), *Testing of Genetically Modified Organisms in Foods*. Food Products Press. An Imprint of The Haworth Press, Inc., New York, London, Oxford, pp. 101–16.

Tritech Research, Inc. 2014. http://www.tritechresearch.com/minj.html (accessed January 16, 2014).

Tung Nguyen, C.T., Son, R., Raha, A.R., Lai, O.M., and Clemente Michael Wong, V.L. 2008. Detection of genetically modified organisms (GMOs) using molecular techniques in food and feed samples from Malaysia and Vietnam. *International Food Research Journal*, 15(2):155–66.

Viljoen, C.D. 2005. Detection of living modified organisms (LMOs) and the need for capacity building. *Asian Biotechnology and Development Review*, 7:55–69.

SUGGESTED REFERENCES

Brandner, D. 2002. Detection of genetically modified food: Has your food been genetically modified? *American Biology Teacher*, 64:433–42.

Fagan, J. 2004. GMO traceability requirements expand. *Food Technology*, 58:124.

Ganguli, A. 2009. *Biotechnology Fundamentals and Applications*. Oxford Book Co., New Delhi, India.
Glenisk for an Organic Ireland. 2014. What is the big deal about GM food? http://www.glenisk.com/why-organic/organic-myth-busting/-what's-the-big-deal-about-GM-food/ (accessed January 20, 2014).
Teixeira, J.A. and A.A. Vicente (Eds.). 2013. *Engineering Aspects of Food Biotechnology*. CRC Press, Boca Raton, FL.
Tiedje, J.M., Colwell, R.K., Grossman, Y.L. et al. 1989. The planned introduction of genetically engineered organisms: Ecological considerations and recommendations. *Ecology*, 70(2):298–315.

4

Applications of Genetic Modification at the Field and Commercial Levels

4.1 INTRODUCTION

As discussed in Chapter 2, biotechnology has many applications including agricultural biotechnology, otherwise referred to as "green biotechnology." Agricultural biotechnology includes plant, animal, and food biotechnologies since agriculture in a proper sense includes both plant and animal production, and food is generally obtained from plant and animal resources. A number of biotechnological techniques have been used in agriculture to augment different traits of plants and animals. These techniques aimed at increasing quantities of plant and animal products, as well as improving their qualitative attributes. The last few decades witnessed many developments in the application of biotechnology in the field of agriculture and food. A number of field crops had their genetic make-up altered to achieve certain desirable characteristics to help improve production in addition to enhancing some quality factors. GM foods are produced from such GM crops, or they include ingredients derived from GMOs or GM plants.

Biofuel production also stems from agricultural resources since the basic materials used in its manufacture originate from plant material. Biofuel comes in two forms: bioethanol and biodiesel. The plant material used to produce each type of biofuel is different. Bioethanol is mainly produced from sugar cane, corn, and some cereals, whereas biodiesel is produced from soybean, rapeseed, and sunflower seed. The use of food crops to produce biofuel has lead to a number of disputes. Critics of this process believe that the world is already short of food products and that the use of such food crops to produce biofuel will further aggravate the world food situation.

Modern agricultural biotechnology includes a wide range of tools which are used to study and manipulate the genetic make-up of plants and animals that are applied in the production and/or processing of agricultural products (GreenFacts 2013). The following sections include discussions on the different types of agricultural biotechnologies in use today and how they are applied at the field and commercial levels.

4.2 PLANT BIOTECHNOLOGY

Plant biotechnology is one field of agricultural biotechnology that encompasses a set of techniques to adapt plants for specific needs or opportunities (USDA-National Institute of Food and Agriculture 2013b). How these techniques are carried out in the laboratory has been discussed in Chapter 3. The aim of this chapter is to explore how they are applied on larger scale at the farm and commercial levels.

One of the main areas that plant biotechnologists focus their research on is the enhancement of some agronomic traits of crops. These include characteristics such as those that improve resistance to pests, reduce the need for pesticides and herbicides, and increase the ability of the plant to survive adverse growing conditions such as drought, cold, and soil salinity (Pew Initiative on Food and Biotechnology 2007).

In applying agricultural biotechnology, it is sometimes common to try and combine a number of needs and/or opportunities in one crop at the same time, for example, herbicide tolerance, pest resistance, drought tolerance, in addition to providing food and healthful nutrition. This helps to manage resources, for example, time and money, in a positive manner.

Although the enhancement of different agronomic traits in a number of crops has been achieved, improvement of products genetically modified to meet food processors or consumers' needs seems to be limited. Efforts in this direction have been going on. More attention has been given to the production of GE crops with enhanced health and nutritional properties.

GM crops are categorized as first, second, or third generation, based on their intended benefit and use (Marx 2010). The main feature of the first-generation GM crops is that they have improved agronomic traits for insect and weed management and were first commercialized in 1994. Second- and third-generation GM crops have been developed during the last 15 years, but their commercial use has not been as extensive as the first-generation GM crops. Some of the intended benefits and uses of second-generation GM crops include the improvement of the nutritional attributes of food crops and reduction of harmful effects or allergenicity of those crops. Other enhanced qualities of second-generation GM crops include improved shelf life and transport stability (Marx 2010). The focus of producing third-generation GMOs is to address industrial application needs, which encompass the production of pharmaceuticals, industrial compounds, or biofuels. It is envisaged that the trend of developing more generations of GM crops would result in many new applications and commercialization of GM crops.

Field food crops such as corn, soybean, as well as some horticultural crops including fruits and vegetables can be modified in different ways to improve their characteristics in a manner that protect them from diseases, weeds, pests, extreme and harsh weather conditions, and possible damage during long-distance travel. Their nutritional quality traits can also be enhanced through GE. It is possible to introduce more than one trait into different crop varieties, for example, combining herbicide tolerance and insect resistance in corn and cotton. Examples of the different ways by which crops can be genetically modified include

- *Herbicide resistance.* Herbicide application only kills weeds without affecting the crop, for example, as used for corn and cotton.
- *Pest resistance.* Crops show less damage caused by insects, for example, as in corn and soybean.
- *Disease resistance.* Crops become more resistant to diseases caused by bacteria, fungi, and viruses, for example, as in corn.
- *Cold temperature resistance.* Crops would withstand and survive freezing and thawing conditions, for example, as for strawberries.
- *Delayed ripening.* Products (particularly perishable ones) experience less tissue damage during transport, for example, as in tomato.

4.2.1 Development of GM Crop Varieties with Improved Resistance to Herbicides Used for Weed Control

Weeds are naturally growing, undesirable, and unwanted plants that accompany the growth of human-planted crops. Weed control systems are essential components of any farming system. Existence of uncontrolled weeds can reduce crop yield by more than 50%, impair crop quality, contaminate the harvest with undesirable weed material, and increase the likelihood that the crop will be attacked by insects or diseases (FAO 2005). These effects result from the weeds' competition with the planted crop for nutrients, water, space, and light (i.e., weeds vs. crops). Weeds can also harbor insect and disease pests, clog irrigation, and drainage systems; undermine crop quality; and deposit weed seeds into crop harvests. Two known classical examples of weeds are the parasitic flowering plant commonly known as "witchweed" (*Striga* spp.) and the parasitic broomrapes (*Orobanche* spp.). These two weeds represent the major pests of staple crops grown in sub-Saharan Africa. Adequate and efficient control of weeds has been found to increase crop yields substantially.

Other harmful and undesirable effects of the presence of weeds include possible damaging of the natural habitat, disturbing and altering the ecosystem processes, and causing displacement of native plant species. Traditionally, weed control has been done using mechanical methods, for example, labor-intensive weeding and hoeing, chemical methods, for example, applying herbicides, and agronomical practices, for example, following crop rotation, or a combination of two or more techniques. The development of herbicide-tolerant crop varieties provides alternative new options for the control of weeds which are negatively affecting agricultural production. Herbicide resistance allows farmers to control weeds with chemicals that would otherwise damage the crop itself (Pew Initiative on Food and Biotechnology 2007).

Although the two terms "herbicide resistance" and "herbicide tolerance" are sometimes used synonymously, a clear distinction of their meanings has been presented by the Weed Science Society of America (WSSA 2013). The following definitions have been cited by the WSSA:

> Herbicide resistance is the inherited ability of a plant to survive and reproduce following exposure to a dose of herbicide normally lethal to the wild type. In a plant, resistance may be naturally occurring or induced by such techniques as genetic engineering or selection of variants produced by tissue culture or mutagenesis.

Herbicide tolerance is the inherent ability of a species to survive and reproduce after herbicide treatment. This implies that there was no selection or genetic manipulation to make the plant tolerant; it is naturally tolerant.

In the present discussion, the term "herbicide resistance" will be used since it is more relevant to what is being presented here.

For centuries, conventional agricultural systems have been using selective herbicides to get rid of weeds that grow and compete for nutrients and water, with field crops. These herbicides kill most, but not all, types of weeds and are not harmful to the crops. Another group of herbicides (referred to as nonselective or broad-spectrum herbicides) have also been used to remove all types of weeds (GMO Compass 2013a). This latter group of herbicides can also kill crops which are sensitive to the chemical herbicide applied. To address this issue, several crops have been genetically modified to be resistant to nonselective herbicides. This group of transgenic crops has been induced with genes that enable them to break down the active chemical ingredient available in the herbicide. This makes the herbicide harmless to the respective crop. The weeds can be controlled throughout the whole growing season by spraying the herbicide at specified times without any noticeable negative effect on the field crops. Examples of the currently followed herbicide-resistant cropping systems include RoundupReady (the active ingredient of which is glyphosate) used for soybean, corn, rapeseed, and cotton, and Liberty Link (the active ingredient of which is glufosinate) used in soybean, corn, canola, cotton, and sugar beet. RoundupReady is produced by Monsanto, while LibertyLink is produced by Bayer, two of the giant biotechnology companies worldwide.

Roundup is a broad-spectrum herbicide used to kill crop weeds since Monsanto introduced it in 1974. It is effective in destroying a wide variety of annual and perennial grasses, sedges, broad-leaf weeds, woody shrubs, and commercial crops (Estes and Watson 2013b). The mechanism by which glyphosate kills these weeds and plants is by inhibiting the creation of EPSP synthase, an enzyme required to synthesize the amino acid phenylalanine (Kleiner 1998).

Glyphosate is not a relatively new herbicide as some people believe. It has in use for over 40 years as a safe, broad-spectrum, and less soil-persistent herbicide for many field crops. When it was used for soybean, it killed the soy as well as the weeds (Institute of Food Science and Technology [IFST] 2008). Since farmers could not use it for soybean, scientists were able to genetically modify soybean in such a way that it is not killed by glyphosate, but rather resist its effect.

4.2.2 Development of GM Crop Varieties with Improved Resistance to Pests

Cropping systems across the world are affected by several thousands of insect pest species, which cause appreciable damage to the crops if not adequately and effectively managed. Each of these species has its own characteristic damage, distribution, and natural enemies. Some plant pests feed on different plants and consume different parts of the plant, for example, leaves, flowers, and stems. Naturally included are the commercially cultivated crops intended for food, feed, or other purposes. This insect attack will damage the crops, reduce the yield, and lower the quality of the product. Pest damage to crops can take place in the field during growth period, during harvest time, or during postharvest stages. It can also occur during crop storage (GMO Compass 2013b). Insects consume a large share of food and fiber destined for humans (pre- and postharvest). The worldwide economic damage caused by insect pests to agricultural and horticultural crops and to orchards stands at 100 billion dollars annually (FAO 2005).

The two pest species that cause most of the crop damage are the *Lepidoptera* and the nematodes. *Lepidoptera* represent a diverse and important group of insects. Nematodes survive saprotrophically or parasitically on plants, animals, and humans. Although most nematodes found in the soil are beneficial, some are pathogens of different plant parts, for example, roots, stems, leaves, or seeds. Plant parasitic nematodes cause huge crop losses worldwide by attacking a wide variety of plant species, such as staple crops, vegetables, and ornamentals.

Different crop protection strategies and approaches to reduce the damage by pests have been tried. The three main strategies which are widely used are

1. The use of chemical pesticides and insecticides. The problems encountered with the application of chemicals include the possible development of pesticide-resistant pests and the potential harmful effects of many chemical pesticides or insecticides for human health and for the environment.
2. The use of natural enemies of the pests. In this approach, alien species are introduced in the environment. These natural enemies help to kill crop pests. One major limitation of this approach is the scarcity of natural pest enemies. In addition to that, this approach might have some risks since it is not easy to predict the possible effects on the surrounding environment.

3. The development of pest-resistant crops through hybridization or GE. Many plant pests have proved either difficult or uneconomical to control with chemical treatment, traditional breeding, or other agricultural technologies and in these instances, in particular, biotechnology has proved to be an effective agronomic tool (Pew Initiative on Food and Biotechnology 2007).

The search for effective, safe, sustainable, and low-cost methods of crop protection has been going on for a long time. One of the advanced techniques, tried and used, is the development of transgenic plants which are pest-resistant. The development of this technique stemmed from an observation that one type of soil bacteria known as *Bacillus thuringiensis* or Bt has the ability to produce a form of protein which is toxic to different types of herbivorous insects. Bt produces the toxin known as Bt toxin in a crystalline inactive form. The active form of the toxin (known as delta endotoxin) is released after the insects consume the protein and it subsequently leads to destruction of the insects' gut. This information has been used by scientists to develop pest-resistant plants. Scientists applied GE techniques to extract the genes responsible for production of the Bt toxins in *B. thuringiensis* and introduced them into plants. These genes would then behave as if they are part of the genetic make-up of the plant and start producing the Bt toxin. Being able to produce the toxin on their own, the plants will be able to defend themselves against specific species of insects. This would help to control insects without the need to use chemical insecticides. Crops which have been genetically modified for insect resistance are designated as Bt crops (Estes and Watson 2013a). Examples of Bt crops being cultivated today include Bt cotton, Bt corn, and Bt potatoes, which are widely grown in the United states, Canada, Argentina, South Africa, France, and Spain. More countries, both in developing and developed regions, are starting to introduce this technology to combat insect attack on cultivated field crops.

In order to adopt the technology aiming at developing pest-resistant crops, it is important to have adequate information on the pest species, the type of damage it causes, the parts of the plant affected, and when the damage might occur. This information helps in monitoring the geographical distribution and life history of the pests and consequently in their effective control.

Biopesticides, also known as biological pesticides, represent a group of pest control interventions which are derived from natural materials such as animals, plants, or microorganisms. Biopesticides are defined differently in the EU and in the United States. In the EU, biopesticides have been defined as "a form of pesticide based on microorganisms or natural products" (European Commission 2008). On the other hand, the EPA in the United States (US EPA 2014) indicates that biopesticides "include naturally occurring substances that control pests (biochemical pesticides), microorganisms that control pests (microbial pesticides), and pesticidal substances produced by plants containing added genetic material (plant-incorporated protectants) or PIPs" (Environmental Protection Agency of the USA 2012). Biopesticides are considered as important components of the integrated pest management (IPM) programs, substituting the synthetic chemical plant protection products (PPP). There are three recognized classes of biopesticides, namely microbial pesticides (including bacteria, viruses, and nematodes), biochemical pesticides (naturally occurring substances used to manage pests and microbial diseases), and plant-incorporated protectants (PIPs) which are GE plants. Examples of commonly used biopesticides include: bacteria-based biopesticides, for example, *B. thuringiensis* or Bt, fungi-based biopesticides, for example, *Beauveria bassiana* (Bb), and virus-based biopesticides, for example, the rod-shaped *Baculoviruses*. The most widely known uses of these pesticides is the one genetically engineered from *Bt* as described earlier, and used to protect crops such as corn, cotton, and potatoes. Biopesticides are usually inherently less toxic than conventional pesticides. On the other hand, they are needed in only small amounts and they mainly affect the target pest and closely related organisms (US EPA 2014).

4.2.3 Development of GM Crop Varieties with Improved Resistance to Bacterial, Fungal, and Viral Diseases

Like all other living organisms, plants can be afflicted by various diseases. Plants can be infected by different types of disease-causing microbes, for example, fungi, bacteria, viruses, by nematodes and other pathogens. Some plant species have a natural ability to fight or resist diseases. The degree of resistance to diseases differs with the plant species. However, some damage to the plants due to infections by these pathogens will eventually occur. If this happens to cultivated crops, it will cause lower yields and produce with lower quality. To address this issue and protect cultivated crops from harmful organisms, different approaches have been used, for example, using chemical materials to destroy the pathogenic organisms. Biotechnology has been used to develop GM crops with resistance to various diseases without the need to use chemicals.

Examples of such crops are papayas and squash which are modified to resist viral diseases.

4.2.3.1 Transgenic Crop Varieties Resistant to Bacterial Diseases
GE has been used to develop transgenic crop varieties which are resistant to diseases caused by bacteria. Examples of these crops include wilt-resistant potato and wheat varieties, blight-resistant Basmati rice variety, and leaf blight-resistant rice, potato, and cabbage. Development of GM plant varieties with bacterial disease resistance has received less attention than diseases caused by fungi and viruses. The lower level of activity on resistance to bacterial diseases compared to other diseases may be due to both a lower perception of the importance of bacterial diseases and the number of crops infected by them when compared to the incidence of viral diseases and to more readily available alternative technologies to combat bacterial diseases (FAO 2005).

4.2.3.2 Transgenic Crop Varieties Resistant to Fungal Diseases
Fungi are known to cause a number of plant diseases. Most of the known plant diseases worldwide are caused by fungal pathogens, for example, the rust fungi are the most widespread plant disease-causing organisms, which cause considerable crop losses each season. For example, the fungal agent of rice blast disease (*Magnaporthe grisea*) destroys 157 million tons of cultivated rice each year, enough rice to feed 60 million people worldwide (Pennisi 2001). Other examples of fungal diseases of crops include cereals' powdery mildew, downy mildew, blight and grey mould, loose smuts Rice Blast, sheath blight, vegetables leaf rot, fruit spot diseases, black scab, silver scurf, and peanuts white mold. Chemical fungicides have been used successfully by many farmers around the world to control plant fungal diseases. Traditional breeding techniques have also been used to develop fungus-resistant cultivars. The high cost of chemical fungicides, their potential effect on the health of farm workers, and the possible effects on the environment lead to the development of transgenic crop varieties with resistance to fungal diseases. The incorporation of plant-derived resistance genes against fungal pathogens into susceptible varieties allowed the development of resistant varieties which can deliver high yields in the absence of chemical fungicide applications.

4.2.3.3 Transgenic Crop Varieties Resistant to Viral Diseases
Viruses are also known as plant disease-causing microorganisms. One of the known viruses that cause plant diseases is the beet necrotic yellow

vein virus (BNYVV) which causes sugar beet to develop smaller, hairier roots which causes yield reduction (GMO Compass 2013c). Experience has shown that the control of plant viral diseases is not an easy job, particularly after the infection commences. GE techniques have been used to generate virus-resistant crop varieties. Biomolecular techniques were established to develop transgene cassette-based approaches for control of most crop viruses (FAO 2005). Other approaches used to control crop viruses include production of virus-free propagation plant material and the control of insects responsible for transmitting virus pests.

4.2.4 Development of GM Crop Varieties with Improved Resistance or Tolerance to Abiotic Factors

Abiotic factors, or stresses, in relation to farming systems refer to conditions of drought, extreme temperatures (very high or very low), soil salinity, frost, and low soil fertility. Abiotic stresses are considered as one of the main factors that limit crop productivity and quality in agricultural areas around the world. Different plants possess different abilities to cope with abiotic stresses, both between species and within members of the same species. Experience has shown that knowledge of the exact mechanism of abiotic stress tolerance would help in improving crop productivity. The development of crop varieties which can tolerate abiotic stresses would be beneficial to agriculture in regions where abiotic stresses are chronic and are leading factors in low crop productivity or poor quality attributes. The abiotic factor which received more attention by agricultural researchers is drought. This is due to the fact that it occurs widely and affects many crops, particularly in developing regions.

Similar to all living organisms, different plants, whether cultivated or wild, need water for their survival. The amount of water needed by different plant species differs and is governed by multiple factors. Some plants need more amounts of water than others. Those which need less water and can survive and flourish under limited amounts of water are described as "drought-tolerant." This group of plants is usually grown in areas with low amounts of rain or where there are inadequate irrigation regiments. Developing crops which are drought-tolerant, or resistant, has been the focus of plant breeders, and recently modern plant biotechnologists, for some time. Both traditional breeding and modern biotechnology approaches have been successfully employed to produce drought-resistant field crops at commercial levels.

Genuity®DroughtGard™ Hybrids are produced by Monsanto to help farmers mitigate the risk of yield loss when experiencing drought stress. The Western Great Plain farmers in the United States had the first opportunity to plant DroughtGard Hybrids on a commercial level (Monsanto 2014). DroughtGard corn produced by Monsanto in 2013 is considered the first drought-tolerant commercial crop. Hybrid seeds carrying this trademark will be produced based on the bacterial cspB gene (DiLeo 2012). It might be released in 2017 in Africa. One concern that has been expressed about developing GM drought-resistant crops is that there is the possibility that they cross with their wild relatives leading to creation of drought-resistant weeds which would be difficult to eradicate (Zubair 2011).

The much debated issue of global warming is expected to have negative effects on many aspects of human and other living organisms' lives. One of these expected effects is that land which is suitable and used for agriculture in developing countries would become less productive, even unusable. Under such conditions, drought-tolerant crops could prove to be successful for cultivation.

As the case with other developments in GE to produce GM crops and foods, using GE to develop drought-resistant crops is not welcomed by everybody. GM Watch (GMWATCH 2013) in their website http://www.gmwatch.org/component/content/article/31-need-gm/12319-drought-resistance believes that there is no need to produce drought-resistant crops through GE. They indicate that drought-resistant crops have been successfully developed through non-GM plant-breeding technologies. Some of the drought-resistant crops produced through these traditional methods, as stated by GM Watch, include maize, beans, cowpea, soybean, rice, chickpea, tomatoes, sorghum, and millet. Information included in this document covers details of when and where these technologies have been developed and used.

4.2.5 Development of GM Crop Varieties with Improved Nutritional Quality Traits

The enhancement of the nutritional quality of food crops has been a main concern and a goal for agriculturalists, food scientists, and nutritionists. Different strategies have been developed and used to achieve this goal. Nutritional enhancement can be done at different stages of the food chain. It can be done at the crop production level and at the food processing level. Plant breeders have succeeded in developing new food crop varieties with enhanced nutritional quality trait profiles. This represents enhancement at the production level. Other strategies used include fortification of food products with various nutrients, representing enhancement at the processing stage. With regard to the involvement of biotechnology, crop GE can be used to generate crop varieties with improved profiles for one or more nutrients. Enhancing the micronutrient (vitamin and mineral content) status of staple crops is considered to be one approach where crop biotechnology could generate crop varieties that could be used to strengthen food security and prevent malnutrition (FAO 2005). Different forms of malnutrition, for example, protein-energy malnutrition (PEM) and micronutrient malnutrition, can be managed through food crop enhancement using biotechnological techniques. Micronutrient malnutrition, especially lack of iron, zinc, and vitamin A, currently affects more than half the world's population. Biotechnology can, thus, have a positive role in improving food and nutrition security worldwide.

Agricultural biotechnology has been used by scientists to develop GE foods with enhanced nutritional properties for the benefit of consumers. A group of food products aimed at consumers are those products with added health benefits, that is, functional foods. Biotechnology has been successfully used to modify and improve the profiles of protein and amino acids, vitamins, minerals, and edible oil and fatty acids.

Seed proteins of a number of major food crops, for example, cereals and legumes, are naturally limiting in some amino acids, particularly the essential amino acids. Cereal grains are usually deficient in the amino acids lysine and tryptophan, while legumes are deficient in the sulfur amino acids methionine and cystine. Plant breeders in Mexico have managed to develop corn varieties which have enhanced levels of the two essential amino acids, lysine and tryptophan. In addition to modifying the genes that encode seed proteins by GE can be used to address the problem of nutritional quality in some staple crop varieties.

Vitamin profile can also be improved through GE. Vitamin A content of staple foods, which are naturally low in this vitamin, has been increased using GE. The highly publicized, high pro-vitamin A transgenic rice, commonly known as "golden rice," is a typical example of the use of biotechnology to enhance vitamin profile in food crops.

Although most staple crops, such as rice, wheat, maize, sorghum, cassava, and beans, are not considered as important resources of mineral elements in human diets, yet minor increases in the mineral content of these staples would have a pronounced effect on human nutrition and health. This is due to the fact that large amounts of these staples are usually consumed in the diet. Both conventional and modern biotechnology techniques can be used to improve the nutritional quality of staple crops in terms of mineral content.

Edible oils and fats have nutritional as well as functional qualities. Nutritionally, they contribute energy when ingested as part of a diet. From a health perspective, the excessive consumption of saturated fatty acids can increase the risk of heart disease as well as other chronic diseases. The fatty acid composition of an edible fat or oil affects the stability, flavor, and texture of fatty foods. This affects the behavior of the fat or oil during processing or storage. Edible oil content, characteristics, and fatty acid composition can be improved through GE. Plant sources of edible oil, that is. seeds, nuts, and fruits, can be modified to achieve this goal. The fatty acid composition of an oil or fat can be modified in a way that suits the purpose for which it is going to be used. The ratio between saturated fatty acids and unsaturated fatty acids can be controlled through GE. An example of a functional food oil developed through biotechnology is Calgene's high lauric acid canola, Laurical™, which is modified to suit confectionery products and chocolates. Some of the reported results on the use of GE in the edible oils area include field trials in Argentina on maize and soybean, and in Mexico on canola expressing high levels of lauric acid (FAO 2005). Work on oil palm with low saturated fatty acids in Indonesia and Malaysia, and on coconut with high lauric acid in Philippines, is also reported. Some of the aspects of edible oils and fats that can be improved through GE in order to enhance their nutritional, health, and functionality characteristics include: reduction of saturated fatty acids, introduction of omega-3 polyunsaturated fatty acids, and enhancing the availability of novel fatty acids such as linoleic acid.

4.2.6 Weighing Risks and Benefits of Applying GE to Crops

One of the important things that needs to be considered carefully with regard to application of GE to crops is to weigh the risks and benefits of this technology. The risks and benefits associated with biotechnology applications have not been adequately assessed by the government agencies responsible for such evaluations. This might be due to the pressure exerted on these agencies to hasten the approval for release of bioengineered crops. Peterson et al. (2000) proposed a different and more detailed approach to assess the risks and benefits of specific types of GM modification, for different species, in different ecological contexts. Some of the possible risks and benefits associated with specific GM crop modifications are shown in Table 4.1. The authors added that compiling such lists only represents the beginning of the comprehensive risk assessment. They believe that there is need to quantitatively assess the risks for specific organisms in different contexts on a case-by-case basis. Further information on the risks and benefits can be obtained by trying to find answers for some questions pertaining to the agricultural, ecological, and social dimensions. Some of the questions that would help to evaluate the relative benefits and risks of a specific GM crop have been suggested by Peterson et al. and are shown in Table 4.2.

4.2.7 Production of Biofertilizers

Like all living organisms, plants need nutrients for nourishment, development, propagation, and all other related activities to sustain their life. These nutrients either come naturally from, or added to, the soil. Fertile soils are those which contain adequate levels of all, or most, nutrients required by the plants growing in them. These nutrients include a wide variety of organic and inorganic substances in forms which are easily available to the plant or that can be converted into available forms. Soil nutrients can be depleted through continuous farming on the same piece of land without addition of nutrients from external sources. They might also be lost through other factors. This is when fertilization of the soil is needed and comes into effect. Phosphorus and nitrogen are two major nutrients needed by plants. They have important roles in plant quality, maturity, and to help the plant to withstand stress. Although these elements exist naturally in the environment, plants have a limited capacity to benefit from them by direct extraction from the environment. Plants are helped by the fungus *Penicillium bilaii* to benefit from the phosphate available in the soil. Through a series of chemical reactions, this fungus increases the phosphate availability to the plant. A biofertilizer produced from this fungus is applied either by inoculating seeds with the fungus or putting it directly in the soil where crops are planted. There are different types of fertilizers and different modes of their application. Traditionally, chemical fertilizers have been used to enrich soil fertility. Details of these are beyond the scope of this book and the focus here will be on biofertilizers produced through modern biotechnologies.

For many centuries, chemical fertilizers (and pesticides) have been used in different cropping systems as agricultural inputs to increase and improve crop production. Concerns about tremendous harm from these chemicals to the environment have been expressed. It is believed that the

Table 4.1 Examples of the Potential Ecological Benefits and Risks of Selected GM Crops

GM Modification	Benefits	Risks
Herbicide resistance in maize, cotton, and other crops	Reduce herbicide use. Increase opportunities for reduced tillage systems	Increase herbicide use. Reduce in-field biodiversity that may reduce the ecological services provided by agricultural ecosystems
Maize with Bt toxin	Reduce pesticide use. Kill fewer nontarget organisms than alternatives such as broad-spectrum pesticides	Promote development of Bt resistance, which will eliminate Bt as a relatively safe pesticide. Kill nontarget caterpillars and butterflies, such as monarchs (Pimentel et al. 2000)
Virus resistance in small grains due to coat proteins	Reduce insecticide use to control insect dispersers of pathogens (Hails 2000)	Facilitate the creation of new viruses (Hails 2000). Move genes into nonagricultural ecosystems where the subsequent increase in fitness of weedy species could eliminate endangered species
Terminator or other sterilizing traits in crops and ornamentals	Prevent the movement of traits to nontarget species. Prevent the movement of introduced species to other ecosystems (Walker and Lonsdale 2000)	Prevent farmers from developing their own seed supplies adapted to local conditions (Conway 2000)
Synthesis of vitamin A or other nutrients	Improve nutrition of people who depend heavily on rice (Conway 2000)	Disrupt local ecosystems if an ecologically limiting nutrient or protein is produced
Nitrogen fixation by nonlegumes	Reduce energy used in fertilizer production and application (Pimentel et al. 2000)	Add to excess N leaching from agriculture, degrading human health and reducing biodiversity

Table 4.2 Questions to Assess the Relative Benefits and Risks of a GM Crop

Type of Impact	Benefit-Related Questions	Risk-Related Questions
Agricultural	Are alternatives available that provide greater agronomic, economic, social, and ecological benefits? Does the GM crop prevent some specific harm to humans or ecosystems, for example, does it reduce pesticide use?	Are risks minimized though good design, for example, is it certain that genes inserted into chloroplast DNA cannot escape through pollen? Has the organism been examined to determine whether genetic modifications to produce a desired trait have not also inadvertently produced risky changes?
Ecological	Does the GM crop help solve an existing environmental problem, for example, does it produce sterile feral animals to control pests (Walker and Lonsdale 2000)?	Does the modified trait have the potential to increase the fitness of the organism outside the managed environment, for example, does it impart herbivore resistance or increase the reproductive rate? In the locale of release, can the trait spread to other species, that is, can the species hybridize with other species nearby?
Social	Will the benefits of this GM organism be widely shared? Does the GM crop provide some specific benefit to humans or ecosystems, for example, does it enhance human nutrition or help restore degraded land?	Is a mechanism in place for surveying for possible negative effects after widespread release has occurred? Who and what are at risk of being negatively affected by this GM crop? Do institutions exist that could mitigate the potential impacts of GM crops?

use of biofertilizers, which are considered to be more friendly to the environment, and are cost effective, is the answer to this problem. Biofertilizers refer to the soil nutrients made available to the plant through the use of living organisms, mainly bacteria and fungi. Biofertilizers advocates indicate that the use of this group of fertilizers will form part of the solution to the problems of increased soil salinity.

Naturally, plants have various relationships with microorganisms, particularly fungi, bacteria, and algae, the most common of which are *Mycorrhiza*, *Rhizobium*, and *Cyanophyceae*. Some of the observed benefits of these microbes include improvement of plant nutrition, disease resistance, and tolerance to biotic and abiotic conditions. These beneficial relationships have been used in a positive manner to develop biofertilizers. Biofertilizers such as *Rhizobium, Azotobacter, Azospirillum*, and blue green algae (BGA) have been developed and used successfully for a long time. *Rhizobium* inoculant is usually applied for leguminous crops. *Azotobacter* can be used with crops such as wheat, maize, mustard, cotton, potato, and other vegetable crops. *Azospirillum* inoculations are recommended mainly for sorghum, millets, maize, sugarcane, and wheat. Blue green algae belonging to a general cyanobacteria genus, *Nostoc* or *Anabaena* or *Tolypothrix* or *Aulosira*, fix atmospheric nitrogen, and are used as inoculations for paddy crops grown both under upland and low-land conditions (Kiguli 2000).

The microorganisms in biofertilizers have the ability to retain and restore the soil's natural nutrient cycle. Both the plant and the soil thus benefit from biofertilzers' application. Sometimes the term "plant-growth promoting rhizobacteria" (PGPR) is used to describe biofertilizers denoting the several roles they play.

Samadhan (2013) reported some of the observed benefits of biofertilizers, for example, increase in crop yield by 20%–30%, replacing chemical nitrogen and phosphorus by 25%, stimulation of plant growth, biological activation of the soil, restoring the soil's natural fertility, and providing protection against drought and some soil-borne diseases.

Some of the documented observations and discoveries related to plant nutrition and soil fertility which have been reported more than 100 years ago include the fixation of atmospheric nitrogen by legumes and the isolation of root nodule bacteria (*Rhizobium*) from root nodules of legumes (Boraste et al. 2009). These discoveries form an integral component of the development and production of biofertilzers.

According to Boraste et al. (2009), strictly speaking, biofertilizers (or more accurately "microbial inoculants") are not fertilizers in that they supply nutrition to plants. Rather they a ganisms contained in a carrier material. The role th nourish is an indirect one and usually comes throug enhancing nutrient availability in the soil. They can rations containing live or latent cells of active types phosphate-solubilizing, or cellulose-digesting microor, can be applied to seeds, soil, or composting areas with the objective of multiplying such microorganisms to enhance the processes that improve nutrient availability to, and assimilation by, the plants.

4.2.8 Global Status of GM Crops

Currently, GM crops are grown in every continent except Antarctica (Freedeman 2009). The International Service for the Acquisition of Agri-Biotech Applications (ISAAA) produces annual reports on the global status of GM crops. These reports document the annual changes and trends in the cultivated GM crops in different countries around the world. It ranks the countries based on the areas cultivated by GM crops, in addition to the biotech crops grown in each country. The 2013 report produced by ISAAA (ISAAA 2014) shows that a record 175.2 million hectares of biotech crops were grown globally in 2013, at an annual growth rate of 3% (as compared to the rate of 6% during 2012), up 5 million from 170 million hectares in 2012 (ISAAA 2013a). The year 2013 was the 18th year of commercialization of biotech crops (1996–2013), when growth continued after a remarkable 16 consecutive years of increases. An unprecedented 100-fold increase in biotech crops planted area from 1.7 million hectares in 1996 to 175 million hectares in 2013 has been recorded. This fact qualifies biotech crops technology to be the fastest adopted technology in recent history. The global area of biotech crops by country in 2013 is reflected in Table 4.3. Figures 4.1 and 4.2 show the biotech crop countries and the global status of commercialized biotech/GM crops in 2013. Figure 4.3 shows the rising trend of area planted with GM crops in both developed and developing countries during the period 1996–2013.

Since the introduction of GM crops in the year 1996 and up to the year 2013, an estimate of millions of farmers in more than 30 countries around the world planted GM crops. These countries come from both developed and developing regions. During the year 2013, reports (ISAAA 2014) indicate that 27 countries planted GM crops. The vast majority of these countries (19) came from developing countries. This information is reflected in Table 4.3 and Figures 4.1 and 4.2. In 2012, Sudan became the fourth country in Africa, after South Africa, Burkina Faso, and Egypt, to commercialize a biotech crop—biotech Bt cotton. For the first time, developing countries grew more (52%) of global GM crops in 2012 than industrial countries. The five leading developing countries in biotech crops are China and India in Asia, Brazil and Argentina in Latin America, and South Africa in the continent of Africa, collectively grew 78.2 million hectares (46% of global) and

Table 4.3 Global Area of Biotech Crops in 2013: By Country (Million Hectares)[a]

Rank	Country	Area (million hectares)	Biotech Crops
1	USA[b]	70.1	Maize, soybean, cotton, canola, sugar beet, alfalfa, papaya, squash
2	Brazil[b]	40.3	Soybean, maize, cotton
3	Argentina[b]	24.4	Soybean, maize, cotton
4	India[b]	11.0	Cotton
5	Canada[b]	10.8	Canola, maize, soybean, sugar beet
6	China[b]	4.2	Cotton, papaya, poplar, tomato, sweet pepper
7	Paraguay[b]	3.6	Soybean, maize, cotton
8	South Africa[b]	2.9	Maize, soybean, cotton
9	Pakistan[b]	2.8	Cotton
10	Uruguay[b]	1.5	Soybean, maize
11	Bolivia[b]	1.0	Soybean
12	Philippines[b]	0.8	Maize
13	Australia[b]	0.6	Cotton, canola
14	Burkina Faso[b]	0.5	Cotton
15	Myanmar[b]	0.3	Cotton
16	Spain[b]	0.1	Maize
17	Mexico[b]	0.1	Cotton, soybean
18	Colombia[b]	0.1	Cotton, maize
19	Sudan[b]	0.1	Cotton
20	Chile	<0.1	Maize, soybean, canola
21	Honduras	<0.1	Maize
22	Portugal	<0.1	Maize
23	Cuba	<0.1	Maize
24	Czech Republic	<0.1	Maize
25	Costa Rica	<0.1	Cotton, soybean
26	Romania	<0.1	Maize
27	Slovakia	<0.1	Maize
	Total	175.2	

Source: Adapted from James, C. 2013. Global status of commercialized biotech/GM crops for 2013. ISAAA Brief 46. ISAAA, Ithaca, NY.
[a]Rounded off to the nearest hundred thousand.
[b]19 biotech mega-countries growing 50,000 hectares, or more, of biotech crops.

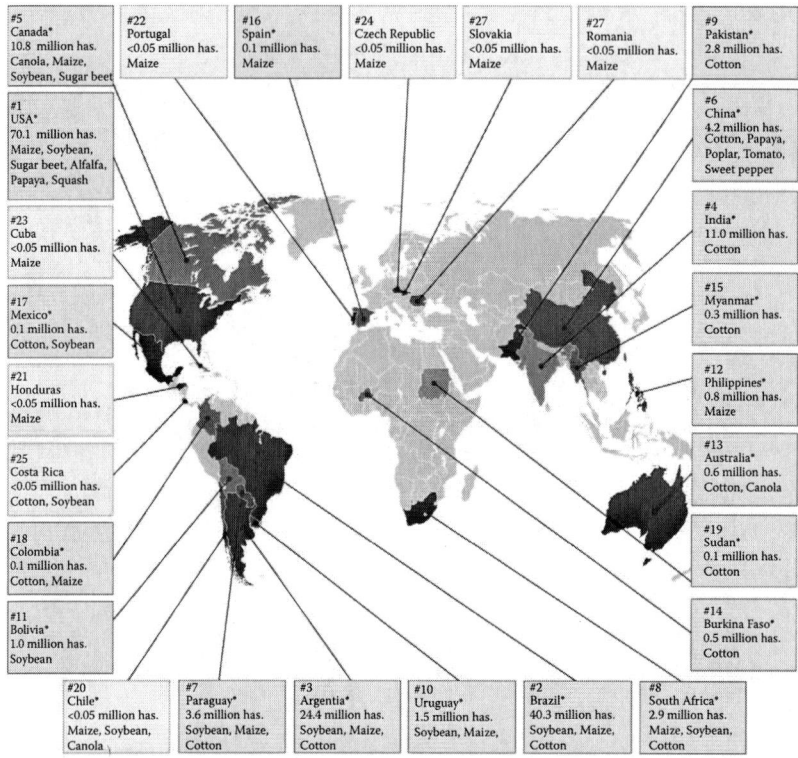

Figure 4.1 Global map of Biotech crop countries and mega-countries, 2013. (Adapted from James, C. 2013. Global status of commercialized biotech/GM crops for 2013. ISAAA Brief 46. ISAAA, Ithaca, NY.)

together represent ~40% of the global population of 7 billion, which could reach 10.1 billion by 2100. Brazil ranks second only to the United States with regard to the area cultivated by biotech crops, with 40.36 million hectares and is emerging as a global leader in biotech crops. The United States continued to be the lead producer of biotech crops globally with 70.1 million hectares, with an average adoption rate of ~90% across all biotech crops. Five EU countries (Spain, Portugal, Czechia, Slovakia, and Romania) planted a record 129,071 hectares of biotech Bt maize, a substantial 13% increase over 2011, with Spain growing 90%, equivalent to 116,307 hectares of the total Bt maize hectarage in the EU. The main GM crops

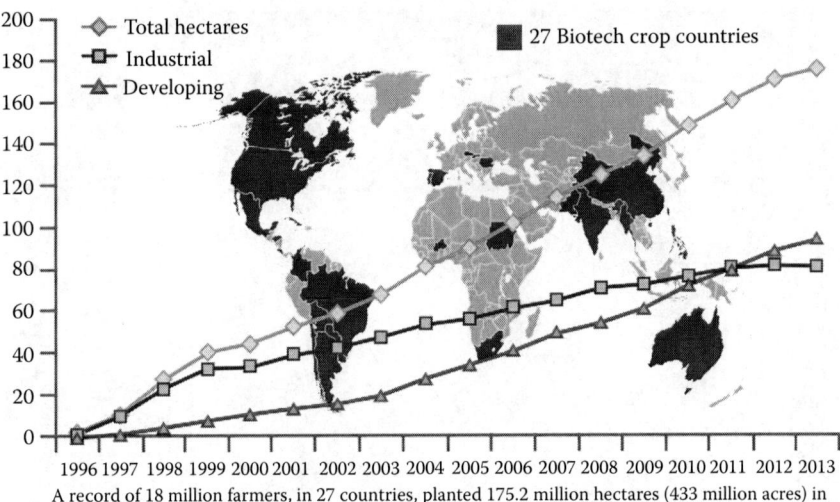

A record of 18 million farmers, in 27 countries, planted 175.2 million hectares (433 million acres) in 2013, a sustained increase of 3% or 5 million hectares (12 million acres) over 2012.

Figure 4.2 Global area of biotech crops: million hectares (1996–2013). (Adapted from James, C. 2013. Global status of commercialized biotech/GM crops for 2013. ISAAA Brief 46. ISAAA, Ithaca, NY.)

grown in 2013 were maize, soybean, cotton, canola, sugar beet, alfalfa, papaya, squash, tomato, and sweet pepper. Between 1997 and 2005, the total area of land planted with GM crops had increased 50 times from 4.2 million acres to 222 million acres.

In the United States and by 2006, 89% of the planted area of soybean, 83% of cotton, and 61% of corn were cultivated with GM varieties. The GM soybean was modified for herbicide tolerance, whereas corn and cotton were genetically modified for both herbicide tolerance and insect protection.

In spite of the fact that most of the GM crops are grown in developed countries, there has been continued large expansion of planting them in developing countries. For example, the largest increase in area planted with GM crops worldwide was in Brazil and India where GM soybean and cotton are grown.

The Monsanto corporation accounts for about 90% of plant traits developed through transgenesis worldwide (Andrews 2013). The approved GM crops in the United States include: corn, soybean, canola, rice beet, flax, and alfalfa (herbicide resistance); corn, cotton, potato, and tomato (insect resistance); corn and chicory (sterile pollen); papaya, squash, and plum

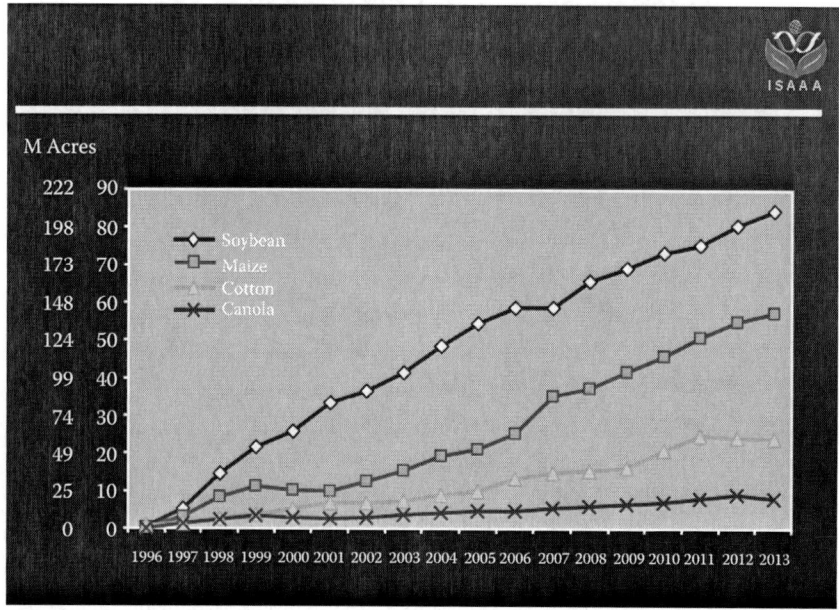

Figure 4.3 Global area of biotech crops, 1996–2013: industrial and developing countries. (Adapted from James, C. 2013. Global status of commercialized biotech/GM crops for 2013. ISAAA Brief 46. ISAAA, Ithaca, NY.)

(virus resistance); tomato (delayed ripening); canola and soybean (altered oil); corn (protein composition); and tobacco (reduced nicotine).

The Center for Environmental Risk Assessment (CERA), ILSI Research Foundation, Washington, DC, has created a website for "Citation for database" related to GM crops as well as to plants with novel traits that may have been produced using some traditional methods such as plant breeding (CERA 2012). It supplies a selection box with multiple value options to choose from, before running the database search. The value options given include event name, crop plant, trait, inserted gene, type of approval, country, original developer, and the year(s) approved.

4.2.9 GM Crops/Foods in Developing Countries with Special Reference to Africa

Herrera-Estrella and Alvarez-Morales (2001) argue that unless GM technologies are adopted in Third World countries to increase food production, more chemical fertilizers, insecticides, and herbicides will be used, further degrading the environment. They also believe that the costs of these agricultural inputs, which have been mainly produced for large mechanized farms in the First World, are not affordable by poor farmers in poor countries. The potential to increase food production at reduced costs and with less negative impact on the environment has been one of the gains modern GM technology has offered. To date, most of these achievements have been reached in developed countries and it is high time that more countries in developing world adopt GM technologies to tap on this potential. Herrera-Estrella (2000) believes that it is unfortunate that most developing countries do not have sufficient resources to implement the necessary biotechnological solutions to the major problems that limit agricultural productivity. It is in the developing world, however, especially in the areas of the world where yields are low due to the lack of technology, that biotechnology could have its greatest impact.

In most developing countries, particularly in the African continent, agriculture is the mainstay and a major component of the local economies. Large sectors of the rural population traditionally practice farming as the main source of income. Globally, people continue to face increased challenges, such as population growth and climate changes, that affect their livelihood. On the other hand, hunger and high levels of poverty continue to persist, calling for sustainable solutions. Agricultural sector can be involved in a positive manner to help in addressing these problems as it contributes about 35% of the African continent GDP and accounts for 70% of its labor force.

A number of developing countries have adopted modern biotechnological techniques and tools to develop GM crops and GM foods. Farmers in countries such as Argentina, Brazil, China, and India have benefited from these developments. Among the African countries which adopted biotechnologies are South Africa, Nigeria, Burkina Faso, Egypt, Uganda, and recently Sudan. There has been some uncertainty and hesitation among policy makers and some farmers in these African countries about the utilization of GM technologies in spite of the documented benefits of the technologies. More information on the potential benefits, safety, and future of the GM technologies would help to address those worries and give a better idea about their use.

Many reported documents (e.g., Falck-Zepeda et al. 2013) indicate that there is a potential in adopting GM technologies in the African context that would have pronounced value on the African agriculture.

The main concern that might hamper these developments is the insufficient level of investment in research and development. The capacity to conduct research for the development of indigenous GM crops could be attributed to the insufficient availability of financial resources and qualified human resources. Technology transfer of GM modification of crops could be tailored to suit local conditions in the respective countries. In general, the development of a focused and suitable strategy for science and technology that would dictate the investment in research and development is considered a useful approach towards benefiting from the appropriate GM technology. Policy makers have a pivotal role toward achieving these goals and it is believed they have the power to do it if they have the will.

As reported by Gouse (2013), based on a strong scientific and technological background, South Africa became the first African country where GM crops were produced on commercial scale when the insect-resistant Bt cotton was produced during the 1997/1998 cropping season. Following that and in 1998/1999, Bt maize was approved for commercial production. Bt yellow maize was planted during the same season, whereas Bt white maize was established in 2001/2002. Both herbicide-tolerant (HT) soybean and HT cotton were released for commercial production during the 2001/2002 planting season. Further, GM crops commercialization in South Africa was approved in the following sequence: in 2003/2004 HT maize, in 2005/2006 GM cotton containing stacked trait (Bt and HT), and in 2007/2008 Bt/HT maize. All the three crops produced through GM technology have been developed and marketed by large multinational corporations (Bothma et al. 2010).

To cater to the biosafety issues related to GM crop adoption and production in South Africa, the South African Committee for Genetic Experimentation (SAGENE) was formed in 1979 with the mandate of monitoring and advising the National Department of Agriculture and industry on the responsible development of GMOs through the provision of guidelines and the approval of research centers and projects. SAGENE became a statutory body in 1992 to advise on modern GM biotechnology (Gouse 2013).

Another African country that adopted GM crop technology is Uganda. Horna et al. (2013) reported that the recognition of the Ugandan government for the need to increase the performance of cotton through the use of GM technologies led to the adoption of the technology. In 2008, the National Biosafety Committee of Uganda approved the guidelines that enabled confined trials on insect-resistant Bt and HT cotton varieties which started in 2009. Another crop of economic significance in Uganda is banana. Although it is a staple crop in this African country, its production has been challenged by a number of constraints, for example, poor agronomic practices, insects and diseases' threat, and soil fertility depletion (Kikulwe et al. 2013). Efforts to address these issues and improve banana production were made and the use of GM technology was called for help. The main objective of using GM was to develop GM cultivars which are pest- and disease-resistant, possess improved agronomic characteristics, and which are acceptable to consumers. The GM banana developed by this technique is the first staple food crop in Uganda produced through modern biotechnology.

According to FAO (2004), potentially useful biotechnologies could be used in developing countries provided that adequate funding and education are used effectively. There is also need for capacity building to better integrate biotechnology in the food science and technology curricula of higher education institutes of learning in developing countries. It is believed that the number of potential areas where biotechnology can be applied to food processing is quite large. Scaling up production of traditional fermented foods is just one example.

FAO and its governing bodies recognize the role biotechnology can play in augmenting agricultural production when properly integrated with other technologies. Information on biotechnology activities in developing countries is scarce and this has prompted FAO to develop an inventory of plant biotechnology products and techniques in use or in the pipeline in developing countries. The inventory has been compiled and organized into a searchable online database called the FAO Biotechnology in Developing Countries Database (FAO-BioDeC) (FAO 2005).

Although a large number of eminent scientists support growing of GM crops and believe that these crops have the potential to improve food security in Africa, Sharife (2009) has a different view. She wonders if the proponents of the "Green Revolution" have the interests of the continent's people and the environment at heart or are more concerned with generating profits for the companies that control the technology.

The adoption and production of GM crops in developing countries is expected to have an impact on many aspects of life in those countries, for example, economic impact. bEcon (Yerramareddy and Zambrano 2011) is a web-based collection of selected peer-reviewed applied economics literature that assesses the impacts of GE crops in developing countries. It presents the experience and lessons learnt by different developing countries which adopted the production of GM crops.

Spielman and Zambrano (2013) summarized the present situation in Africa with respect to GM technology adoption and use. In Table 4.4, they reflect the number of GM products in Africa South of the Sahara in the year 2011. They stated that agricultural biotechnology and GE crop research in Africa is advancing, though slowly. Some indicators show that new crops, traits, and advanced technologies have some potential in the near future. Some progress has also been noted in the introduction of biosafety regulations in many African countries. The observed progress does not seem to match the potential opportunities presented by the new GM technologies. This means that more effort is needed in that direction. It is also suggested that more and close collaboration between the private sector and the public research institutions need to be strengthened so that the critical assets and competencies of the private sector be utilized appropriately.

Table 4.4 Number of GM Products in Africa South of the Sahara, 2011

Country	Number of Products
South Africa	40
Kenya	9
Nigeria	5
Uganda	3
Zimbabwe	2
Burkina Faso	2
Tanzania	1
Cameroon	1
Ghana	1
Malawi	1
Mauritius	1
Total	66

Source: Spielman, D.J. and Zambrano, P. 2013. Policy, investment, and partnerships for agricultural biotechnology research in Africa: Emerging evidence, in Falck-Zepeda, J., Guillaum, G., and Sithole-Niang, I. (Eds.), *Genetically Modified Crops in Africa: Economic and Policy Lessons from Countries South of the Sahara.* Table 1. International Food Policy Research Institute, Washington, DC.

4.3 ANIMAL BIOTECHNOLOGY

Like other types of biotechnology, animal biotechnology is based on the same principles and uses similar techniques as other types of biotechnology, for example, plant biotechnology. With this understanding, animal biotechnology can be defined as the application of scientific and engineering principles to the processing or production of materials by animals or aquatic species to provide goods and services (USDA-National Institute of Food and Agriculture 2013a).

Animal biotechnology encompasses the use of animal cells to generate valuable products through recombinant DNA technology, and the application of biotechnological approaches to rapidly multiply animals of desired genotype or to introduce specific changes in their genotype to achieve certain goals (Animal Biotechnology 2013). It includes both old and modern types of biotechnologies, as detailed below.

4.3.1 Animal Genetic Modification Techniques

4.3.1.1 Techniques That Do Not Use GE

4.3.1.1.1 Domestication and Artificial Selection

Domestication of wild animals has been practiced thousands of years ago by early civilizations. Today, different breeds of livestock are very much different from their ancestors with regard to different traits and behavior. Examples for these differences include amount of milk produced by cows, number, and size of eggs produced by chicken. These observed differences resulted through spontaneous mutations or through intervention of human by artificial selection or breeding. Traditional breeding techniques that date back to 5000 B.C. are still in use at present. Such techniques include crossing diverse strains of animals (known as hybridizing) to produce greater genetic variety.

4.3.1.1.2 Assisted Reproductive Procedures

An example of these procedures used to alter animal characteristics is what is known as "artificial insemination" (AI). AI involves collection of semen from select bulls with desired characteristics, for example, the milk that their daughters produce, and using the semen to fertilize cows artificially. The select bulls are tested for fertility to ensure that they carry their genes during the AI process. It is anticipated that the AI technique will continue to be an integral component of the animal production systems.

4.3.1.2 Techniques That Involve GE
4.3.1.2.1 *Transgenics*
This technique is based on the same transgenic principles discussed earlier for plants, that is, transfer of genes through different species. Methods to develop transgenic animals and transgenic aquatic species, for example fish, came into use since the early 1980s. Among transgenic livestock traits such as increased growth rate, enhanced lean muscle mass, and increased resistance to diseases, enhanced milk production have been achieved. Fish have been genetically engineered to grow faster and have increased muscle mass. Transgenic poultry, goats, swine, and cattle have been produced to generate large amounts of human proteins in their products, for example, eggs, milk, and blood. The goal here is to use these products as human pharmaceuticals, for example, antibodies, enzymes, and clotting factors.

4.3.1.2.2 *Gene Knock-Out Technology*
The term "knockout" used to describe this type of animal biotechnology refers to the inactivation of a specific gene in an animal. The main purpose of this technology is to create a possible source of replacement organs for humans. The process of transplanting cells, tissues, or organs from one species to another is known as "xenotransplantation." Among the different animals suitable for this kind of technology, the pig has been found to be the most viable organ donor to humans. The main challenge facing the process of xenotransplantation is the lack of immunological compatibility between pig cells and human cells, which would result in rejection of the donated organ. The presence of markers on the cells of the pig makes it easy for the human immune system to recognize them as foreign and consequently reject them. GE is applied to knockout or inactivate the pig gene responsible for the protein that forms the marker associated with the pig cells.

Another example of the "knockout" technology in animals is the inactivation of the "prion-related peptide (PRP) gene that may produce animals resistant to some diseases linked to prions, for example, bovine spongiform encephalopathy (BSE) and Creutzfeldt-Jakob disease (CJD) (USDA-National Institute of Food and Agriculture 2013a).

4.3.1.2.3 *Somatic Cell Nuclear Transfer (Cloning)*
Somatic cell nuclear transfer (SCNT), alternatively known as "cloning," is another example of application of animal biotechnologies. It involves the use of somatic cell nuclear transfer to develop multiple copies of mammals which are nearly identical copies of other animals, for example, transgenic animals, genetically superior animals, or animals that possess some other desirable

characteristics. Some of the animals cloned using this technique include cattle, sheep, goats, pigs, horses, mules, cats, rats, and mice. Dolly, the sheep, was the first cloned animal. SCNT begins with culturing somatic cells from an appropriate tissue (fibroblast) taken from the animal to be cloned (the donor). Nuclei from the cultured somatic cells are then microinjected into an enucleated oocyte (egg cell from which nucleus has been removed) obtained from another animal of the same or a closely related species. A number of steps of further culturing and *in vitro* development follow. The embryos are transferred and implanted into the uterus of a recipient surrogate female where it develops into a fetus resulting in the birth of live offspring. Experience has shown that the degree of success for propagating animals through cloning is very low and is usually less than 10%. It is affected by multiple factors, for example, the animal species, source of the recipient ova, cell type of the donor nuclei, treatment of donor cells before nuclear transfer, and the technique followed for nuclear transfer (USDA-National Institute of Food and Agriculture 2013a; North Carolina Association for Biomedical Research 2013).

4.3.1.2.4 Production of Infertile Aquatic Species
The aquaculture industry is considered the fastest growing among the animal food-producing sectors (Rasmussen and Morrissey 2007). Some of the main developments in transgenic fish research include the use of growth hormones, the use of antifreeze proteins, the use of metabolic genes, and genetic modification to induce sterility.

The technique of producing infertile species is specific to aquatic organisms, for example, fish and mollusks (aquatic invertebrates which include animals such as squid, octopuses, cuttlefish, snails, slugs, clams, oysters, and scallops). It has been developed to address one of the problems encountered in aquaculture production systems. This problem involves some species which are not indigenous to a certain area posing an ecological hazard to native species in case the foreign species escape confinement and enter the natural ecosystem. The technique is based on modifying the chromosome complement of fish or mollusks in a way that render them infertile. This can be achieved through producing triploid individuals, that is, carrying three sets of chromosomes instead of the normal two sets. Triploid organisms have been developed by using different procedures that interfere with the final steps in the development step of meiosis. Examples of these procedures applied to newly fertilized eggs include controlled application of high or low temperatures, treatment with various chemical substances, or subject to high hydrostatic pressure. Later on, in the process suppression of the first cell, division of the zygote can be achieved in order to produce individuals with four sets of chromosomes (tetraploid). Infertile triploid individuals can subsequently be produced through controlled mating of the developed tetraploids with normal diploids. In addition to addressing the issue of confinement, sterility in fish allows for reproductive energy to be diverted toward somatic growth, resulting in higher growth rates for some triploid individuals.

In practice, it has been a difficult task to achieve a 100% level of sterilization. This called for searching for alternative methods to fulfill the issue of reproductive confinement of fish.

As with any new technology, animal biotechnology is confronted with a number of challenges, faces a variety of uncertainties, and raises some concerns. The most important among these concerns are the issue of food safety, effects on human and animal health, possible environmental degradation, and most importantly animal welfare.

4.3.2 Biotechnology for Animal Feed Production

The feed of farm animals usually contains ingredients made from plants developed through GE. Additives and enzymes, for example, Vitamin B12, biotin, amino acids, beta carotene, which are sometimes incorporated in animal feed are also produced using GE microorganisms. In countries where animal feed must be labeled, for example, some EU countries, the products obtained from the animals, for example, meat, milk, and eggs do not require labeling. Different crops incorporated in the development of animal feed include soybean, maize, cottonseed, wheat, rye, and oats, some of which are genetically engineered.

The end products of animals fed GE ingredients, for example, meat, milk, and eggs, have been evaluated using advanced sensitive test methods to find out whether there is any difference resulting from the type of feed, that is, GE or no-GE. At present, it is not possible to tell whether an animal was fed GM soy just by testing its resulting products. The only possible way to establish the presence or absence of GMOs in animal feed is to analyze the origin of the feed itself (GMO Compass 2013d).

4.4 FOOD BIOTECHNOLOGY

4.4.1 Introduction—Genetically Modified Foods

There is a continuous and pressing need to increase food production and to improve its quality and safety. This is necessitated by the ever growing number of world human population coupled with increasing numbers of food insecure and malnourished individuals. Food technology, as a discipline, has largely contributed to address some of the world food problems through increase in quantity of food produced and enhancing its quality attributes. The general aims of food technology are to exploit natural food resources as efficiently and profitably as possible. Adequate and economically sound processing, prolongation of shelf life by preservation and optimization of storage and handling, improvement of safety and nutritive value, adequate and appropriate packaging, and maximum consumer appeal are key prerequisites to achieving these aims (Board on Science and Technology for International Development 1992). Biotechnology has played many roles in applying the principles of science in various fields of human life. Some of the areas in which biotechnology has been involved, and which lead to positive outcomes, include development of hybrid plants, recombinant DNA technology, and the development of vaccines. Biotechnology has also played important roles in food processing, which resulted in improving the quality and safety of foods.

The application of modern biotechnology, including GE, in agriculture to improve quality and increase production of crops, livestock, fishery, and food is considered one of the most significant technological advances of modern era. Food biotechnology is the application of biological techniques to food crops, animals, and microorganisms to improve the quality, safety, and ease of processing and production economics of food (Institute of Food Science and Technology [IFST] 2008). In this sense, food biotechnology encompasses traditional food processing methods such as those used to produce bread, cheese and other fermented dairy products, and beer, in addition to modern biotechnologies including GE.

The desire to produce improved plants and animals to be used as food sources has existed since human has been producing food. Food biotechnology will no doubt play important roles in the future of the global food supply, quality, and safety. It will also have positive roles in food manufacturing practices to produce value-added, nutritious, and wholesome foods with health-enhancing characteristics. The International Food Information Council Foundation (IFICF) (2014) published a comprehensive up-to-date guide on food biotechnology. This resource book entitled *Food Biotechnology: A Communicator's Guide to Improving Understanding* (3rd edition) contains latest science and consumer-friendly information in the form of talking points, handouts, a glossary, a PowerPoint presentation, and tips for engaging with media.

The roots of modern food biotechnology (that can be referred to as "basic food biotechnology") have been practiced for thousands of years. Fermentation and malting are two classical examples of these basic technologies. Many developments in these technologies have taken place.

Ghosh and Williams (2010) have cited the following general categories for foods produced through modern biotechnology:

- Foods consisting of, or containing, living/viable organisms, for example, corn
- Foods derived from, or containing, ingredients derived from GMOs, for example, corn meal containing proteins or oil from GM soy
- Foods containing single ingredients or additives produced by GM microorganisms, for example, colors, vitamins, or essential amino acids
- Foods containing ingredients processed by enzyme produced through GM microorganisms, for example, high-fructose corn syrup produced from starch using the enzyme glucose isomerase or cheese produced using the enzyme chymosin

In the following sections, further details of GM foods will be presented.

4.4.2 Composition of GE Foods

Information about the occurrence of any compositional changes in the GE food crops is important. It helps to determine whether these changes are beneficial or harmful with regard to human nutrition and health. The resulting changes may be regarded as intended or unintended, as discussed in the following sections.

4.4.2.1 Intended Changes in Composition

Golden rice enriched with provitamin A: an improved GE rice cultivar with increased levels of carotenoids which are precursors of vitamin A. As part of its natural composition, rice contains very little amounts of vitamin A. Eye diseases occur widely in regions of the world where rice is a staple food. This has been attributed to vitamin A deficiency in the staple rice. Consumption of golden rice is expected to address this nutritional and health problem.

Improved fatty acid profile of soybean, maize, rapeseed, and other oil crops: GE is used to modify the fatty acid composition, for example, enhanced polyunsaturated fatty acid composition. Edible oil extracted from these altered crops could be used positively to manage some chronic diseases like cardiovascular diseases, some types of cancer, and obesity.

The composition of some crops has been genetically modified using GE techniques to improve protein or amino acid profile in a way that render these crops nutritionally superior to their conventional counterparts. For example, one GE potato variety has been modified to have higher good quality protein. The gene responsible for this improvement has been extracted from the Amaranth plant, which is known to have high-quality protein. Other examples of compositional changes include production of gluten-free wheat to cater for the physiological needs of people who are gluten-intolerant and crops with increased levels of antioxidants which act as free radical scavengers and prevent against some chronic diseases. Changes in GE food crops composition can also take the shape of reduction or elimination of some harmful or undesirable components such as allergens, caffeine, and nicotine.

4.4.2.2 Unintended Changes in Composition

The academic review.org website (http://academicsreview.org/reviewed-content/genetic-roulette/section-2/2-10-chemical-compositions) (Academic Reviews, 2014) states that there have been many studies on GM crops which confirm that only minor variations in chemical composition occur. These differences are said to be within the normal range of chemical content observed among conventional crop varieties of a given food crop. Compositional changes arising as a result of GE are fewer and smaller than those observed in conventional varieties, as the source reports.

4.4.3 Biotechnology in Food Processing and the Uses of GMOs in the Food Industry

4.4.3.1 GM Crops (Products) Used as Food

A large number of GM crops have been used to develop GM foods. The DNA Hot Science website (DNA Hot Science 2014) gives information of some of these crops. Some of the most common GM crops used as food are presented in the following sections.

Apples. One of the problems encountered with apples (in addition to other plant products, e.g., potato) is browning when cut and exposed to air. In the year 2012, an apple variety has been genetically modified to resist this type of browning. The gene responsible for the production of the enzyme polyphenol oxidase which helps the browning process has been modified in such a way that less amounts of the enzyme are produced in the apple. This modified apple variety is known as "nonbrowning Arctic apple" and has been produced by "Okanagan Speciality Fruits" (Arctic Apples 2014).

Corn (Maize). GE corn (Figure 4.4) has been developed through introducing a gene from a soil bacteria (*Bacillus thuringeinsis*) which is able to produce a toxin that acts as a pesticide and kills certain insects. The GE corn is known as Bt corn as related to the bacteria donating the gene and it is rendered pest-resistant. Different products are obtained from this GE corn, for example, grits, meal, and corn flour, corn oil, sugar, and syrup. Grits are the coarsest products resulting from corn dry milling. They are generally used to produce breakfast cereals and snack foods. Corn meal is used in the production of cornbread, muffins, fritters, bakery mixes, and pancake mixes. Corn flour is the finest product of the corn dry-milling process. It is usually used to produce pancake mixes, muffins, doughnuts, breadings, and batters in addition to baby foods, meat products, and some fermented foods. These products are consequently used to process foods like snack foods, baked goods, fried foods, edible oil products, confectionery special purpose foods, and soft drinks.

Figure 4.4 GM corn.

Cotton (Seed Oil). Bt cotton is GE cotton used in the same way as Bt corn, consequently making it pest-resistant. Cotton seed oil is extracted from cotton seeds and is used as an edible oil to cook, fry, or bake foods. It is also incorporated in snack foods and is blended with other edible oils.

Papaya. GE papayas have been developed to resist the ringspot virus. The first virus-resistant papayas were produced in Hawaii in 1999. Before this, the Hawaii's papaya industry was threatened by the deadly ringspot virus. The development of a virus-resistant papaya crop helped to address the problem.

Peas. GM peas developed by inserting a gene from kidney beans, which creates a protein that acts as a pesticide, have shown possible allergens in mice. This observation suggested that the same allergic reaction may happen to people consuming these GM peas.

Potato. At present, there are no transgenic potatoes marketed for human consumption. This followed the development of the "New Leaf Potato" by Monsanto in the late 1990s, which targeted the fast food market. It was eventually withdrawn from the market in 2001 after it failed to pick up in the retail market. This has not stopped efforts to develop GE potato with different desired traits, for example, resistance to late blight disease, bruise prevention, and production of less acrylamide during frying. Pollacknov (2014) reported that a potato genetically engineered to reduce the amounts of acrylamide, a potentially harmful ingredient in French fries and potato chips, has been approved for commercial planting by the United States Department of Agriculture on November 7, 2014. The new potato also resists bruising, a characteristic long sought by potato growers and processors for financial reasons. Potatoes bruised during harvesting, shipping, or storage can lose value or become unusable. The biotech tubers were developed by the J. R. Simplot Company, a privately held company based in Boise, Idaho, USA.

Rapeseed or Canola (Oil). Herbicide (Roundup)-resistant cultivars of rapeseed were produced by Monsanto. Rapeseed is cultivated mainly for edible oil production, which is regarded as the third largest source of vegetable oil in the world. The oil is used in various edible oil products including fried foods, baked products, and snacks. In Canada, where "double zero" rapeseed was developed, the crop was given the name "canola" (Canadian oil).

Rice. Naturally, rice is deficient in vitamin A or its precursor beta carotene. On the other hand, rice constitutes the staple food for a high percentage of world population particularly in developing regions. Vitamin A deficiency has been a chronic problem for people in developing countries, especially affecting women and children. One approach sought as a solution to this problem is the development of rice which contains beta carotene. A transgenic strain of rice known as "Golden Rice" has been engineered. It contains beta carotene that would be converted into vitamin A when consumed by humans. Another development reported for GM rice is a variety containing human genes not targeted for human consumption but for human treatment of infant diarrhea, particularly in the developing world.

Soybean. GE soybean (Roundup Ready Soybean; Figure 4.5) has been developed by Monsanto to be resistant to the Roundup WeatherMAX herbicide which could be sprayed to kill weeds with no harm on the soy plant itself. Products obtained from soy include soybean oil, soy flour, soy protein concentrate, soy protein isolate, lecithin, and textured soy protein. Soybean-based foods include soy milk and other soy beverages, tofu, breads, pastries, snack foods, baked or fried foods, and special purpose foods.

Squash. Some zucchini and yellow crookneck squash have been genetically modified but they are not well received by farmers.

Figure 4.5 GM soybean.

Sugar Cane. Some sugar cane cultivars have been genetically modified for pesticide resistance. GM sugar cane has not met the expected success in the market due to the low level of acceptance by consumers.

Sweet Corn. Insecticide-resistant GM sweet corn has been developed. Although it was mainly intended for animal feed, it made its way into the human food supply and has been incorporated in foods marketed for human consumption.

Tomato. The first commercially grown GM whole food crop was tomato. In 1994, Calgene, a biotechnology company from California, USA, genetically engineered a tomato variety named "FlavrSavr." This pioneer transgenic crop was modified to ripen without softening and attain a longer shelf life by delaying its natural tendency to rot and degrade quickly. It was welcomed by consumers who purchased the fruit at a substantial premium over the price of regular tomatoes. In 1997 and due to economic difficulties, Calgene was forced to withdraw the FlavrSavr tomato from the grocery shelves. In spite of these problems, research for new modifications on tomato continued. In 2001, research scientists announced research results for the development of salt-resistant tomato. This would help to utilize areas with high salt content which are otherwise uncultivable with other crops. It also helps to lower salt content of those soils.

Honey. Honey can be produced from flowers of GM crops. In Canada, some types of honey are produced by bees collecting nectar from GM canola plants. Some European countries do not allow imports of honey produced from GM crops, thus the ban of this type of Canadian honey.

4.4.3.2 Food Companies Using GMO Ingredients

A large number of food companies worldwide use GM ingredients in producing some of their food brands. A list of those companies has been compiled (Table 4.5), based on the information presented in the two references cited in Table 4.5. This list is not exhaustive or static. Some companies included here may opt to shift to using non-GM components, while other companies, not included in the list, may venture into including GM components for the first time. It is thus important to update the list as things change.

4.4.3.3 Examples of GM Foods

It might be impractical to come up with a full list that includes all genetically modified foods produced worldwide. This is because there are no regulations guiding labeling of GM foods in the United States, the major producer of GM foods globally. Butcher (2003) reports that some estimates

GENETICALLY MODIFIED FOODS

Table 4.5 Companies Using GMO Products

Company	Reference
Aunt Jemima	1 and 2
Betty Crocker	1 and 2
Campbells	1 and 2
Coca Cola	1 and 2
Duncan Hines	1 and 2
Frito-Lay/Pepsi	1 and 2
Heinz	1 and 2
Hershey's Nestle	1 and 2
Hormel	1 and 2
Kellogs	1 and 2
Kraft/Philip Morris	1 and 2
Morningstar Farms	1 and 2
Nabisco	1 and 2
Nature Valley	1 and 2
Ocean Spray	1 and 2
Pepperidge Farms	1 and 2
Pillsbury	1 and 2
Procter and Gamble	1 and 2
Quaker	1 and 2
Stouffers	1 and 2
Aurora Foods	1
Banquet	1
Best foods	1
Bisquick	1
Cadbury/Sweppes	1
Capri Sun	1
Carnation	1
Chef Boyardee	1
ConAgra	1
Kool-Aid	1
Delicious Brand cookies	1
Famous Amos	1
General Mills	1
Green Giant	1
Healthy Choice	1

(*Continued*)

Table 4.5 (*Continued*) Companies Using GMO Products

Company	Reference
Hellman's	1
Hungry Jack	1
Hunt's	1
Holsum	1
Interstate Bakeries	1
Jiffy	1
KC Masterpiece	1
Keebler/Flowers Industries	1
Kid Cuisine	1
Knorr	1
Lean Cuisine	1
Lipton	1
Loma Linda	1
Marie Callender's	1
Minute Maid	1
Mrs. Butterworth's	1
Ore-Ida	1
Orville Redenbacher	1
Pepsi	1
Pop Secret	1
Post Cereals	1
Power Bar Brand	1
Prego Pasta Sauce	1
Pringles	1
Ragu sauce	1
Rice-A-Roni/Pasta-Roni	1
Smart Ones	1
Tombstone Pizza	1
Totino's	1
Uncle Ben's	1
Unilever	1
V-8	1
Beech-Nut	2
Blue Sky	2

(*Continued*)

Table 4.5 (*Continued*)   Companies using GMO Products

Company	Reference
Boca	2
Crisco	2
Dannon	2
Enfamil	2
Eggo	2
Good Start	2
Hansen	2
Hostess	2
Isomil	2
Kashi	2
Keebler	2
Land O'Lakes	2
Libby's	2
Lifesaver	2
Marie Callender's	2
Nestle	2
Peter Pan	2
Progresso	2
Similac	2
Skippy	2
Smucker's	2
Yoplait	2

[1] Miami-water.com (http://miami-water.com/blog/2217/product-list-of-gmo-genetically-modified-foods/).
[2] Celestialhealing.net (http://www.celestialhealing.net/monsanto/GMO_FoodCo.htm).

say as many as 30,000 different food products on grocery stores shelves are modified. GMO Compass produced a number of documents describing the different categories of GM foods available in the market (GMO Compass 2014a,b,c,d,e,f).

Foods, in general, come from plant or animal sources. Aquatic organisms consumed as food, for example, fish, are grouped under animal sources. The following sections include information about foods which are derived from GE sources. At present, there are many GM crops that are used as food sources, but there are no GM animals, with the exception of fish, that are used for food production. With regard to foods derived from GM plant sources, the plant product is either directly consumed as food, for example, some fruits and vegetables, or the GM crop is used as a commodity for further processing into food ingredients, which are subsequently incorporated in processed foods, for example, milled corn products and milled soy products. The most common GM food crops are: apples, corn (maize), cotton (seed oil), flax, papaya, peas, potato, rapeseed or canola, red-hearted chicory (radicchio), rice, soy, squash, sugar beets, sugar cane, sweet corn, and tomato.

Lanphier (2014) cites the following food crops as the top foods containing GMOs (in the United States): Sugar beets (95% of U.S. crop)—approximately 50% of white sugar sold in the United States is made from GMO sugar beets; soy (94% of U.S. crop); Canola (90% of U.S. crop); Cotton (90% of U.S. crop); and Corn (88% of U.S. crop).

4.4.3.4 Processed Foods and Ingredients Based on GM Products

Cheese is the classical example for this group of foods. Cheese processing is based on coagulation or curdling of milk, a step traditionally accomplished using rennet which is a mixture of enzymes produced in mammalian stomach. Rennet was originally obtained from the stomach of calves or from microbial sources. Both sources were not sustainable due to scarcity, high cost, or bad taste. The active enzyme in rennet is known as chymosin or rennin. Chymosin can now be produced through GE. The rennet-producing genes from animal's stomach are extracted and inserted into certain bacteria, fungi, or yeast to induce them to produce chymosin. After fermentation, the microbes are killed and chymosin (known as fermentation-produced chymosin—FPC) is isolated and used for cheese production. FPC was the first artificially produced enzyme to be registered and its use permitted by the U.S. Food and Drug Administration (USFDA). FPC products have been on the market since 1990 and have been considered the ideal milk clotting enzyme (Law, 2010).

Some of the most common processed foods that contain GM components and the companies producing them have been presented by Derbick (1999) and are shown in the following list. It is indicated that these processed foods tested positive for being GM and that these tests do not represent any "safety tests." These foods were not labeled to show the presence of GM components.

Examples of processed foods containing GM components

- Aunt Jemima Pancake Mix
- Ball Park Franks

GENETICALLY MODIFIED FOODS

- Betty Crocker Bac-O's Bacon Flavor Bits
- Boca Burger Chef Max's Favorite
- Bravos Tortilla Chips
- Duncan Hines Cake Mix
- Enfamil ProSobee Soy Formula
- Frito-Lay Fritos Corn Chips
- Gardenburger
- General Mills Total Corn Flakes Cereal
- Green Giant Harvest Burgers (now called Morningstar Farms)
- Heinz 2 Baby Food
- Jiffy Corn Muffin Mix
- Kellogg's Corn Flakes
- Light Life Gimme Lean
- McDonald's McVeggie Burgers
- Morning Star Farms Better'n Burgers
- Nabisco Snackwell's Granola Bars
- Nestle Carnation Alsoy Infant Formula
- Old El Paso Taco Shells
- Ovaltine Malt Powdered Beverage Mix
- Post Blueberry Morning Cereal
- Quaker Chewy Granola Bars
- Quaker Yellow Corn Meal
- Quick Loaf Bread Mix
- Similac Isomil Soy Formula
- Ultra Slim Fast

4.4.3.5 Foods Produced from Animals Consuming GM Crops or Treated with Bovine Growth Hormone (BGH)

Animal feed given to farm animals is based on the meals resulting from the processing of crops, for example, and oilseed crops like soybean and canola, some of which are GE. These meals are high in protein and are regarded as nutritious feed for livestock and poultry. Animal products, for example, meat, milk, and eggs of animals fed on meals based on GM crops cannot be distinguished from those of animals fed non-GM products. The only way to confirm the presence of GMOs in animal feed is to analyze the origin of the feed itself (Staff, GMO Compass 2014).

Some countries approve and allow the use of the recombinant bovine somatotropin (also known as rBST or bovine growth hormone—BGH) to dairy cows in order to increase milk production. About 22% of cows in the United States are injected with recombinant bovine growth hormone (Bio Elite Wellness 2014). Though milk from rBST-treated cows may contain traces of rBST, it has no direct effect on humans. The World Health Organization, the USFDA, the American Medical Association, the American Dietetic Association, and the national Institute of Health have independently stated that dairy produce and meat from BST-treated animals are safe for human consumption (Brennand 2014).

4.4.3.6 Non-GMO Companies and Food Products

As shown in Table 4.5, there are many food processing companies which use GM products or ingredients to process their food products. Nevertheless, there are also other food processing companies that rely on non-GM products to produce their food commodities. The Non-GMO Project has spearheaded the efforts to educate the public on the food companies that do not use GM products or ingredients to process foods. Schillinger Genetics, Inc. (eMerge 2014) is a U.S.-based body dedicated to helping farmers, feed, and food companies, and consumers realize benefits and value from different novel products. It provides all sectors throughout the entire chain a better alternative than the typical commodity markets by improving the bottom line, nutritional benefits, or substitution of a more expensive or limited resource product. They advocate and support non-GM products.

Sarich (2013) produced a list of over 400 companies which are not using GMOs in their products. This list has been compiled courtesy of the "Non-GMO Project" (nongmoproject.org).

The website REALfarmacy.com (2013) has also listed 400 companies that do not use GMOs in their products.

The companies included in these lists are

365, 479 Degrees

A
A. Vogel, Adams Vegetable Oils, Agrana, Agricor, Inc., Ah!Laska, Alexia, Alter Eco, Alverado Street Bakery, Amande, Amella, Among Friends, Amy's Kitchen, Andalou Naturals, Angie's Artisan Treats, Ariven Planet, Arrowhead Mills, Artisan Bistro, Artisan Bistro Home Direct, Atlantic Organic, Atlantic Rose, Attune Foods, Autumn Sky Wild

B
Back to Nature, Bainter Extra Virgin Sunflower Oil, Bakery On Main, Barbara's, Barlean's Organic Oils, Barnana, Barney Butter, Basic Food Flavors, Inc., Beach Bum Foods, Beanfields, Beanitos, Bearitos, Berlin

GENETICALLY MODIFIED FOODS

Natural Bakery, Better Bean, BetterStevia, Bhakti Chai, Biad Chili Products, Bites of Bliss, Blue Diamond, Blue Lotus Chai, Blue Print, Bold Organics, Bora Bora, Boulder Canyon Natural Foods, Brad's Leafy Kale, Brad's Raw 4 Paws, Brad's Raw Chips, Brad's Raw Crackers, Brad's Raw Onion Rings, Braga Organic Farms, Bragg, Brand Aromatics, Bridgewell Resources, Bubbies, Buenatural

C
Cabo Chips, Cadia, Cal-Organic Farms, Califia Farms, California Olive Ranch, Canfo Natural Products, Canyon Bakehouse, Cape Cod Select, Catania, CaveChick, Cedar's, Central Market Organics, Chappaqua Crunch Granola, cheweco organics, Chez Marie, Inc., ChiaRezza! OMG Foods, Inc., Choice Organic Teas, Chosen Foods, CHS Oilseed Processing, Chunks O' Fruti, Ciao Bella Gelato, Ciranda, CleanVia, Coconut Secret, Cocozia, Cool Cups, Coral LLC, Country Choice Organic, Crispy Cat, Crofters, cruncha ma•me, Crunch Master, Curties Juice

D
Dave's Gourmet, David's Unforgettables, Deli-catessen, Della, Desert Essence, Doctor In The Kitchen, Doctor Kracker, Dr. Arenander's BrainGain & Oral Care, Dr. Bronner's Magic, Dream, Drew's LLC, Dulsweet

E
Earth Balance, Earth's Best, EatPastry, Eatsmart, EcoTeas, Edazen, Eden, Edward & Sons, Eighth Wonder, Emerald Cove, Emile Noel, Emmy's Organics, Emperor's Kitchen, Endangered Species Chocolate, Ener-G Foods, Engine 2, Enjoy Life Foods, Envirokidz, EO, Erewhon, Essential Living Foods, Inc, Everyday Superfoods

F
Fairfield Specialty Eggs, familia, FanciFood, Fantastic World Foods, Farm to Table Foods, Farmer's Market, Farmhouse Culture, Field Day, Field Roast Grain Meat Company, Fillmore Farms, Fiordifrutta, Flamous Organics, Flax USA, Flora, Follow Your Heart, Freekeh Foods, Freekehlicious, Freeline Organic Foods, Fresh & Easy, Frey, Frontier, froovie, Fruit Bliss, FruitChia, Fry Group Foods, Fungi Perfecti LLC, Funky Monkey Snacks

G
Garden Bar, Garden of Eatin', Garden of Life, gimMe, Gin Gins, Gingras XO, Giving Nature, GL Soybeans, Global River, GlucoLift, Gluten Free Pantry by Glutino, Glutino, Gnu Foods, Go Raw, Golazo, GoMacro, Inc.,

GoodBelly, GoodBelly+, Good Health Natural Foods, Good Karma, GoOrganic/GoNaturally, GoPicnic, Grain Place Foods, Grains of Wellness, Green Gem, Green Island Rice, Green Mountain Gringo, Green Mustache, Grimmway Farms, Growing Naturals, Guayaki, Guiltless Gourmet

H
Haig's Delicacies, Haiku, Hail Merry, Hapi Foods Group, Inc., HAPPYBABY Pouches, HAPPYTOT, Harvest Bay, Haute Cuisine, Health is Wealth, Health Warrior, Heavenly Organics, Herbal Zap, High Country Kombucha, Hiland Naturals, Hilary's Eat Well, Himalania, HimalaRose, HimalaSalt, Hodgson Mill, Hol-Grain, Home Appetit, HomeFree, House Foods, Houweling's Tomatoes

I
Ian's, Imagine, Immaculate Baking, Immortality Alchemy, Imperial Gourmet, Indianlife, Intiyan, It Tastes Raaw

J
Jaali Bean, JaynRoss Creations LLC, Jeff's Naturals, Jessica's Natural Foods, Jolly Llama, JustFruit

K
KAMUT, Keller Crafted Meats, Kettle Foods, Kettlepop, Kiji, KIND Healthy Snacks, Kiwa, Konriko, Koyo, Kur Organic Superfoods

L
La Reina, La Spagnola, La Tolteca, La Tourangelle, Lafiya Foods, Lassens, Laughing Giraffe Organics, Laurel Hill, Lekithos, LesserEvil, Let's Do, Licious Organics, Lillabee Allergy Friendly Baking, Little Duck Organics, livingNOW gluten-free, Loeb's, Lotus Foods, Lucy's, Luna & Larry's Coconut Bliss, Lundberg Family Farms

M
Mac-n-Mo's, Made In Nature, Madhava, Mamma Chia, Manitoba Harvest, Marconi Naturals, Maria & Ricardo, Marinelli's True Italian Pasta Sauce, Mariner Biscuit Company, Martha's All Natural, Marukan, Marukome USA, Mary's Chicken, Mary's Gone Crackers, Mary's Little Garden, Mary's Organic Chicken, Mary's Organic Turkey, Mary's Pasture Raised Chicken, Maui Maid, Mediterranean Organic, Mediterranean Snacks, MegaFood, Melt Organic, Metabolic Response Modifiers (MRM), Mighty Mustard,

GENETICALLY MODIFIED FOODS

Mighty Rice, Mighty-O Donuts, Mindful Meats, Minsa, Minsley, Miracle Noodle, Miso Master, Modesto WholeSoy Co., Momo's, Montana Specialty Mills, LLC, Mori-Nu, Mt Vikos, Muesli Munch, Multiple Organics, MXO GLOBAL INC., My Chi Delights

N
Naked Coconuts, Naosap Harvest, Napa Valley Naturals, Nasoya, Nathan's, Native Forest Distributed by Edward & Sons, Natural Directions, Natural Habitats, Natural Nectar, Natural Sea, Natural Tides, Natural Vitality, Naturally Splendid Enterprises Ltd, Nature Built, Nature Fed, Nature Way, Nature's Express, Nature's Path, Navitas Naturals, Nejaime's, NestFresh, New Chapter, New England Naturals, New Organics, NewYork Superfoods, Nexcel Natural Ingredients, Nexsoy, Niagara Natural, NibMor, Nordic Naturals, North Coast, NOW Foods, NOW Healthy Foods, NOW Real Food, NOW Real Tea, Nu Life Market, Numi Organic Tea, Nummy Tum Tum, NurturMe, Nutiva, Nutrigold®, Nuts About Granola

O
Oh Baby Foods, Old Wessex, Oleicus/Oleico, Once Again, One Degree Organic Foods, One World, Organic Baby, Organic Planet, Organic Valley, Organicville, Oriya Organics, Ozery Bakery

P
Pacific Natural Foods, Pacific Northwest Farmers Cooperative, Paisley Tea Co, Palo Root Tea, Pampas Rice/Organic Latin America, Pan De Oro, Pascha, Pastorelli Food Products Inc., Peace Cereal, PEACOCK, Peanut Butter & Co., Peeled Snacks, Peggy's Premium, PJ's Organics, Planet Rice, Plum Organics, Popcorn, Indiana, Popcornopolis, powbab, President's Choice, PROBAR, PuraSource, Pure, Pure Country Pork, Pure Eire, Purely Decadent, Purely Elizabeth, Pyure Brands

Q
Q.bel, Qrunch Foods, Quinn Popcorn

R
R.W. Knudsen, Rainbow Light Nutritional Systems, Rapunzel, RAU, Red Hat Co-operative Ltd, Reese, Revive, RiceSelect, Righteously Raw, Rigoni di Asiago Honey, Rishi Tea, Rising Moon Organics, Risodipasta, Rivara, ROBE and Riverina Natural Oils LLC, Roots, Route 11, Royal Hawaiian Orchards, Rumiano Family Cheese, Runa, Ruth's Foods, RW Garcia

S
Sacha, Saffron Road, Sage V Foods, Sainthood Herbs, Salba Smart, Sally's Smart Foods, Salute Santé!, Sambazon, San-J, Santa Cruz Organic, Scratch and Peck, SeaSnax, Secret Squirrel, Seven Stars Farm, Sharkies, Shortstacks, Silk, Silver Hills Sprouted Bakery, Simple Origins, Simply Soy Yogurt, Simply Suzanne, Sir Kensington's, SK Food, Skout, Snyder's of Hanover, So Delicious Dairy Free, Sol Cuisine, Somersault Snack Co., Sophie's Kitchen, Source, Soyatoo, Spectrum, Spectrum Ingredients, Spicely, Squarebar, Stahlbush Island Farms, Stahlbush Island Farms Ingredients, Stark Sisters Granola, Stash Teas, Stiebrs Farms Go-Organic Eggs, Stone Buhr Flour Company, Straus Family Creamery, Stretch Island Fruit Co, Suja Juice, Sun Cups, SunE900, Sunfood Superfoods, SunRidge Farms, Sunset, Sunset Kidz, Sunshine Burger, Superberries, Surf Sweets, Sushi Sonic, Sweet Sass Foods, Sweet Tree, Sweet Leaf

T
TAMBOR, Taste of Nature, Tasty Brand, TeaPops, That's it. The Better Chip, The Chia Co, The Fresh Market, The Ginger People, The Pure Wraps, The Republic of Tea, The Scoular Company, The Simply Bar, The Solio Family, Theo Chocolate, Third Street, Inc., Three Farmers, Tiny But Mighty TOMMYS, Tonnino, Trace Minerals Research, Traditional Medicinals, Tree of Life, Tropical Traditions, Tru Joy Sweets, truRoots, truwhip, Turtle Island Foods, Two Leaves Tea Company, Two Moms in the Raw, twofold

U
Udi's, Uncle Sam, Union Market, Upfront Foods

V
Van's Natural Foods, Vega, Veggie-Go's, Venus, VerMints, Veronica Foods, Viana, Victoria, Vigilant Eats, VitaV, Viterra

W
Watts Brothers, Way Better Snacks, Wayfare, Weetabix, Western Foods, WestSoy, Whole Alternatives, Whole Earth, Whole Harvest, Whole Pantry, Wholesome Chow, Wholesome Sweeteners, WholeSoy & Co., Wild Veggie, Wildbrine, Wildwood, Willamette Valley, Wingfoot, Wisdom of the Ancients, Woodstock

X, Y, Z
XO Baking Co., Yamasa, Yoga, Yogavivie, Zema's Madhouse Foods, Ziggy Marley Coco'Mon, Ziggy Marley Hemp Rules, Zing Bars, Zulka

4.4.3.7 Non-GMO Seed Companies

Major seed companies are in the process of phasing out production of non-GMO soybean seed and substituting it with the GE varieties. One company, Schillinger Genetics, Inc., emerged with the aim of addressing the need of farmers who want to grow non-GMO seeds. This company is a specialty soybean seed company based in West Des Moines, Iowa, USA, has developed and introduced high-quality non-GMO food-grade soybean seed varieties under its brand name eMerge Genetics (eMerge, 2014). eMerge is expected to fill the expanding market demand for food-grade, non-GMO soybeans from both local and overseas food companies.

4.4.3.8 Non-GMO Project Verified

The Non-GMO Project is a nonprofit organization committed to preserving and building the non-GMO food supply, educating consumers, and providing verified non-GMO choices. The Project is governed by a Board of Directors which works with a collaborative network of technical and communications advisers from different backgrounds and sectors. The Board of Directors believe that everyone deserves an informed choice about whether or not to consume genetically modified organisms (NON-GMO Project 2014). The headquarters of the project are in Bellingham, Washington, USA. The project has designed the seal shown below to designate foods of non-GM origin. The "Non-GMO Project Verified" Seal is used to distinguish non-GMO products. It appears as part of the food product label (see Figures 4.6 and 4.7).

4.4.4 Biotechnology in Functional Foods and Nutraceuticals

The role of diets in health promotion, disease prevention, and slowing progression of chronic diseases has been well recognized. Consumers are increasingly becoming health conscious and concerned about the type of food they purchase and consume. More people started to demand foods that promote and maintain health. A general consumer trend to look for healthier foods, especially those which use ingredients that claim to have health benefits, has been observed. As consumers become more aware of the need for preventive measures to improve their health and well-being, market opportunities are created for new nutraceuticals in the form of functional food products or dietary supplements. The use of food as a pharmaceutical agent to manage and improve health and to control some diseases has been recognized centuries back. The pharmaceutical use of food forms the basis of the concept of nutraceuticals. The recognized health benefits of

Figure 4.6 Non-GMO Project Verified seal. (With permission from the NON-GMO Project. 2014. http://www.nongmoproject.org/ (accessed February 20, 2014).)

foodstuffs lead to current interest in nutraceuticals and functional foods. Nutraceutical biotechnology has evolved into a billion-dollar industry, with the promise of producing foods that provide functions beyond the basic nutrients they contain (BioSpectrum 2012). Pew Initiative on Food and Biotechnology (2007) published a report entitled "Application of Biotechnology for Functional Foods." The report reviews the potential that

Figure 4.7 Food products showing the Non-GMO Project seal. (With permission from the NON-GMO Project. 2014. http://www.nongmoproject.org/ (accessed February 20, 2014).)

modern biotechnology has in developing functional foods. It discusses the various applications of biotechnology used in production of functional foods as well as the regulatory and legal aspects related to them.

Although the two terms, that is, nutraceuticals and functional foods, are sometimes used synonymously, they have different meanings. The term "nutraceutical," which combines the two words nutrition and pharmaceutical, has been defined by Ramaa et al. (2006) as "food or part of food that provides medicinal and health benefits including the prevention and/or treatment of a disease." On the other hand, a broad definition of "functional foods" has been given by Clydesdale (1997) as

> Foods similar in appearance to conventional foods, which are consumed as part of a normal diet and have demonstrated physiological benefits and/or reduce the risk of chronic disease beyond basic nutritional functions.

Usually changes in consumer trends toward certain commodities drive the direction toward which manufacturing industries move. This has been the case in the food industry where food manufacturers start to focus on production of health promoting food commodities including nutraceuticals and functional foods. The nutraceutical industry started to grow as an important component of the food industry and at the same time benefiting from the new biotechnologies. In a related article, Cosgrove (2010) examines the increasingly intertwined relationship between biotech and new food/nutraceutical product development. The article highlights insights of Dr. Cheryl Barton in a report entitled "Biotech for Wellness: Driving successful R&D and licensing in nutraceuticals through new business models and collaboration." Dr. Barton indicated that

> A number of factors are leading to the convergence of the food and biotech sectors including the increasing use of ingredients that claim to have health benefits and which have been studied in clinical trials to demonstrate these benefits; and the increasing scientific evidence for a link between diet and the cause or treatment of a number of diseases.

Some examples of the common nutraceutical ingredients of interest between the two sectors, cited in the report include omega 3 fatty acids, phytosterols and stanols, probiotics and prebiotics, with the aim to support health or target risk factors for chronic diseases such as cholesterol. A number of manufacturing companies started to follow a more biotech-linked approach to nutraceutical development. Examples of such companies are B.R.A.I.N., InterMed Discovery, Matis, Medisyn Technologies, Redpoint Bio, and Senomyx (Cosgrove 2010).

Modern biotechnologies have recognized roles in production of both plant-based and animal-based functional foods. Hsieh and Ofori (2010) outlined these roles as follows.

4.4.4.1 Production of Plant-Based Functional Foods
- *Biofortification with essential micronutrients*: the intention here is to apply GE to produce varieties of food products with higher amounts of health-promoting nutrients. Options other than biofortification are found to be less effective and more expensive. Examples of biofortified crops include rice, corn, wheat, staples biofortified with vitamin A, zinc, and iron.
- *Biofortification with phytochemicals (phytonutrients)*: phytochemicals describe a group of plant-based chemical substances which are found to be active in disease prevention and control. They are mainly present in some fruits and vegetables. Examples of phytochemicals include carotenoids, polyphenolics (anthocyanins, isoflavones), and antioxidants. GE can be used to induce or improve the level of these health-promoting substances.
- Modification of macronutrients, for example, oils with healthier fatty acid profile and more stable cooking oils. The fatty acid profiles of vegetable oils as well as their other characteristics, for example, heat or storage stability can be enhanced through application of GE.
- Production of hypoallergenic foods, that is, foods that produce less or no allergic reactions for sensitive consumers. The level of allergy-causing components found in some food materials can be reduced to a level that does not harm sensitive people. Application of GE techniques can help in achieving this goal.
- Reduction of antinutrients naturally present in some foods of plant origin. GE can be used to lower the level of naturally occurring antinutrients available in some food products of plant origin. This means that the availability of nutrients will be improved.

4.4.4.2 Production of Animal-Based Functional Foods
This group of foods received less attention than foods of plant origin. A limited number of animal-based foods have been produced so far, for example, meat with modified fatty acid profile, milk suitable for lactose-intolerant people, and milk with improved fatty acid profile. There is the potential of more work to be done using GE to benefit from functional foods of animal origin.

4.4.4.3 Application of Biotechnology for Production of Foods and Food Ingredients

4.4.4.3.1 Probiotics and Prebiotics

Probiotics and prebiotics are components which can be found in some foods and food supplements, or that can be incorporated into foods, and which yield health benefits related to their interactions with the human gastrointestinal tract (GI). The term "probiotics" refers to live microorganisms that have a health benefit when consumed in adequate amounts. They are usually bacteria selected from species naturally found in the human intestinal tract. Probiotic microorganisms may be concentrated and added directly to a food. The most common foods having probiotic organisms are fermented dairy products such as yogurt (Pew Initiative on Food and Biotechnology 2007).

According to the Joint FAO/WHO Working Group (2002), probiotics are defined as "live microorganisms which, when administered in adequate amounts, confer a health benefit on the host." Another definition (Ashwell 2002) states that a probiotic is "a live microbial food ingredient that, when ingested in sufficient quantities, exerts health benefits." On the other hand, prebiotics have been defined as "nondigestible food ingredients that beneficially affect the host by selectively stimulating the growth of one or a limited number of bacterial species in the colon, such as *Bifidobacteria* and *Lactobacilli*, which have the potential to improve host health" (Ashwell 2002). The relationship between probiotics and prebiotics can be simply stated as one of them (prebiotics) is the food of beneficial bacteria (probiotics).

The use of microorganisms to produce fermented foods is not a new concept. Microorganisms have been used to ferment foods for thousands of years. It has been found that not all bacteria present in fermented foods exhibit a probiotic effect. Strains of microorganisms may only be considered probiotic if they show clinically established health benefits.

Probiotic-containing foods are produced and consumed in many parts of the world. They are very common in Japan and Europe and more recently have been introduced in the United States.

Prebiotics can be found naturally in many foods, for example, whole grains, onions, bananas, garlic, honey, leeks, artichokes, fortified foods, beverages, dietary supplements, and other food applications (International Food Information Council Foundation [IFICF] 2009). They can also be synthesized artificially, for example, enzymatically from sucrose. Examples of prebiotic substances include inulin, fructo-oligosaccharides (FOS), polydextrose, arabinogalactan, polyols, lactulose, and lactitol. The relationship between probiotics and prebiotics is that the probiotic microbes use the prebiotic components as feed. Taken together, the probiotic bacteria and the prebiotics that feed them are known as "synbiotics" indicating the synergistic association between them.

A recognized health effect of probiotics is their antidiarrheal effect and helping stool regularity. There is also evidence that yogurt containing sufficient amounts of live and active cultures of *Streptococcus thermophilus* and *Lactobacillus bulgaricus* can help reduce symptoms associated with lactose intolerance (Gibson 1999; Sanders 1999). Prebiotics consumption in the diet will promote the growth and multiplication of beneficial bacteria in the gut, consequently enhance the positive effect of probiotic bacteria. Additionally, prebiotics have been shown to promote the absorption of some mineral elements, for example, calcium and magnesium (Adolfsson 2004).

Some of the benefits of probiotics as outlined in UK Essays (2014) include

- Lowering the cholesterol level by break down the bile in the gut.
- Lowering the blood pressure by the production of ACE-inhibitor-like peptides.
- Protect from infections by competitive inhibition and improving immune system of the host.
- *L. bulgaricus* shows anticarcinogenic activity by decreasing the activity of β-glucuronidase.
- Block the adhesion site for the pathogens.
- Probiotics provide the antagonistic environment for the pathogens.

Today, there are many probiotic food products marketed as dairy products, tablets, and sachets with prescribed composition of probiotic microorganisms.

Probiotics biotechnology aims to address the different problems arising during their formulation into usable form, for example, stress during formulation, adapting to new environment, and competing with existing bacteria in the body.

The branch of biotechnology that deals with the pathogen modifications is called patho-biotechnology. This field has been developed by two scientists Roy Sleator and Colin Hill (Sleator and Hill 2006, 2007, 2008). They described the exploitation of pathogenic stress factors in biotechnology, medicine, and food. Among other advances, their approach promises the design of more technologically sustainable and effective probiotic cultures with improved biotechnological food and other applications.

Biotechnology aims first to improve the external stress tolerance of microorganisms, for example, to water and temperature. Microorganisms have the natural ability to accumulate compatible solute, for example, betaine and trehalose. These solutes help to minimize water loss through stabilization of proteins. The role played by biotechnology to address this issue is production of compatible solute-accumulating strains of microorganisms. Transgenesis is used to accomplish this procedure. BetL system (one of three natural uptake systems in the bacteria *Listeria monocytogenes*) is selected and a gene, which is betaine transporter, is transferred into the bacteria *Lactobacillus salivarius*. This process helps the bacteria to increase its salt tolerance and resistance and increase its betaine uptake.

Another goal of biotechnology in probiotics is to improve the process of colonization in the GI tract. This is achieved by inserting BetL gene into *Bifidobacterium (B) breve* and trehalose gene taken from *Escherichia coli* into *Lactococcus (L) lactis*. The modified bacteria strains exhibited improved tolerance to gastric juices, bile, and intestinal colonization.

Biotechnological procedures are also applied to help probiotic bacteria to compete with pathogens. Pathogens are found to use oligosaccharides expressed on host cells as receptors to adhere themselves. Biotechnology is used to produce receptor-mimic structure express with which pathogens bind. This approach renders the pathogens harmless to the probiotic strains as well as the host.

Probiotics have been used to treat inflammatory bowel disease by creating more host-friendly gut flora. Selective use of probiotic bacteria can create an environment where stimulation of the immune system is restrained and intestinal inflammation reduced (Pew Initiative on Food and Biotechnology 2007).

One biotechnology company in the United States has recently acquired its first probiotic product.

Enterologics, Inc. (Saint Paul, MN, USA) completed its acquisition of The BioBalance Corporation and its subsidiary BioBalance LLC from New York Health Care, Inc. With this purchase, the biotechnology company obtained its first probiotic biotherapeutic product, *E. coli* M17, or Probactrix. *E. coli* M17 has an active Investigational New Drug application (IND) with the U.S. Food and Drug Administration (USFDA) for the indication chronic pouchitis (OWM 2011).

4.4.4.3.2 Fermented Foods
Fermentation is the process during which bioconversion of organic substances by microorganisms and/or enzymes of microbial, plant, or animal

origin takes place. It results in the production of a diversity of metabolites, for example, enzymes which are capable of breaking down carbohydrates, proteins, and lipids present within the substrate and/or fermentation medium, vitamins, antimicrobial compounds, texture-forming agents, amino acids, organic acids, and some flavor compounds. Improvements in the production, quality, and yields of these metabolites have been researched using different microbial biotechnologies. Fermentation is one of the oldest forms of food preservation which has been used in most parts of the world. It is estimated that fermented foods contribute to about one-third of the diet worldwide (Ghosh 2013).

Microorganisms including bacteria, yeasts, and mold form a key and integral component of the fermentation process. They have thus received much attention in the research activities to improve the fermentation process and its resulting products. Microbial cultures used in fermentation have been improved through both traditional and modern biotechnologies. Traits which have been considered for commercial food applications in both developed and developing countries include sensory and nutritional qualities, virus resistance in the case of dairy fermentations, and the ability to produce antimicrobial compounds for the inhibition of undesirable microorganisms.

A variety of foods are produced by solid-state fermentation. Examples of such foods are yoghurt (using *L. bulgaricus* and *S. thermophilus* in the ratio of 1:1), buttermilk (using *Streptococcus lactis*, *Streptococcus cremoris*, and *Lecuconostoc cremoris*), cheese (using *S. lactis*, *S. cremoris*, *Lactobacillus lactis* for curd formation and *Penicillium roquefortii* and *P. cammebertii* for ripening) (Ghosh 2013).

Biotechnology in the food processing sector targets the selection and improvement of microorganisms with the objectives of improving process control, yields, and efficiency as well as the quality, safety, and consistency of bioprocessed products (FAO 2004). Biotechnological processes used for the improvement of microbial cultures for use in food-processing applications include traditional biotechnologies such as classical mutagenesis and conjugation, which mainly focus on improving the quality of microorganisms and the yields of their metabolites. Another traditional biotechnology applied is hybridization, which is used for the improvement of baker's and brewer's yeasts and yeast involved in beverage production. *Saccharomyces cerevisiae* strains have, for example, been researched for improved fermentation, processing, and biopreservation abilities and for capacities to increase the wholesomeness and sensory quality of wine (FAO International Technical Conference 2010).

Before the advent of the technique of electroporation (explained in Chapter 2), genetic exchange in bacteria used to occur naturally by three different mechanisms, namely transduction, conjugation, and transformation. Transduction involves genetic exchange mediated by a bacterial virus (bacteriophage). Conjugation, or bacterial mating, is a natural gene transfer system that requires close physical contact between donors and recipients and is responsible for the dissemination of plasmids in nature. In transformation, certain microorganisms are able to take up naked DNA present in the surrounding medium. This process is limited to strains that are naturally competent.

GE has also been employed in fermentation development and improvements. Baker's yeast was engineered to improve the rate at which bread dough rises by increasing the efficiency with which maltose is broken down. This modification was done by using genes from yeast and placing them under a strong constitutive promoter. Another example of GE application in fermentation is the modification of brewer's yeast enzyme system to utilize carbohydrates more efficiently resulting in improved alcohol yield, consequently producing a full-strength low-carbohydrate beer. The wine industry has also benefited from GE of yeast when two genetically modified yeast strains were authorized for use (Bauer et al. 2007).

The Office of International Affairs National Research Council (1992) published a report dealing with the applications of biotechnology to traditional fermented foods. The report aimed at creating greater awareness of the opportunities to reduce hunger and improve nutrition in developing countries through the application of biotechnology to commonly used methods of food preparation and preservation. It mainly focused on the application of biotechnology to traditional fermented foods. Scientists from developed and developing countries participated in the discussion and described their research in this field. Some recommendations on priorities for future research were included in the report.

4.4.4.3.3 Sweeteners
The term "sweetener" as used in relation to food products means a sugar (sucrose) substitute. It can also be used to describe sugar alcohols (e.g., xylitol, sorbitol), honey, syrup, and unrefined sweeteners. The main purpose of using a sugar substitute in food is to give the sweet taste of sugar while providing less energy. Sugar substitutes approved for food use can come from natural sources or can be synthesized (artificial sweeteners) with the majority being the artificial sweeteners.

Some sugar substitutes belong to the chemical group of sugar alcohols known as "polyols." Polyols are less sweet than sucrose but they have similar bulk properties and are widely used in the food industry. An example of polyols is erythritol, which is a four-carbon compound used as biological sweetener. It is also used as a functional sugar substitute in special foods for people with diabetes and obesity. Erythritol is produced by microbial techniques using osmophilic yeast and has been produced at commercial level using mutant strains of *Aureeobasidium* sp. and *Pseudozyma tuskubaensis*. Due to its high yield, low cost, and productivity at industrial level, erythritol is used as a source for the production of other sugars (Moon et al. 2010).

Another sweetener is D-psicose. Chemically this is a hexoketose monosaccharide C-3 epimer of D-fructose. It is rarely found naturally (sometimes known as "rare sugar") and is present in small amounts in agricultural products. It has a very low metabolic energy level estimated at 0.3% that of sucrose and its relative sweetness stands at 70%, making it an ideal substitute for sucrose in food products (Matsuo et al. 2002; Mu et al. 2012). It can be isolated from the antibiotic psicofuranine, from which it derives its name. D-psicose has shown useful physiological functions such as combating hyperglycemia, hyperlipidemia, and obesity. In food processing, it is used to improve the gelling properties and impart desirable flavor. In 2012, the U.S. Food and Drug Administration (USFDA) approved D-psicose as a food additive and designated it as generally recognized as safe (GRAS) (USFDA 2012). In the United States, six intensely sweet sugar substitutes have been approved for use. These are stevia, aspartame, sucralose, neotame, acesulfame potassium, and saccharin.

Another sweetener is D-tagatose. This is a naturally occurring ketohexose sweetener that can be used as a novel functional sweetener in foods, beverages, and dietary supplements. It is often found in dairy products, fruits, and cacao. It is similar in texture to sucrose and is 92% as sweet, but with only 38% of the calories. Tagatose at commercial level can be produced from the monosaccharide galactose which is obtained through enzymatic hydrolysis of lactose. Galactose is then isomerized under alkaline conditions, using calcium hydroxide, to D-tagatose. The resulting mixture is consequently purified and solid tagatose is obtained through crystallization. Xu et al. (2012) described a method for production of D-tagatose utilizing alginate immobilized *Lactobacillus fermentum* CGMCC2921 cells. Tagatose has been included in the FAO/WHO GRAS list since 2001.

4.4.4.3.4 Biocolors

Food color is an important sensory attribute that affect consumers' degree of acceptability to the food. It is also regarded as the most important factor in determining whether a consumer would purchase and/or consume a certain food since it is the first sensory food property perceived by the consumer. Food colors add to the esthetic value of food and it has been found to have a direct effect on stimulating or suppressing an individual's appetite.

Food colors can be produced from natural sources or can be manufactured artificially, that is, synthetic colors. A few decades back synthetic colors, for example, azo dyes were widely used in various preparations. However, their use has been negatively affected after safety issues were raised about their use and consumption in various foods. Results of toxicological tests carried out for them indicated that they have negative physiological effects. That led to their removal of these colors from the list of permitted food colors.

Biocolors is a term used to refer to any dyes obtained from biological sources, that is, living organisms such as plants, animals, and microorganisms. As such, they come under colors from natural sources which have the potential to be used in place of synthetic colors in coloring foods, drugs, or cosmetics. The food industry has benefited from the advent of biocolors in different food commodities including bakery products, dairy products, confectionery, and beverages. Biocolors can be obtained using one of the following approaches (Sharma 2014):

- Using biotechnology
- Using microorganisms
- Using enzymes

Biotechnology has been applied in the production of natural food colors and has been successful in many directions including genetic modification for pigment production, improvement of traditional pigment extraction methods, microbial production of pigments, and pigment production by using tissue culture (Sharma 2014). Details of how each of these biotechnological techniques is used to produce food colors are presented by Sharma (2014) in his article "Understanding Biocolour—A Review."

4.4.5 The Future of GM Food

The future of any human activity builds on, and is affected by, its past and its present conditions. GM foods is not an exception. Multiple factors including research and development, political, socioeconomic, ethical, media, and geographical factors are expected to affect the future of GM foods. However, the controversy and the debate about their safety and effect on human health in addition to their impact on the environment will prevail and continue in the future. This is because research results supporting both views, that is, that they are safe or that they are unsafe, continue to come out every day. The fact of the matter is that new types of GM food are developed and marketed regardless of the objections of those who oppose GE. This trend is expected to continue in the future. Despite consumers' concern about GM food, more of it is being developed and marketed every year. There is little doubt that in the near future that plants and animals will be genetically modified in new and different ways, for example, scientists are experimenting with GE plants to produce plastic (Freedeman 2009). The bottom line is that people need to make sure that they gain a thorough understanding of the effects that raising and consuming GMOs have on both the environment and their bodies.

The future of GM foods is viewed differently in different parts of the world, for example, in the United States versus Europe and in developing countries. The present trends in producing, regulating, labeling, and consuming GM foods will have a direct effect on the future of GM foods. Production of new GM foods intended for different purposes and targeted at different groups in the society is on the rise. This trend will continue in the future.

In the United States, production and consumption of GM foods is expected to continue growing in the future. This growth will be reflected both in food varieties and quantities produced. This is based on the observed trends in GM food production and consumption since the introduction of the first GM foods in the markets. The raging battle over GM food labeling, which has not been resolved yet, will continue for some time before reaching a decision as to whether labeling should be mandatory or not. If mandatory labeling of GM foods becomes a requirement for GM foods, it is anticipated that consumption of GM foods will drop. Some present GM food consumers might feel that introduction of this additional labeling means people need to be cautious and the food might be unsafe bearing in mind the strong anti-GM campaigns.

The future of GM foods in Europe has been discussed by Burr (2013) who raises the question "what is the future of GM foods in Europe?" As alluded to earlier, the future would be influenced by past practices and trends. Most of the EU countries have a history of being cautious about GMOs and GM foods particularly in relation to potential health and environmental risks. Objections by environmental groups and the widespread apprehension about GMOs and GM foods among European consumers have been going on for decades. This conservative approach by European countries lead to the situation that there is not much GMO agriculture in the EU. It is reported that there is only one GM crop, namely Monsanto's MON810 corn, which is currently cultivated in the EU (Casert 2013). Another GM crop, DuPont-Pioneer Maize 1507, is expected to be approved for cultivation by the EU's 28 member nations. The future of GM foods in Europe may also be tainted by the relocation of the operations of the two agricultural biotech giants BASF (Germany-based) and Syngenta (Swiss-based) to the United States (Burr 2013). Although the future of GM agriculture in Europe is moving toward a GMO-free agriculture, its food system would not necessarily be GMO-free. This is due to the fact that trade regulations in Europe do not restrict importation of GM crops from different parts of the world. Forty-nine GM crops are approved for import in Europe. On the other hand, the strict labeling laws and the negative attitude of European consumers towards GM technology might result in less demand for GM foods in the EU countries. A 2010 European Barometer study (European Commission. Eurobarometer 2010) showed that less than a quarter of Europeans feel that GM foods are either safe to eat or safe for the environment. The same report also shows that the majority of Europeans believe that GM foods could have benefits as well as some risks. The more recent European Barometer study of 2013 (European Commission 2013) indicated that three-fourths of Europeans believe that science and technology have a positive impact on the society. It is anticipated that the future of GMOs and GM foods in Europe will not be good unless Europeans can be convinced that GMOs are safe for both the environment and consumers' health.

The future of GM foods is also presented in a documentary (Roberts 2014) which states that GM foods will be the eventual cure for the world hunger, as seen by GM technology proponents. On the other side, the GM opponents speculate that in future the GM food production will increase the problem of world hunger. Other views presented in this documentary express concerns that future risks might extend beyond personal health and that there might be a global pandemic and shortage of crops in the future. Some legal issues might also be encountered in the future as people are becoming involved in lawsuits in connection with GM technology and GM foods all over the world, the documentary predicts.

Development and production of functional foods and functional ingredients is one area in which biotechnology has a promising future potential. At present, a number of advances that indicate this trend are reported. Hsieh and Ofori (2010) gave examples of plant-based functional foods with production potential in the future. These involve using GE in biofortification with essential micronutrients, for example, vitamin A, iron, and zinc and with phytonutrients, for example, carotenoids and polyphenolic compounds with antioxidant properties. The bioavailability of nutrients can also be improved through genetic manipulation. GE also has the potential to modify macronutrients in a way that render them healthier, for example, edible oils with an improved and healthier fatty acid profile. Another nutritional aspect that can be achieved through GE is the production of food crops with reduced antinutrient levels. Other future prospects of using GE to improve the nutritional profile of foods are described by Takase (2010). These include improving the amino acid and fatty profiles of food crops, enhancing the bioavailability of minerals and vitamins, eliminating allergens and other harmful or toxic substances and the production of an edible vaccine.

Takeuchi (2010) describes an example of the application of biotechnology to produce healthy cooking oils which suppress fat accumulation in the body. He was able to develop a dietary oil with higher amounts of medium-chain fatty acids, which are known to be readily converted into energy and less likely to accumulate as body fat in animal studies. The oil was certified as a food specified for health use in 2002 and marketed as a dietary oil "less likely to cause body fat accumulation," and a product name "healthy Resseta" by Nisshin Oillio, Japan.

Large sectors of the population in developing countries rely on pulse crops as the main source of protein due to the high cost of animal protein. GE technology can be used as an effective tool to improve pulse crop yields by manipulating the physiological processes and the ability to withstand biotic and abiotic stresses (Eapaen 2010). In future, GE can be used to produce transgenic pulse crops which are high yielding and have improved nutritional qualities. This would help improve the food and nutrition security of the vulnerable groups particularly in developing regions of the world. Efforts in GM will continue to address two main aspects related to food, firstly to enhance the nutritional quality of food and secondly to increase the relevance of GM technology to developing countries in a way that helps to address their specific problems and needs.

4.5 BIOTECHNOLOGY FOR BIOFUEL PRODUCTION

As discussed in Chapter 2, biotechnology has been used in many industries. Based on the purpose for which it is used, it has been designated a specific color. One of its uses is the production of biofuels. Although this is not a food commodity, it is presented here because the source of material used for production, that is, soybean, corn, palm, and rapeseed falls in the food domain. On the other hand, biofuel production can have a direct effect on the sustainable availability of food worldwide, since the main crops used in biofuel production are food crops.

Biofuels is the term used to describe fuels other than fossil fuels and which are produced from plant material. At present, two types of biofuels are produced, namely bioethanol and biodiesel (Figure 4.8). They mainly differ in the source from which they are derived: bioethanol is produced through fermentation of the carbohydrate material of crops such as corn, sweet sorghum, sugar beet, cassava, and sugar cane, whereas biodiesel is produced from oil crops such as rapeseed, soybean, sunflower, and palm (ISAAA 2013b). Bioethanol is mainly used to substitute petrol for road transport vehicles. In some parts of the United States (e.g., in the State of Iowa), it is added to the gas pumped at petrol stations. Biofuels are produced in many countries around the world, including the United States, Brazil, Canada, China, Germany, France, and Thailand. Some of these countries, for example, United States, India, and EU countries have set clear goals to boost the proportion of biofuels used for transport (ISAAA 2013b).

Figure 4.8 Biofuel isolated from rape seeds and oil cakes.

The main reason for resolving to produce biofuels is that an alternate energy source to fossil fuels is needed at present and in the future. Fossil fuels have been described as limited, nonrenewable, and environment pollutant. In addition to this, it is anticipated that a large increase in demand for energy, mainly coming from developing countries, will soon become a reality. Some of the expected benefits of producing and using biofuels include less dependency on fossil fuel, cleaner environment, additional source of income to farmers cultivating crops used for biofuel production, and a possible boost for rural development.

It is believed that biofuel production and use have positive impacts on the environment. Biofuels are regarded as "carbon-neutral" in the sense that the carbon dioxide they release upon combustion is initially extracted from the atmosphere during biomass production, resulting in zero net greenhouse gas emissions (ISAAA 2014). The addition of ethanol to gasoline helps to oxygenate the fuel mixture consequently leading to more complete burning of the fuel. This mechanism reduces the release of volatile organic substances in the atmosphere hence leading to a cleaner environment. Adding ethanol to vehicle fuel will also eliminate the need to add lead, as the practice now in some fuel types. Biofuels are also biodegradable and nontoxic, which means they are more environment-friendly.

The recent shift in trend towards using some food crops, for example, corn and soybean for biofuel production has fueled some controversy. Some people believe that there are still millions of people who are food insecure and are not getting enough food and that this shift towards using food crops to produce fuel will exacerbate the problem. On the other hand, some people believe that biofuels can replace 30% of current transportation energy needs in an environmentally responsible way without affecting global food production with plausible technology developments (Ragauskas et al. 2006). Regardless of these concerns and opposition, biofuel production continues to flourish.

In order to optimize biofuel production and utilization, a number of issues still need to be carefully addressed. There is need to develop specific high-yielding energy crops in order to maximize production. It is also important to enhance the infrastructure needed for biofuel production, transport, and distribution (ISAAA 2014). Any possible negative impact on food production and availability need to be taken into consideration and carefully addressed. This needs continuous evaluation of the impact of biofuel production on the food availability and the worldwide food security situation.

Studies describing technological and operational advances in the production of biofuels from biomass are being conducted, the results of which appear in an online peer-reviewed journal *Biotechnology for Biofuels* (*Biotechnology for Biofuels* 2014). Biotechnology for biofuels emphasizes understanding and advancing the application of biotechnology and synergistic operations to improve plants and biological conversion systems for the production of fuels from lignocellulosic biomass and any related economic, environmental, and policy issues. The journal focuses on the following areas:

- Development of plants for biofuels production
- Plant deconstruction
- Pretreatment and fractionation
- Enzyme production and enzymatic conversion
- Fermentation and bioconversion
- Integrated systems
- Bio-based chemical products of fermentation
- Process design and economics
- Life-cycle studies

REFERENCES

Academic Reviews. 2014. Chemical composition of crops is highly variable. http://academicsreview.org/reviewed-content/genetic-roulette/section-2/2-10-chemical-compositions (accessed January 19, 2014).

Adolfsson, O. 2004. Yogurt and gut function. *The American Journal of Clinical Nutrition*, 80:245–56.

Andrews, R. 2013. "All about" series. All about genetically modified foods. http://www.precisionnutrition.com/all-about-gm-foods (accessed December 11, 2013).

Animal Biotechnology. 2013. http://www.whatisthebiotechnology.com/pages/animal_biotechnology.html (accessed December 19, 2013).

Arctic Apples. 2014. How'd we do that? http://www.arcticapples.com/arctic-apples-story/how-we-keep-apples-from-turning-brown (accessed January 11, 2014).

Ashwell, M. 2002. Concepts of functional foods, in Walker, R. (Ed.), *ILSI Europe Concise Monograph Series*. http://www.ilsina.org/file/ILSIFuncFoods.pdf (accessed February 20, 2014).

Bauer, F.F., Næs, T., Esbensen, K., Young, P.R., du Toit, M., and Vivier, M.A. 2007. Functional wine-omics, in Blair, R.J. and Pretorius, I.S. (Eds.), *Proceedings of the Thirteenth Australian Wine Industry Technical Conference*, July 29–August 2, 2007, Adelaide, South Australia, pp. 178–83.

Bio Elite Wellness. 2014. Examples of genetically modified food. http://www.getfreenutritiontips.com/blog/examples-of-genetically-modified-food/ (accessed January 19, 2014).

BioSpectrum. 2012. Biotech nutraceuticals is the way forward. http://www.biospectrumasia.com/biospectrum/opinion/1682/biotech-nutraceuticals-forward#.Uwg-3dKsiSo (accessed February 20, 2014).

Biotechnology for Biofuels. 2014. About biotechnology for biofuels. *Online Journal*. http://www.biotechnologyforbiofuels.com/about (accessed January 20, 2014).

Board on Science and Technology for International Development, Office of International Affairs National Research Council. 1992. Application of biotechnology to traditional fermented foods. National Academic Press, Washington, DC.

Boraste, A., Vamsi, K.K., Jhadav, A. et al. 2009. Biofertilizers: A novel tool for agriculture. *International Journal of Microbiology Research*, 1(2):23–33.

Bothma, G., Mashaba, C., Mkhonza, N., Chakauya, E., and Chikwamba, R. 2010. GMOs in Africa: Opportunities and challenges in South Africa. *GM Crops and Food. Commentary*, 1(4):175–80. Landes Bioscience. Previously published online: www.landesbioscience.com/journals/gmcrops/article/13533. CSIR Biosciences; Pretoria, South Africa.

Brennand, C.P. 2014. Bovine somatotropin in milk. Food safety fact sheet. Utah State University Cooperative Extension. https://extension.usu.edu/files/publications/factsheet/FN-250_6.pdf (accessed February 20, 2014).

Burr, M. 2013. The future of GM food in Europe. http://www.linkedin.com/today/post/article/20131213094514-13968697-the-future-of-gm-food-in-europe (accessed May 1, 2014).

Butcher, M. 2003. Genetically modified food—GM foods list and information. http://www.disabled-world.com/fitness/gm-foods.php (accessed January 19, 2014).

Casert, R. 2013. EU execs back for genetically modified corn. http://news.yahoo.com/eu-execs-back-ok-genetically-modified-corn-172414115—finance.html (accessed May 1, 2014).

Celestialhealing.net. 2014. Companies who use GMO ingredients. http://www.celestialhealing.net/monsanto/GMO_FoodCo.htm (accessed February 14, 2014).

CERA (Center for Environmental Risk Assessment). 2012. GM crop database. http://cera-gmc.org/index.php?action=gm_crop_database (accessed December 15, 2013).

Clydesdale, F.M. 1997. A proposal for the establishment of scientific criteria for health claims for functional foods. *Nutrition Reviews*, 55(12):413–22.

Conway, G. 2000. Genetically modified crops: Risks and promise. *Conservation Ecology*, 4(1):2. http://www.consecol.org/vol4/iss1/art2 (accessed November 12, 2014).

Cosgrove, J. 2010. Biotech for wellness. Nutraceuticals world. http://www.nutraceuticalsworld.com/contents/view_online-exclusives/2010-07-12/biotech-for-wellness/ (accessed February 20, 2014).

Derbick, D. 1999. Genetically-modified processed and whole foods list. Sightings. http://rense.com/politics6/gms.htm (accessed January 19, 2014).

DiLeo, M. 2012. Monsanto's GM drought tolerant corn. http://www.biofortified.org/2012/08/monsantos-gm-drought-tolerant-corn/ (accessed December 10, 2013).

DNA Hot Science. 2014. Gallery of genetic modification. http://www.pbs.org/wnet/dna/pop_genetic_gallery/index.html (accessed January 19, 2014).

Eapaen, S. 2010. Recent advances in the development of transgenic pulse crops, Chapter 10, in Bagchi, D., Lau, F.C., and Ghosh, D.K. (Eds.), *Biotechnology in Functional Foods and Nutraceuticals*. CRC Press, Boca Raton, FL, pp. 139–55.

eMerge. 2014. Non-GMO perspective. http://www.emergegenetics.com/ (accessed February 20, 2014).

Environmental Protection Agency of the USA. 2012. Regulating biopesticides. http://www.epa.gov/opp00001/biopesticides/ (accessed April 10, 2014).

Estes, L. and Watson, G. 2013a. Bt crops. http://www.bio.davidson.edu/people/kabernd/seminar/2002/resist/bt.htm (accessed December 10, 2013).

Estes, L. and Watson, G. 2013b. Roundup-ready crops. http://www.bio.davidson.edu/people/kabernd/seminar/2002/resist/roundup.htm (accessed December 10, 2013).

European Commission. 2008. Encouraging innovation in biopesticide development. http://ec.europa.eu/environment/integration/research/newsalert/pdf/134na5.pdf (accessed April 10, 2014).

European Commission. 2013. EU-wide poll shows public support for responsible research and innovation. http://europa.eu/rapid/press-release_IP-13-1075_en.htm (accessed May 1, 2014).

European Commission. Eurobarometer. 2010. Special Eurobarometer. Biotechnology. http://ec.europa.eu/public_opinion/archives/ebs/ebs_341_en.pdf (accessed May 1, 2014).

Falck-Zepeda, J., Gruère, G., and Sithole-Niang, I., eds. 2013. *Genetically Modified Crops in Africa Economic and Policy Lessons from Countries South of the Sahara*. International Food Policy Research Institute, Washington, DC.

FAO. 2004. Biotechnology applications in food processing: Can developing countries benefit? http://www.fao.org/biotech/logs/C11/summary.htm (accessed December 13, 2013).

FAO. 2005. Status of research and application of crop biotechnologies in developing countries (FAO-BioDeC). FAO Corporate Document Repository. http://www.fao.org/docrep/008/y5800e/Y5800E02.htm (accessed December 15, 2013).

FAO International Technical Conference. 2010. Agricultural biotechnologies in developing countries: Options and opportunities in crops, forestry, livestock, fisheries and agro-industry to face the challenges of food insecurity and climate change (ABDC-10), Guadalajara, Mexico, March 1–4, 2010. Current status and options for biotechnologies in food processing and in food safety in developing countries. http://www.fao.org/docrep/meeting/019/k6993e.pdf (accessed February 20, 2014).

Freedeman, J. 2009. *Genetically Modified Food: How Biotechnology is Changing What We Eat*. First Edition. The Rosen Publishing Group Inc., New York, NY.

Ghosh, S. 2013. Biotechnology in food processing industry. Biotechnology forums. http://www.biotechnologyforums.com/thread-1692.html (accessed February 20, 2014).

Ghosh, D.K. and Williams, P. 2010. Global food biotechnology regulations and urgency for harmonization, Chapter 29, in Bagchi, D., Lau, F.C., and Ghosh, D.K. (Eds.), *Biotechnology in Functional Foods and Nutraceuticals*. CRC Press, Boca Raton, FL, pp. 531–41.

Gibson, G.R. 1999. Dietary modulation of the human gut microflora using the prebiotics oligofructose and inulin. *American Society of Nutrition Science*, 129:1438S–41S.

GMO Compass. 2013a. Herbicide resistant crops. http://www.gmo-compass.org/eng/agri_biotechnology/breeding_aims/146.herbicide_resistant_crops.html (accessed December 11, 2013).

GMO Compass. 2013b. Pest resistant crops. http://www.gmo-compass.org/eng/agri_biotechnology/breeding_aims/147.pest_resistant_crops.html (accessed December 11, 2013).

GMO Compass. 2013c. Disease resistance. http://www.gmo-compass.org/eng/agri_biotechnology/breeding_aims/148.disease_resistant_crops.html (accessed December 11, 2013).

GMO Compass. 2013d. Genetic engineering: Feeding the EU's livestock. Feed. http://www.gmo-compass.org/eng/grocery_shopping/processed_foods/153.animal_feed_genetic_engineering.html (accessed December 19, 2013).

GMO Compass. 2014a. Food products. Overview: Foods, genetic engineering, and labelling. http://www.gmo-compass.org/eng/grocery_shopping/processed_foods/26.overview_foods_genetic_engineering_labelling.html (accessed January 19, 2014).

GMO Compass. 2014b. Processed foods. Bread and baked goods. http://www.gmo-compass.org/eng/grocery_shopping/processed_foods/28.gmos_bread_baked_goods.html (accessed January 19, 2014).

GMO Compass. 2014c. Processed foods. Sweets, chocolate, and ice cream. http://www.gmo-compass.org/eng/grocery_shopping/processed_foods/31.sweets_chocolate_ice_cream_genetic_engineering.html (accessed January 19, 2014).

GMO Compass. 2014d. Processed foods. Meats and sausage. http://www.gmo-compass.org/eng/grocery_shopping/processed_foods/32.genetic_engineering_meats_sausage.html (accessed January 19, 2014).

GMO Compass. 2014e. Processed foods. Dairy products and eggs. http://www.gmo-compass.org/eng/grocery_shopping/processed_foods/29.dairy_products_eggs_genetic_engineering.html (accessed January 19, 2014).

GMO Compass. 2014f. Processed foods. Beverages. Juice, soft drinks, wine, and beer. http://www.gmo-compass.org/eng/grocery_shopping/processed_foods/30.beverages_genetic_engineering.html (accessed January 19, 2014).

GMWATCH. Non-GM successes: Drought tolerance. http://www.gmwatch.org/component/content/article/31-need-gm/12319-drought-resistance (accessed December 11, 2013).

Gouse, M. 2013. Socioeconomic and farm-level effects of genetically modified crops: The case of Bt crops in South Africa, Chapter 1, in Falck-Zepeda, J., Gruère, G., and Sithole-Niang, I. (Eds.), *Genetically Modified Crops in Africa Economic and Policy Lessons from Countries South of the Sahara*. International Food Policy Research Institute, Washington, DC, pp. 25–41.

GreenFacts. 2013. What is agricultural biotechnology? http://www.greenfacts.org/en/gmo/3-genetically-engineered-food/1-agricultural-biotechnology.htm (accessed December 10, 2013).

Hails, R.S. 2000. Genetically modified plants—The debate continues. *Trends in Evolution and Ecology*, 15(1):14–8.

Herrera-Estrella, L. 2000. Genetically modified crops and developing countries. *Plant Physiology*, 124(3):923–6. American Society of Plant Biologists. http://www.plantphysiol.org/content/124/3/923.full.pdf+html (accessed December 15, 2013).

Herrera-Estrella, L. and Alvarez-Morales, A. 2001. Genetically modified crops: Hope for developing countries? The current GM debate widely ignores the specific problems of farmers and consumers in the developing world. *EMBO Reports*, 2(4):256–8.

Horna, D., Zambrano, P., Falck-Zepeda, J., Sengooba, T., and Kyotalimye, M. 2013. Genetically modified cotton in Uganda: An ex ante evaluation, Chapter 3, in Falck-Zepeda, J., Gruère, G., and Sithole-Niang, I. (Eds.), *Genetically Modified Crops in Africa Economic and Policy Lessons from Countries South of the Sahara*. International Food Policy Research Institute, Washington, DC, pp. 61–97.

Hsieh, Y.P. and Ofori, J.A. 2010. Advances in biotechnology for the production of functional foods, Chapter 1, in Bagchi, D., Lau, F.C., and Ghosh, D.K. (Eds.), *Biotechnology in Functional Foods and Nutraceuticals*. CRC Press, Boca Raton, FL, pp. 3–28.

Institute of Food Science and Technology (IFST). 2008. Information statement genetic modification and food. www.ifst.org (accessed April 20, 2014).

International Food Information Council Foundation (IFICF). 2009. Food insight: Your nutrition and food safety resource. Functional foods fact sheet: Probiotics and prebiotics. http://www.foodinsight.org/Resources/Detail.aspx?topic=Functional_Foods_Fact_Sheet_Probiotics_and_Prebiotics (accessed February 21, 2014).

International Food Information Council Foundation (IFICF). 2014. Food biotechnology: A communicator's guide to improving understanding. 3rd Edition. http://www.foodinsight.org/education/food-biotechnology-communicator%E2%80%99s-guide-improving-understanding (accessed February 10, 2014).

ISAAA. 2013a. Global status of commercialized biotech/GM crops. ISAAA Brief 44-2012. http://www.isaaa.org/resources/publications/briefs/44/executivesummary/default.asp (accessed December 10, 2013).

ISAAA. 2013b. Pocket K No. 24 Biotechnology for green energy: Biofuels. http://www.isaaa.org/resources/publications/pocketk/24/default.asp (accessed February 22, 2014).
ISAAA. 2014. Global status of commercialized biotech/GM crops. ISAAA Brief 46-2013. http://www.isaaa.org/resources/publications/briefs/46/executivesummary/default.asp (accessed March 18, 2014).
James, C. 2013. Global status of commercialized biotech/GM crops for 2013. ISAAA Brief 46. ISAAA, Ithaca, NY.
Joint FAO/WHO Working Group. 2002. Report on drafting guidelines for the evaluation of probiotics in food. London, Ontario, Canada. ftp://ftp.fao.org/es/esn/food/wgreport2.pdf (accessed February 20, 2014).
Kiguli, L.N. 2000. The utilisation of Azolla filiculoides Lam. as a biofertiliser under dryland conditions. Master thesis, University of Rhodes, South Africa.
Kikulwe, E.M., Birol, E., Wesseler, J., and Falck-Zepeda, J. 2013. Benefits, costs, and consumer perceptions of the potential introduction of a fungus-resistant banana in Uganda and policy implications, Chapter 4, in Falck-Zepeda, J., Gruère, G., and Sithole-Niang, I. (Eds.), *Genetically Modified Crops in Africa Economic and Policy Lessons from Countries South of the Sahara*. International Food Policy Research Institute, Washington, DC, pp. 99–142.
Kleiner, K. 1998. Let us spray. *New Scientist*, 158(2128):16.
Lanphier, L. 2014. More GMO info—What you must know exhibit health—A better way to health and wellness. http://www.exhibithealth.com/health-education/more-gmo-info-what-you-must-know-1518/ (accessed May 12, 2014).
Law, B.A. 2010. *Technology of Cheesemaking*. Wiley-Blackwell, UK, pp. 100–1.
Marx, G.M. 2010. Monitoring of genetically modified food products in South Africa. PhD dissertation, Faculty of Health Sciences Department of Haematology, University of the Free State, Bloemfontein, South Africa.
Matsuo, T., Suzuki, H., Hashiguchi, M., and Izumori, K. 2002. D-psicose is a rare sugar that provides no energy to growing rats. *Journal of Nutritional Science and Vitaminology (Tokyo)*, 48(1):77–80.
Miami-water.com. 2014. Product list of GMO genetically modified foods. http://miami-water.com/blog/2217/product-list-of-gmo-genetically-modified-foods/ (accessed February 14, 2014).
Monsanto. 2014. Genuity DroughtGard hybrids (by Monsanto). http://www.monsanto.com/products/pages/droughtgard-hybrids.aspx (accessed June 10, 2014).
Moon, H.J., Jeya, M., Kim, I.W., and Lee, J.K. 2010. Biotechnological production of erythritol and its applications. *Applied Microbiology and Biotechnology*, 86(4):1017–25.
Mu, W., Zhang, W., Feng, Y., Jiang, B., and Zhou, L. 2012. Recent advances on applications and biotechnological production of D-psicose and its applications. *Applied Microbiology and Biotechnology*, 94(6):1461–7.
NON-GMO Project. 2014. http://www.nongmoproject.org (accessed February 20, 2014).

North Carolina Association for Biomedical Research. http://www.aboutbioscience.org/topics/animalbiotechnology (accessed December 19, 2013).

Office of International Affairs National Research Council. 1992. Applications of biotechnology to traditional fermented foods. Report of an Ad Hoc Panel of the Board on Science and Technology for International Development National Academy Press, Washington, DC. http://pdf.usaid.gov/pdf_docs/PNABZ036.pdf

OWM. 2011. Biotechnology company acquires first probiotic product. http://www.o-wm.com/content/biotechnology-company-acquires-first-probiotic-product (accessed February 21, 2014).

Pennisi, E. 2001. The push to pit genomics against fungal pathogens. *Science*, 292:2273–4.

Peterson, G., Cunningham, S., Deutsch, L. et al. 2000. The risks and benefits of genetically modified crops: A multidisciplinary perspective. *Conservation Ecology*, 4(1):13. http://www.consecol.org/vol4/iss1/art13/ (accessed November 10, 2014).

Pew Initiative on Food and Biotechnology. 2007. Application of biotechnology for functional foods. http://www.pewtrusts.org/uploadedFiles/wwwpewtrustsorg/Reports/Food_and_Biotechnology/PIFB_Functional_Foods.pdf (accessed February 20, 2014).

Pimentel, D., Lach, L., Zuniga, R., and Morrison, D. 2000. Environmental and economic costs of nonindigenous species in the United States. *BioScience*, 50(1):53–64.

Pollacknov, A. 2014. U.S.D.A. approves modified Potato. Next up: French fry fans. The New York Times. http://www.nytimes.com/2014/11/08/business/genetically-modified-potato-from-simplot-approved-by-usda.html?_r=0 (accessed November 17, 2014).

Ragauskas, A.J., Williams, C.K., Davison, B.H. et al. 2006. The path forward for biofuels and biomaterials. *Science*, 311(5760):484–9. Biotechnology for biofuels. http://www.biotechnologyforbiofuels.com/about (accessed February 22, 2014).

Ramaa, C.S., Shirode, A.R., Mundada, A.S., and Kadam, V.J. 2006. Nutraceuticals—An emerging era in the treatment and prevention of cardiovascular diseases. *Current Pharmaceutical Biotechnology*, 7(1):15–23.

Rasmussen, R.S. and Morrissey, M.T. 2007. Biotechnology in aquaculture: Transgenics and polyploidy. *Comprehensive Reviews in Food Science and Food Safety*, 6:2–16.

REALfarmacy.com. 2013. 400 companies that do not use GMOs in their products. http://www.realfarmacy.com/400-companies-no-gmos/ (accessed October 10, 2013).

Roberts, J.J. 2014. Genetically modified food: Future of food. http://ezinearticles.com/?Genetically-Modified-Food:-Future-of-Food&id=5858335 (accessed May 1, 2014).

Samadhan. 2013. A complete business solution. BioFertilizers. http://www.samadhan.net/project/agriculture/rice-mills/bio-fertilizers/ (accessed December 15, 2013).

Sanders, M.E. 1999. Probiotics: A publication of the Institute of Food Technologists Expert Panel on Food Safety and Nutrition. *Food Technology*, 53:67–77.

Sarich, C. 2013. Over 400 companies who aren't using GMOs in their products. Natural society. http://naturalsociety.com/over-400-companies-who-arent-using-gmos-in-their-products/ (accessed February 20, 2014).

Sharife, K. 2009. GM: The food of the future. Pambazuka News. Issue 459. http://pambazuka.org/en/category/features/60523 (accessed May 1, 2014).

Sharma, D. 2014. Understanding biocolour—A review. *International Journal of Scientific and Technology Research*, 3(1):294–9.

Sleator, R.D. and Hill, C. 2006. Patho-biotechnology: Using bad bugs to do good things. *Current Opinion in Biotechnology*, 17(2):211–16. 10.1016/j.copbio.2006.01.006 (accessed February 20, 2014).

Sleator, R.D. and Hill, C. 2007. Patho-biotechnology: Using bad bugs to make good bugs better. *Science Progress*, 90(1):1–14. 10.3184/003685007780440530 (accessed February 20, 2014).

Sleator, R.D. and Hill, C. 2008. Bioengineered bugs—A patho-biotechnology approach to probiotic research and applications. *Medical Hypotheses*, 70(1):167–9. 10.1016/j.mehy.2007.03.008 (accessed February 20, 2014).

Spielman, D.J. and Zambrano, P. 2013. Policy, investment, and partnerships for agricultural biotechnology research in Africa: Emerging evidence, in Falck-Zepeda, J., Guillaum, G., and Sithole-Niang, I. (Eds.), *Genetically Modified Crops in Africa Economic and Policy Lessons from Countries South of the Sahara*. Table 1. International Food Policy Research Institute, Washington, DC.

Staff, GMO Compass. 2014. Genetic engineering: Feeding the EU's livestock. http://www.gmo-compass.org/eng/grocery_shopping/processed_foods/153.animal_feed_genetic_engineering.html (accessed January 19, 2014).

Takase, H. 2010. The improvement and enhancement of phyto-ingredients using the new technology of genetic recombination, Chapter 11, in Bagchi, D., Lau, F.C., and Ghosh, D.K. (Eds.), *Biotechnology in Functional Foods and Nutraceuticals*. CRC Press, Boca Raton, FL, pp. 157–80.

Takeuchi, H. 2010. Application of biotechnology in the development of a healthy oil capable of suppressing fat accumulation in the body, Chapter 7, in Bagchi, D., Lau, F.C., and Ghosh, D.K. (Eds.), *Biotechnology in Functional Foods and Nutraceuticals*. CRC Press, Boca Raton, FL, pp. 103–12.

UK Essays. 2014. Probiotics and biotechnology. http://www.ukessays.com/essays/biology/probiotics-and-biotechnology.php (accessed February 20, 2014).

US EPA. 2014. What are biopesticides? http://www.epa.gov/pesticides/biopesticides/whatarebiopesticides.htm (accessed March 10, 2014).

US Food and Drug Administration (USFDA). 2012. *GRAS Notice Inventory*. GRN # 400. http://www.accessdata.fda.gov/scripts/fcn/fcnDetailNavigation.cfm?rpt=grasListing&id=400 (accessed February 20, 2014).

USDA-National Institute of Food and Agriculture. 2013a. Animal biotechnology. http://www.csrees.usda.gov/nea/biotech/in_focus/biotechnology_if_animal.html (accessed December 19, 2013).

USDA-National Institute of Food and Agriculture. 2013b. Plant biotechnology. http://www.csrees.usda.gov/nea/biotech/in_focus/biotechnology_if_plant.html (accessed December 10, 2013).

Walker, B. and M. Lonsdale. 2000. Genetically modified organisms at the crossroads: Comments on "Genetically modified crops: Risks and promise" by Gordon Conway. *Conservation Ecology*, 4(1):12. http://www.consecol.org/vol4/iss1/art12 (accessed November 12, 2014).

Weed Science Society of America (WSSA). 2013. Herbicide resistance and herbicide tolerance definitions. http://wssa.net/weed/resistance/herbicide-resistance-and-herbicide-tolerance-definitions/ (accessed December 10, 2013).

Xu, Z., Li, S., Fu, F. et al. 2012. Production of D-tagatose, a functional sweetener, utilizing alginate immobilized *Lactobacillus fermentum* CGMCC2921 Cells. *Applied Biochemistry and Biotechnology*, 166(4):961–73. http://www.ncbi.nlm.nih.gov/pubmed/22203394 (accessed February 10, 2014).

Yerramareddy, I. and Zambrano, P. 2011. bEcon: Economics literature about the impacts of genetically engineered crops in developing economies. www.mendeley.com/groups/1296883/becon/ (accessed December 15, 2013).

Zubair, U. 2011. http://boltakarachi.blogspot.com/2011/06/drought-resistant-gm-crops.html (accessed December 11, 2013).

SUGGESTED REFERENCES

Copping, L.G. (Ed.). 2010. *The GM Crop Manual: A World Compendium*. British Crop Production Council, Alton, Hampshire.

Cummings, C.H. 2008. *Uncertain Peril: Genetic Engineering and the Future of Seeds*. Beacon Press, Boston.

Glover, J., Mewett, O., Cunningham, D., and Ritman, K. 2005. Genetically modified crops in Australia: The next generation. Summary Report. Department of Agriculture, Fisheries and Forestry, Australian Government, Bureau of Rural Sciences. http://data.daff.gov.au/brs/brsShop/data/biotech_5__2_.pdf (accessed October 28, 2014).

Halford, N.G. 2003. *Genetically Modified Crops*. Imperial College Press, London, UK.

Kuyek, D. 2002. Genetically modified crops in Africa: Implications for small farmers. www.grain.org/docs/Africa-gmo-2002-en.pdf (accessed October 28, 2014).

Macaulay, J. 2003. Biopharming: Growing medicine crops. *Food Technology*, 57:20.

McGregor, R. 2004. Taste modification in the biotech era. *Food Technology*, 58:24–30.

Miller, M. (Ed.). 2011. *Genetically Modified Crops in Africa*. United Nations Environmental Programme.

Mousdale, D.M. 2008. *Biofuels: Biotechnology, Chemistry, and Sustainable Development*. CRC Press, Boca Raton, FL.

Panesar, P.S. and Marwaha, S.S. (Ed.) 2013. *Biotechnology in Agriculture and Food processing: Opportunities and Challenges*. CRC Press, Boca Raton, FL.

Parker, I.M. and Kareiva, P. 1996. Assessing the risks of invasion for genetically engineered plants: Acceptable evidence and reasonable doubt. *Biological Conservation*, 78:193–203.

Pearce, F. 1999. Crops without profit. *New Scientist*, 164(2217). http://www.newscientist.com/19991218/newsstory4.html (accessed May 6, 2014).

Ruttan, V.W. 1999. The transition to agricultural sustainability. *Proceedings of the National Academy of Sciences USA*, 96(11):5960–7.

Shetty, K., Paliyath, G., Pometto, A., and Levin, R.E. 2006. *Functional Foods and Biotechnology*. CRC Press, Boca Raton, FL.

5

Laws, Regulations, and Labeling for GM Foods

5.1 INTRODUCTION

Consumers have been eating food products with GM ingredients for decades, yet many consumers are unaware of this. Many studies tried to ascertain consumers' views about GM foods and whether GM foods should be labeled or not. Some consumers say that they need to know whether their food contains GM ingredients, just as some want to know whether their food is natural or organic. Informing consumers on food ingredients has been the main motivation for labeling.

Regulation of GM crops and GM foods proved to be a complex multidimensional issue that needs to be addressed carefully. It has been an issue of concern to the public, scientists, policy makers, as well as the food industry. Weirich (2007) edited a book that surveys GM food labeling policies and the cases that support them. It is considered the first comprehensive, interdisciplinary treatment of the controversy on the issue of labeling GM foods. The contributors are drawn from a wide group of professionals such as food and agricultural scientists, economists, attorneys/legal scholars, bioethicists, and philosophers.

Governments around the globe have considered GM regulation and labeling as an urgent priority and are working hard to set regulatory processes addressing GM foods. Different governments have taken different approaches based specifically on political, social, and economic conditions prevailing around the region or in the country. Differences in regulations among countries are clear and the most marked differences occur between the United States and Europe. In general, most governments have set "the protection of public health as affected by food consumption" as the first priority to be achieved through promoting and enforcing high standards of safety throughout the food chain.

Some of the pertinent questions related to regulation of GM foods and the exact responsibilities of different parties, that is, the government and the food industry, are frequently asked and need to be satisfactorily answered for the benefit of the consumer. Also what sort of agency or body that would be most appropriate to handle the regulation of GM foods needs to be identified. Other questions that need to be answered and clarified and that the regulators face include: How prescriptive should the regulations be? Should there be any penalty or sanctions for non-compliance? What monitoring and enforcing procedures should be followed? Are new regulations needed for gene technology or would already existing regulations suffice? To what extent can governments go in order to promote trade and commerce at regional and international levels? Would national governments be able to streamline and handle any conflict that might arise between international obligations and local needs? If so, what modalities can be followed?

A number of challenges face governments in the process of drafting regulatory framework for GM foods. Governments need to ensure consumer protection, as well as trust and confidence in what the government sets as regulations. Consumers also need to be part of the decision-making process with regard to regulation development. Governments should have processes that will not impede development of the technology at all levels, for example, research and commercial levels. A reasonable balance between international regulatory systems and foreign trade and the national sovereignty of states needs to be observed by governments.

The regulatory process is further complicated by the need for specific and different regulations for the different stages of the use of gene technology to produce foods, that is, in the laboratory and at the industrial and marketing levels.

The development of a general regulatory framework on GE began in 1975 shortly after the first application of recombinant DNA (rDNA) new technology. The need for regulations resulted from the belief that this new technology may have some risks in addition to the identified benefits. Since the GE technology was then used to produce different types of foods, identified as GM or GE foods, it was necessary to regulate this category of foods. The main focus of the GM foods regulation has been safety assessment. The basic concepts for safety assessment of foods derived from GMOs have been developed through the collaborative work between the Organization for Economic Cooperation and Development (OECD) and the United Nations' WHO and the FAO. An outcome of that collaboration came in the year 2003 when the Codex Alimentarius Commission (CAC) of the FAO/WHO produced the document "Principles and Guidelines on foods derived from biotechnology" (Codex Alimentarius Commission 2003). The aim of this publication was to help countries establish, coordinate, and standardize regulations on GM foods to ensure consumer safety and at the same time harmonize and facilitate international trade. An additional related document updating the guidelines for import and export of food was released by CAC in 2004 (Codex Alimentarius Commission 2004).

5.2 GENERAL ASPECTS OF GM FOOD REGULATIONS AND LABELING

5.2.1 The Principle of Substantial Equivalence

"Substantial equivalence," or SE, is an internationally recognized standard that measures whether a biotech food or crop shares similar health and nutritional characteristics with its conventional counterpart (Council for Biotechnology Information 2001). It is considered as the starting point in the assessment of safety of GM foods and is widely used by national and international bodies dealing with GM foods, for example, FAO, WHO, OECD, USFDA, and by some countries such as Canada (through the Canadian Food Inspection Agency) and Japan (through Japan's Ministry of Health and Welfare).

SE considers that the most practical method to establish food safety of a novel food or one of its components is to check whether it is substantially equivalent to an analogous conventional food product, if one exists. SE embodies the concept that if a new food or food component is found to be substantially equivalent to an existing food or food component, it can be treated in the same manner with respect to safety (i.e., the food or food component can be concluded to be as safe as the conventional food or food component) (FAO/WHO 1996).

GM foods which are considered substantially equivalent are regarded to be as safe as their conventional counterparts. On the contrary, food products that are not substantially equivalent may still be safe on the condition that they are subjected to a range of other tests to confirm their safety, before they can be marketed.

SE takes into consideration the occurrence of natural harmful or toxic substances and antinutrients in foods of plant origin. Such foods can still be consumed safely. In order to find out whether a modified food is substantially equivalent or not, it is tested by the manufacturer for the presence of any changes in some defined components such as toxins and allergens which were not present in the conventional counterpart. The results of such tests are further evaluated by an official regulatory body, for example, the USFDA. Data and comments resulting from the initial test and that of the regulatory agency are then submitted to regulators who would be required to determine whether significant differences occur between the modified and unmodified products. In case no differences are found, regulators will not usually recommend further food safety testing. Regulators may require further safety testing under certain circumstances, for example, if the tested product does not have a natural equivalent or if it exhibits significant differences from the unmodified food (OECD 1993).

SE originated from an OECD working group report. The group was mandated to draft a document on the safety implications of modern food biotechnology. Background documents used in the report were drawn from previous examples of how the safety of novel foods and food components had been evaluated. The report included recommendations on some concepts and principles that address the safety assessment of foods developed through modern biotechnology. Those principles are in line with recommendations set by FAO and WHO and have received wide international acceptance.

When applying the SE principle, some aspects need to be taken into consideration. Such aspects include the composition and properties of the counterpart food as well as those of the novel food, the method used to alter the nature of the new food, and the way in which the new genetic material is expressed.

If the new food is considered as substantially equivalent to an existing food, then further safety or nutritional concerns are considered insignificant. This food would then be treated in a similar manner to its natural counterpart to which it was compared. Regulators might be faced with a situation where a totally new class of foods is introduced. In such a situation, it would not be practical to apply the SE principle, and previous experience and judgment used in the evaluation of materials of similar nature is taken into account. If a product is judged not to be substantially equivalent to an existing one, then further tests would be recommended.

5.2.2 GM Food Labeling

Labeling of GM products, including GM foods, has been one of the main issues which concern regulators. Some of the debatable issues that lead to a lot of controversy, the same way as GMOs, in general are: Should GM products and GM foods be labeled or not? Should labeling be mandatory or voluntary?

There is broad scientific consensus that food on the market derived from GM crops poses no greater risk than conventional food (Charles et al. 2010; Ronald 2011; Pinholster 2012). There are claims that there is no evidence supporting the view that consumption of GM foods has a negative health effect on humans (American Medical Association (AMA) 2012; Key et al. 2008). On the other hand, some opponents of GM technology are calling for additional and more rigorous testing for GM foods.

There is a lot of controversy over GM food labeling. People in the agribusiness industry believe that labeling should be voluntary and that the demands of the free market and consumer preferences would influence the labeling process. In case consumers prefer labeled foods over unlabeled ones, the industry would regulate itself or run the risk of losing consumers. On the other side of the controversy stand the consumer interest groups who believe in mandatory labeling of GM foods stating that people have the right to know what they are eating.

5.2.3 Issues with GM Food Labeling

Whitman (2000) states that there are a number of questions that need to be answered before regulating a mandatory nature for GM foods. One important question relates to the added cost of production incurred, and the economic impact if labeling of GM foods becomes mandatory. This extra cost comes as a result of constructing two separate production lines at the industrial level as well as a result of practices at the farm level where farmers will try to keep GM crops separate from non-GM crops during planting, harvesting, and shipping. Testing and detailed record-keeping need to be conducted at various stages of the food supply chain. Estimates of the cost of mandatory labeling vary from a few dollars per person per year to about 10% of a consumer's food bill (Gruère and Rao 2007). It is expected that the industry will pass on the extra cost burden to the consumer in the form of higher food prices. The question here is: Will consumers be ready to absorb the added cost of this initiative?

Another question is about the acceptable limits of GM contamination (if any) in non-GM products. At what level can these limits be set? Who would be responsible to monitor food-producing companies for compliance? What kind of penalty would be imposed? As with other issues related to GM technology, there is a divided difference of opinion among different nations, for example, the EU has determined that 0.9% is an acceptable level of cross-contamination, Australia and new Zealand decided a threshold value of 1%, while Japan has specified a 5% threshold (Hansen 2001). Many consumer interest groups believe that only 0% would be acceptable. On the other hand, some food-producing companies such as Gerber Baby Foods and Frito Lay have pledged not to use any GM ingredients in manufacture of any of their products.

Byrne et al. (2010) describes two methods by which regulators can verify whether a food is, or is not, genetically engineered. The first one is content-based in which foods are tested for the physical presence of foreign DNA or protein. Auer (2003) discusses the methods used to detect GE components in crops and processed foods. The second way to verify the presence of GE components is process-based. It involves detailed record keeping of seed source, field location, harvest, transport, and storage. Details of this approach are explained by Sundstrom et al. (2002). Other issues that cause concern about GM foods labeling include the level of detectability of GM food cross-contamination and public awareness about GM food labels. With regard to detecting GMO contamination, it has been found that current techniques are not able to detect minute quantities of contamination. This means that ensuring a 0% level of contamination will not be possible. There is disagreement among scientists as to what level of contamination is practically possible. With regard to public awareness and to what extent would consumers understand the meaning of food label information, the question is who would be responsible to educate the public about GM food labels? Would it be of any benefit to label GM foods if consumers are not able to interpret the meaning of the information on the label?

In January 2000, an international trade agreement for labeling GM foods was established, with more than 130 countries signing the agreement. The policy states that exporters must be required to label all GM foods and that importing countries have the right to judge for themselves the potential risks and reject GM foods, if they so choose (Helmuth 2000).

5.2.4 Pro- and Anti-Labeling Arguments

There is no universal agreement on whether to label GM foods or not, and should labeling be mandatory or voluntary. Some people are pro-GM labeling and present their arguments and supporting views. Others believe that labeling, if enacted, should be voluntary and they also support their views. The next section states and discusses views and arguments of both camps.

5.2.4.1 Pro-Labeling Arguments
The argument presented by pro-labeling groups focuses mainly on consumer rights. Raab and Grobe (2003) and Byrne et al. (2010) state that pro-labeling lobby believes that consumers have a right to know what is in their food, especially concerning products for which health and environmental concerns have been raised. CCHU9005 Group 3 (2014) believe that consumer rights and the concept of free informed choice is a crucial factor in the deliberation whether or not to impose a labeling system differentiating GM food from non-GM foods. It is an undeniable fact that consumers hold enormous power in market decisions since they are the target group that industry aims at. If consumers refrain themselves from buying any goods, there would be no need for the manufacture of such goods. Respect of consumer's choices is thus a main issue in the argument of the mandatory GM food labeling movement. Consumers are expected to make an informed choice provided they have all the information about what they are buying, what is in it, and what they are consuming. Another concern raised by the pro-labeling of GM foods advocates is the religious factor. Religious and moral views of consumers with regard to food need to be respected and taken into consideration, which can only be achieved if they know what their food contains.

Consumer surveys and polls show that a high majority of American consumers prefer a labeling policy (Byrne et al. 2010; CCHU9005 Group 3, 2014). On the other hand, Gruère and Rao (2007) and Phillips and McNeill (2000) indicate that at least 21 countries and the European Union have established some form of mandatory labeling. This is an indication of the wide international support for GM food labeling.

The issue of GM food safety and the potential health risks that might be experienced by consuming GM food is an additional argument that supporters of mandatory GM food labeling raise. Their argument stems from the claim that it is hard to prove there are no risks associated with GM consumption. This is because the GM industry is still young and the database of information to the scientific effects of biotechnology is little. The precautionary principle needs to be applied to GM foods since they might have unknown effects. In such situations, it is better to minimize risk than to take action under misinformation, pro-labeling groups argue.

5.2.4.2 Anti-Labeling Arguments

Concern for consumers' choices and rights seems to be on both sides of the argument. As pro-labeling advocates raised a number of concerns related to consumers that would support their case for labeling, anti-labeling supporters are also including the welfare of consumers in the heart of their argument.

Labeling of GM foods might indicate that they are different from non-GM foods. This might leave a false impression on consumers and give them a wrong idea that GM foods are unsafe for human consumption. Although no solid proof exists that GM foods are different from traditional foods, labeling them would imply a warning about negative health consequences. Labeling is expected to mislead consumers, scare them instead of helping them make informed decisions. This might result in a stigma for GM foods, and more consumers would tend to turn away from purchasing and consuming them. Consumers who do not wish to eat GM foods have the option of turning to organic foods, which are already labeled. In this case, labeling of GM foods will not have any benefit with regard to consumer choices and it will only be an additional cost of production.

Labeling might not be desired by all consumers. The extra cost of production that might be caused by labeling is expected to be paid by all consumers. Experience with mandatory labeling in the European Union, Japan, and New Zealand has not resulted in consumer choice. Rather retailers have removed GM products from their shelves due to perceived consumer aversion to GE products (Byrne et al. 2010). The possible extra cost incurred by labeling might be very little or negligible. Daniells (2014) reports that the cost to consumers of requiring labeling of GM food is $2.30 per person annually, or less than a penny a day, according to a new analysis commissioned by Consumers Union, the policy arm of Consumer Reports. This might indicate that the extra cost due to GM food labeling is minimal and possibly not felt by consumers and that they might be ready to pay that extra cost.

Anti-labeling supporters raise the issue of the practicability of GM food labeling. They argue that it would be extremely difficult and possibly impractical to label GM foods due to the fact that there are too many GM foods in the market. Different foods are modified to different levels and in different ways. It would be practically impossible to keep track of all of them. Also what is the exact definition of a GM food? If a single GM component is indirectly included in a food, for example, in the case of non-GM livestock consuming GM feed, or cheese processed using GM chymosin would it be considered GM?

5.2.5 Global GM Food Labeling

Since the advent of GE with its great potential, particularly in relation to GM foods, the need for a regulatory framework became evident. At international level, the two outstanding instruments are the Cartagena Protocol on Biosafety and the Codex Alimentarius. At country level, national laws that govern GMOs and GM foods are set. The Cartagena Protocol on Biosafety, which was adopted in Montreal, Canada, in the year 2000, provided an international regulatory framework to manage GMOs with regard to their safe transfer, handling, and use in relation to their effect on human health and the environment (Convention on Biological Diversity 2014). A major challenge with the effectiveness of this Protocol is that the major GM crop-producing countries, for example, the United States, Argentina, and Canada, are not part of it and consequently it is not binding them in any way. This makes international trade harmonization on GM products somewhat difficult. Codex Allimantarius develops food standards, and guidelines, which help to protect consumers' health and at the same time ensure international fair trade in food. In addition to this, it aims to coordinate international food standards. With regard to GMOs and GM foods, Codex set guidelines for risk assessment and for standards of food labeling that cover all pre-packaged foods including GM foods, particularly for nutritional and health claims (Codex Allimantarius. International Food Standards 2014).

The Center for Food Safety (CFS) 2014 has compiled a list of countries which require mandatory GE foods labeling. As of 2013, 64 countries are reported to require GMO labeling. This list is periodically updated by the Center. Table 5.1 shows the countries with mandatory labeling of GE foods, as reported by the CFS.

Table 5.1 Countries with Mandatory Labeling of GE Foods

Australia	Latvia
Austria	Lithuania
Belarus	Luxembourg
Belgium	Malaysia
Bolivia	Mali
Bosnia and Herzegovina	Malta
Brazil	Mauritius
Bulgaria	Netherlands
Cameroon	New Zealand
China	Norway
Croatia	Peru
Cyprus	Poland
Czech Republic	Portugal
Denmark	Romania
Ecuador	Russia
El Salvador	Saudi Arabia
Estonia	Senegal
Ethiopia	Slovakia
Finland	Slovenia
France	South Africa
Germany	South Korea
Greece	Spain
Hungary	Sri Lanka
Iceland	Sweden
India	Switzerland
Indonesia	Taiwan
Ireland	Thailand
Italy	Tunisia
Japan	Turkey
Jordan	Ukraine
Kazakhstan	United Kingdom
Kenya	Vietnam

Source: From Center for Food Safety. 2014a. Label genetically engineered foods. http://www.centerforfoodsafety.org/issues/976/ge-food-labeling/ (accessed June 23, 2014); Center for Food Safety. 2014b. International labeling laws. http://www.centerforfoodsafety.org/issues/976/ge-food-labeling/international-labeling-laws (accessed May 3, 2014). With permission.

5.3 GM FOODS REGULATIONS AND LABELING IN SELECTED COUNTRIES

5.3.1 Introduction

Usually the introduction of new safety regulations for food products follows the occurrence of a problem that was not anticipated, for example, the restrictions on the processing and sale of beef once the nature of BSE was appreciated (The NCBE Guide 2014). However, the regulations related to GM products (including GM foods) followed a different trend. They were, actually, developed in advance of the occurrence of any hazards to human health or to the environment.

Governments of different countries around the globe have followed different strategies and taken various approaches with regard to GE and its applications to develop GMOs and products made from them, particularly GM foods. The assessment and management of potential risks associated with application of GE techniques have been addressed in different manners. Like any other technology, GE has also been regulated in some countries. Other countries are still in the process of developing relevant regulations. There are marked differences in the regulation of GMOs and GM foods among different countries, with some of the most marked differences observed between the United States and Europe. Table 5.2 reflects some characteristics of GM regulations in different country groups, while Table 5.3 shows GM food labeling requirements in different countries. The main aspects taken into consideration while drafting GMOs and GM food regulations are safety to human health and the environment. One of the key issues facing GM food regulators is whether they should be labeled or not. Labeling can be mandatory or voluntary. The threshold of GM content in mandatory labeling differs among different countries. Details of these levels are discussed under sections dealing with labeling in different countries.

The debate on GM food labeling emanates from the different views on whether GM foods in the market pose any greater risk than conventional foods. There are conflicting scientific views on this issue. There seems to be a broad scientific consensus which supports the idea that GM foods are as safe as their conventional counterparts and that there is no clear evidence to support the belief that the consumption of GM foods has a negative effect on human health (Key et al. 2008; Ronald 2011; American Medical Association (AMA) 2012; Pinholster 2012). On the contrary, some scientists and anti-GM groups believe that the issue of GM food safety is not settled yet and that further rigorous testing is required for GM foods.

Table 5.2 Characteristics of GM Regulations in Different Country Groups

Group	Food Safety Approval Regulation	Labeling Regulation	Specificity	Countries
1	Process-based, mandatory	Stringent mandatory Includes derived products	Traceability requirements, 0.9% threshold	EU, East Europe
2	Process-based, mandatory	Stringent mandatory Includes derived products	No traceability, low threshold	Brazil, China, Russia, Switzerland, Norway
3	Process-based, mandatory	Product-based, mandatory	Labeling exceptions, 1%–5% threshold	Australia, Japan, Korea, Saudi Arabia, Thailand, Taiwan
4	Substantial equivalence, mandatory	Voluntary for substantially equivalent	5% threshold for labeling	United States, Canada, Argentina, Hong Kong, Philippines, South Africa
5	Mandatory (in place or pending)	Mandatory	"Pragmatic" product-based labeling	Indonesia, Chile, Ecuador, Vietnam
6	Mandatory (in place or pending)	Intention to require labeling	Slow regulatory process	India, Kenya
7	Considering mandatory	No clear position	Wait-and-see approach	Bangladesh, most African countries
8	No	No	GM free	Zimbabwe, Zambia

Source: Adapted from Ghosh, D.K. and Williams, P. 2010. Global food biotechnology regulations and urgency for harmonization, Chapter 29, Table 29.1, *Biotechnology in Functional Foods and Nutraceuticals*. CRC Press, Boca Raton, FL, pp. 531–41.

Table 5.3 GM Food Labeling Requirements in Different Countries

Country	Labeling of GM Food Ingredients Required for:			
	GM Food (Unprocessed)	Products of Animals Fed with GM Feed	Processed GM Products (More/Less)	GM Additives, Vitamins Processing Aids
European Union (all the 25 member countries)	Yes	No	Yes	No
Australia	Yes	—	—	—
New Zealand	Yes	—	—	—
Saudi Arabia	Yes	—	Yes	—
Indonesia	Yes	—	—	—
Japan	Yes	—	—	—
South Korea	Yes	—	—	—
China	Yes	—	Yes	—
Brazil	Yes	—	—	—
Russia	Yes	—	—	—
United States	No	No	No	No
Canada	No	No	No	No
Switzerland	Yes	No	Yes	Yes
Norway	Yes	—	Yes	—
Iceland	No	No	No	No

Source: From Gene-modified blogspot.com. 2011. Genetically modified food and related issues. Labeling of GM food. http://gene-modified.blogspot.com/2011/06/labelling-of-gm-food.html (accessed May 15, 2014). With permission.

Note: This table was compiled in 2011.

5.3.2 Australia–New Zealand

Food Standards Australia New Zealand (FSANZ), formerly known as Australia New Zealand Food Authority (ANZFA), is the official government body responsible for developing food standards for Australia and New Zealand (Food Standards Australia New Zealand 2014). Relevant government agencies and other stakeholders are consulted by the FSANZ during different stages of food standards development. Standards are usually developed after rigorous scientific assessment of risk to public safety and health. Decisions made by the FSANZ must be approved by a joint Australia and New Zealand Food Regulation Ministerial Council. This Council includes health ministers of all Australian states and territories as well as health minister of New Zealand in addition to other ministers nominated by each country.

FSANZ assesses the safety of GM foods in accordance with internationally established scientific principles and guidelines developed through the work of the OECD, the FAO, the WHO, and the CAC.

The safety assessment in Australia and New Zealand is characterized by

1. Case-by-case consideration of GM foods
2. Consideration of the intended and unintended effects of the genetic modification
3. Comparisons with conventional foods having an acceptable standard of safety (Food Standards Australia New Zealand 2007)

In 2001, FSANZ and the Office of Gene Technology Regulator (OGTR) took over the responsibilities and activities of GE in Australia from the Genetic Manipulation Advisory Committee, which was assigned this responsibility since 1987. The OGTR is a Commonwealth Government Authority within the Department of Health and Ageing and reports to Parliament through the Ministerial Council on Gene Technology.

FSANZ evaluates the following aspects for GM foods or ingredients: nutritional content, any nutritional change, toxicity, allergenicity, and stability or any unintended effect of the inserted genetic material (Better Health Channel 2014).

FSANZ has the legal responsibility of evaluating and reporting the safety of all GM foods and ingredients marketed in Australia and New Zealand. Mandatory pre-market safety assessment for all GM foods and ingredients is conducted by FSANZ before these products enter the country and before they are used in foods for human consumption. No food will be approved for sale unless it is proved to be safe to eat. GM food labeling is mandatory in Australia and New Zealand. These regulations became law in December 2001, with the intention of helping consumers make informed decisions as to buy or consume a GM food or not.

A food produced from GM material, or manufactured using GE enzymes must be approved by FSANZ before it is released for sale in Australia or New Zealand. Lists of foods which meet criteria for such approvals are periodically updated and published on the FSANZ website (Record of GM Product Dealings 2014).

All foods marketed in Australia, including products containing GM ingredients, must be in compliance with the Australia New Zealand Food Standards Code (referred to as "the Code"). GM foods and ingredients are regulated in the code under "Standard 1.5.2—Food produced using gene technology" (Australia New Zealand Food Standards Code 2014).

As of August 2008, FSANZ had approved 34 GM food commodities, which include varieties of GM corn, cotton, canola, sugar beet, soybean, and potato. Although there is approval for the commercial growing of GM canola, only GM cotton from which cottonseed oil is extracted is currently grown commercially in Australia (Ghosh and Williams 2010).

Following the announcement on December 22, 2008, by the Minister for Agriculture and Food that the State Government had approved limited commercial-size trials of GM canola, the State Government of Western Australia established an interdepartmental committee to investigate the labeling of GM food (Western Australian Agriculture Authority 2011). It aimed to answer the following questions:

- Does current GM food labeling support informed consumer choice?
- What are the limitations to requiring additional GM food labeling?
- What can the State Government do to educate and support consumer choice about GM food?

The main recommendations of the committee emphasized the need for public education programs in order to improve understanding about GM foods and their labeling. Increased mandatory labeling of GM foods to indicate which GM DNA and/or proteins have been removed through processing has also been recommended. The committee also recommended that the industry should be encouraged to adopt more informative labeling of GM foods and provide adequate information on GM foods to consumers on a voluntary basis.

5.3.3 Canada

5.3.3.1 Regulations
In Canada, GM foods are regulated by Health Canada (HC), the Canadian Food Inspection Agency (CFIA), Environment Canada (EC), Agriculture and Agri-Food Canada (AAFC), and the Department of Fisheries and Oceans (DFO). HC, under the Food and Drugs Act, and the CFIA are assigned the responsibility of safety and nutritional value evaluation of GM foods. Through science-based regulation, guidelines,

and public health policy, in addition to health risk assessment of the food supply, HC works to protect the health and safety of Canadians (Health Canada 2014).

The Plant Biosafety Office (PBO) working under the umbrella of the CFIA is responsible for the environmental assessment of biotechnology-derived plants (Canadian Food Inspection Agency 2014). The Canadian regulatory system relies on whether a product has "novel" features or not regardless of method of origin. Novel foods, as defined in Canada, include the following categories: Foods resulting from a process not previously used for food; products that have never been used as a food; and foods that have been modified by genetic manipulation (Ghosh and Williams 2010), thus GM foods fall under the umbrella of novel foods. HC is responsible for regulation of novel foods in Canada through a pre-market notification requirement, which is specified under Division 28 of part B of the "Food and Drugs Regulations" (Novel Foods). This notification is required for, and helps, assurance of food safety when prepared or consumed according to its intended use.

The Canadian regulation of foods derived from biotechnology is a long process which involves a 7–10 years period of research, development, testing, and evaluating the safety of a new GM food. There are eight steps that need to be followed in order to complete the process (Health Canada 2014). These steps are

1. Pre-submission consultation
2. Pre-market notification
3. Scientific assessment
4. Request for additional information
5. Summary of report of findings
6. Preparation of food rulings proposal
7. Letter of no objection
8. Decision document on Health Canada website

5.3.3.2 Labeling
The Government of Canada believes that labeling of foods derived from biotechnology is an important issue that affects consumer preference or choice. The Canadian General Standards Board formed a committee to develop a standard for voluntary labeling of GE foods. The stakeholders represented in the committee included consumer groups, food companies, universities, government, and general interest groups. The standard is known as "Voluntary Labeling and Advertising of Foods That Are and

Are Not Products of Genetic Engineering." The Standards Council of Canada adopted this national standard in April 2004.

HC and the CFIA work together in preparing food labeling policies under the Food and Drug Act. The role played by HC in food labeling is to safeguard health and safety, whereas the CFIA has the responsibility of leading the program to develop the general food labeling policies and regulations. The CFIA responsibilities include consumer protection from fraud and misrepresentation with respect to food labeling, packaging, and advertising, in addition to setting food labeling and advertising guidelines. Currently Canada has a mandatory food labeling policy if a health or safety issue is encountered with a food, which might be mitigated through labeling, for example, presence of a food allergen or a change in food composition or nutritional value (Health Canada 2014). In such situations, special labeling to alert consumers or susceptible groups is required. This is a general requirement that applies to all foods including GM foods.

The Canadian Federation of Agriculture does not support implementation of mandatory labeling saying that it would incur huge losses to the industry. Their fear is that consumers may regard food labels as a warning and avoid such foods. Food processors might opt to avoid inclusion of GM components in their products rather than labeling them, hence a need for reformulation arises. This is expected to increase food prices and reflect negatively on food markets.

5.3.4 The European Union

In Europe, anti-GM food groups have been very active in protesting against production, importing, marketing, or consumption of GM foods. These groups have used two major food scares that took place in Europe, namely bovine spongiform encephalopathy (commonly known as mad cow disease) in Great Britain, and the dioxin-tainted foods which originated from Belgium. In general, the consumers' confidence about the European food supply was shaken due to these unfortunate health incidences. Europeans started to lose trust in what their governments tell them about GM foods. This public position led European governments to respond in a way that might regain some public trust. This was reflected in the requirement for mandatory food labeling of GM foods in stores. At the same time, the European Commission (EC) has established a 1% threshold for contamination of unmodified foods with GM products.

The general legislative approach in the EU throughout its existence has been that "if anything can conceivably be regulated, regulate it." Consequently, it had built a comprehensive system and machinery for considering and approving (or otherwise) applications for approval of specific lines of GMOs and for controlling the release of GMOs into the environment (IFST 2008).

Until the 1990s, Europe used to have regulations governing GM products which were less strict than in the United States. This situation has changed a lot and at present the EU has possibly the most stringent and strictest regulations in the world for the presence of GMOs in food and feed (Davison 2010). All GMOs in addition to irradiated foods are regarded as "new foods" and are required to be subjected to extensive, case-by-case, science-based food evaluation by the European Food Safety Authority (EFSA 2012). Foods are evaluated for safety, labeling, traceability, and freedom of choice (GMO Compass 2014).

The regulatory system and labeling of food products in the EU has witnessed a series of developments and changes which started during the early 1990s. Continuous updates are announced as need arises or change of conditions is encountered. The general principle followed is that before GM foods are approved for sale in the EU, they must be rigorously evaluated for safety to comply with the "EC Novel Foods Regulation (258/97)" which came into effect on May 15, 1997. This is an EU-wide system which indicates that a "novel food" needs to be assessed for safety before it is approved for marketing. A "novel food" in this sense denotes a food which has not been consumed by humans in a wide extent in the EU before and it includes GM foods.

The EC "Novel Foods and Novel Food Ingredients Regulation (EC 258/97)" which applies labeling requirements to GM foods which will be approved in future includes labeling if there are any health or ethical concerns or if it contains a live GMO. This regulation was approved in 1997 after many years of deliberations. The Novel Foods Regulation is required to be applied under any of the following conditions: Foods and ingredients derived from GM, modified or new molecules, any product that have not previously been eaten by humans in the EU to a significant degree, and if novel processing methods are used. In addition to these, the Novel Food Regulation requires that labeling should be included if

- The novel food differs from the equivalent familiar food due to a change in composition or nutritional value

- Consumption of the novel food has health implications, for example, an allergen is present that is not present in the existing equivalent food
- The novel food creates ethical considerations, for example, a food plant containing DNA of animal origin
- The novel food is or contains a viable genetically modified organism

A labeling regulation (EC 1831/97) specifically designed for GM soybean, corn, and other GM foods followed the Novel Foods Regulation. An amendment to this labeling regulation (EC 1139/98) was adopted and used by the different EU States in September 1998. Article 8 of the EU Novel Food Regulation requires an ingredient which is no longer equivalent to be labeled. One major requirement of the regulations is that products should be clearly labeled if they contain genes that may result in toxicity or allergenicity, particularly if such genes would not normally be expected to occur in the food.

Two directives (Directive 90/220/EEC, dealing with release and marketing, and Directive 90/219/EEC, dealing with use in containment) released by the EU regulate the production, release, and marketing of GM plants. They were required to be implemented by EU member states using individual state regulations. In April 2004, the EU Regulation EC 1829/2003, referred to as "Labeling of Genetically Modified Foods and Feed" has been implemented in Europe. The main objectives of this regulation are to

- Protect human and animal health through stringent safety assessment of GM food and feed before it can be sold
- Ensure common procedures for risk assessment and authorization are efficient, transparent, and do not take too long
- Ensure clear labeling that responds to the concerns of consumers (including farmers buying feed) and enables them to make informed choices (European Commission 2014)

Regulation EC 1829/2003 covers all GM food and animal feed, regardless of the presence of any GM material in the final product. Consequently, products such as edible oils, flour, and glucose syrups need to be labeled as GM if their original source is genetically modified. On the other hand, foods produced through GM technology, for example, cheese produced with GM enzymes, do not have to be labeled. Similarly, food products obtained from animals fed on GM animal feed, for example, meat, milk, and eggs, do not need to be labeled. Table 5.4 shows some examples of labeling rules set in this regulation.

This regulation was not welcomed by food and feed manufacturers in Europe as well as their overseas suppliers and has been a great issue of concern for them.

In the EU, if a food contains or consists of GMOs, or contains ingredients produced from GMOs, this must be indicated on the label. For GM products sold "loose," information must be displayed immediately next to the food to indicate that it is GM (Food Standards Agency 2013).

5.3.5 Japan

In Japan, the Ministry of Health, Labor, and Welfare (MHLW) is responsible for carrying out scientific reviews to evaluate and confirm the safety of new GM crops. On the other hand, the Ministry of Agriculture, Forestry, and Fisheries (MAFF) is assigned the responsibility of assessing the safety of the environment and animal feed as affected by GM cropping (The Organic and non-GMO Report 2014). The MAFF also has the responsibility of approving new GM crops for feed use. With regard to GM crops to be used for food use, Japan has approved 44 GM crops for food use. These include 15 canola, 12 corn, 7 cotton, 5 potato, 4 soybean, and 1 sugar beet varieties. The complete list of products that have undergone safety assessment and been announced in the Official Gazette as of April 2014 is published in the website of the Department of Food Safety, MHLW (http://www.mhlw.go.jp/english/topics/food/pdf/sec01-2.pdf).

The MAFF and MHLW have implemented labeling requirements under the Food Sanitation Law and Japanese Agricultural Standards (JAS) for GM crops approved in Japan. Some food products have been selected for JAS labeling requirements. The selection of these foods is based on the fact that they are developed from ingredients that could include GM products and that GM DNA or protein can be identified in the foods. Selected foods are soy- and corn-based products, including tofu, natto, soymilk, miso, other products made with soy ingredients, corn snacks, cornstarch, popcorn, in addition to other products made with corn ingredients (The Organic and non-GMO Report 2014). The food labeling policy followed in Japan indicates that if the GM content of the food is more than 5%, the food must be labeled as "GM Ingredients Used" or "GM Ingredients Not Segregated." For a food product to be labeled as "non-GM," the GM

Table 5.4 Examples of Labeling Requirements under EC Regulation No. 1829/2003 for Authorized GMOs (Updated April 2008)

GMO Type	Hypothetical Example	Labeling Required?
GM plant	Chicory	Yes
GM seed	Maize seeds	Yes
GM food	Maize, soybean, tomato	Yes
Food produced from GMO	Maize flour, highly refined soya oil, glucose syrup from maize starch	Yes
Food produced from animal fed GM animal feed	Meat, milk, eggs	No
Food produced with help from a GM enzyme	Cheese, bakery products produced with help of amylase	No
Food additive/flavoring produced from GMOs	Highly filtered lecithin extracted from GM soybeans used in chocolate	Yes
Feed additive produced from a GMO	Vitamin B2 (riboflavin)	No
GMM used as a food ingredient	Yeast extract	Yes
Alcoholic beverages which contain a GM ingredient	Wine with GM grapes	Yes
Products containing GM enzymes where the enzyme is acting as an additive or performing a technical function	—	Yes
GM feed	Maize	Yes
Feed produced from a GMO	Corn gluten feed, soybean meal	Yes
Food containing GM ingredients that are sold in catering establishments	—	Yes (the FSA's legal view is that labeling is required across EU Member States under EC Regulation 1829/2003)

Source: From Food Standards Agency. 2013. GM labeling. http://www.food.gov.uk/policyadvice/gm/gm_labelling#.U0sVJtKsi1J (accessed May 2, 2014). With permission.

Note: GM, genetically modified; GMM, genetically modified microorganism.

content must be less than 5%. It is the responsibility of the food processor to show that all non-GM ingredients were identity preserved during production and processing. Documentation of this process must be provided by the supplier of the food product. Labels allowed to be used for these products include "non-GM product segregated" or "not genetically modified." Monitoring of compliance with these regulations rests with the MAFF and the MHLW. They randomly and periodically check samples for the presence of GM content. If a food sample was found to contain more than 5% of GM material in a product labeled "non-GM," then the local Japanese manufacturer or food importer will be required to change labeling to read "GM ingredients."

Japanese regulations are very strict with regard to unapproved GM varieties that are detected in foods. In this respect, a "zero tolerance" policy is applied. In order to assure full compliance with this policy, processed as well as imported foods present in the market are sampled and tested. If any unapproved GM component is detected in a product at the port of entry, the product will be re-exported, destroyed, or used in no-food products. In case the violation is detected at the retail level, the product manufacturer will be ordered for an immediate recall. The main products being tested include corn, soybeans, papaya, and potatoes. The MHLW, which oversees the testing, has so far found one unapproved potato variety, an unapproved papaya on two occasions, and one unapproved corn: StarLink (The Organic and non-GMO Report 2014).

In Japan, grocery stores sell both GM and non-GM foods. That allows consumers to choose the type of food they prefer to consume. It has been observed that customers are beginning to show a strong preference for unmodified fruits and vegetables.

5.3.6 South Africa

Although South Africa was the first country in the African continent to grow GM crops and is one of the few African countries to adopt the GM technology so far, consumer knowledge and awareness of biotechnology are limited. Regulations controlling GM labeling regarding consumer preferences are lacking (Botha and Viljoen 2009). Voluntary GM labeling, which aimed at detecting and quantifying GM components in foods, has been the only option used as a marketing strategy by the government. The food labeling was only meant to indicate the absence of GM in food products. Labels used on food products include: "GMO-free," "non-GM," and "organic." Botha and Viljoen (2009) concluded that in the absence of specific regulations, voluntary GM labeling does not provide discerning consumers with the choice intended and that a regulated approach for GM labeling needs to be in place in order to protect consumers.

5.3.7 United States

Currently, the United States is the largest commercial producer of GM crops and GM foods in the world (ISAAA 2014).

Foods regulation in the United States is carried out through three agencies working in cooperation. These are the U.S. Food and Drug Administration (FDA), the U.S. Department of Agriculture (USDA), and the U.S. Environmental Protection Agency (EPA). The FDA is concerned with food safety relating to new plant varieties, dairy products, seafood, food additives, and processing aids. It regulates GM food products which are sold to consumers or which are mixed into processed foods and sold to consumers. The USDA is responsible for regulation of meat and poultry products in addition to overseeing the release of GM plants and animals into the environment for field testing or commercial use. The EPA is responsible for regulation of pesticides and pesticidal agents including plants with pesticidal properties, which may affect plants and animals in the environment. It may, therefore, have to approve new plant varieties resistant to attack by pests. In other words, the EPA evaluates GM plants for environmental safety, the USDA evaluates whether the plant is safe to grow, and the FDA evaluates whether the plant is safe to eat. This regulatory process implemented in the United States is sometimes criticized as being confusing since three different government agencies have jurisdiction over GM foods. Some overlaps of responsibilities can take place and separation of powers at the end might be difficult.

The FDA regulates food labeling using an approach designed to provide consumers with information relative to health, nutrition, and safety. The Federal Food and Drug Cosmetic Act (FD&C Act) lays out the FDA's science-based labeling policy; all foods, whether or not they are derived from transgenic crops or animals, are subject to the policy (Council on Science and Public Health 2012). The FDA has the authority to initiate regulatory action if a product fails to meet the requirements of the FD&C Act 3.

The framework set by the FDA which addresses labeling policy of bioengineered foods is underpinned by a law that covers the following points (Council on Science and Public Health 2012):

- All food labels should include a name that accurately describes the basic nature of the food. With regard to bioengineered foods, name changes only apply if the food is significantly different from its traditional counterpart, such that the common or usual name no longer adequately describes the new food.
- If food production processes result in major differences, for example, in nutritional profile or lead to any safety concerns, the food label must disclose these differences.
- Voluntary labeling by food producers and processors about production methods is allowed by the FDA on the condition that the labeling is not misleading or carries false information.
- The FDA discourages the use of acronyms, which might be confusing or misleading consumers, for example, "GMO-free." Some consumers may not understand what the acronym stands for, while others might interpret "genetically modified" as referring to conventional techniques.
- In general, the FDA believes that its current labeling policies, combined with pre-market safety assessments, are sufficient to ensure the safety of bioengineered foods.

The current FDA policy was developed in 1992 (USFDA 1992) and states that agri-biotech companies may voluntarily request the FDA for a consultation. Companies planning to develop new GM foods are not required to consult the FDA, nor are they required to follow the FDA's recommendation after consultation (Whitman 2000). The principle of "substantial equivalency" was adopted by the FDA in 1992 and the policy issue stated that GE food is substantially equivalent to conventional varieties of food unless evidence showed otherwise. In 2006, the FDA put into practice a number of guidelines for the evaluation of plants which are genetically modified to produce new proteins such as those with pharmaceutical applications.

The "Food, Drug, and Cosmetic Act" governs the FDA's position on food labeling. This Act only deals with food additives and not whole foods or food products that are considered "GRAS" (generally recognized as safe). The relationship between the regulations of biotechnology products GRAS and health claims has been discussed by Simon et al. (2010). The FDA regards the key factors in reviewing safety to be the characteristics of the food and its intended use, rather than the fact that new methods have been used in its production (The NCBE Guide 2014). This is a basic difference between the United States and the EU regulations, which creates a major issue in the current controversy regarding GM foods.

The FDA requires labeling of GM foods in the following situations: 1, if the food has a significantly different nutritional property; 2, if a newly developed food has an allergen that consumers would not expect to be present; and 3, if a food contains a toxicant beyond acceptable limits (Byrne et al. 2010).

In 2001, the FDA proposed voluntary food labeling draft guidelines for foods that do or do not contain GM ingredients. This draft guidance represents FDA's views on voluntary labeling of foods indicating whether foods have or have not been developed using bioengineering (USFDA 2014). The FDA advised that labeling requirements that apply to foods in general should also apply to foods produced using biotechnology. Additionally, the agency indicated that it is required that the label on the food must reveal all material facts about the food. Some examples cited by the FDA to substantiate that statement include

1. Bioengineered foods which are significantly different from their traditional counterparts should show that and should be labeled in a way to describe the difference
2. If it is not clear how the food or one of its constituents should be used, the label must include a statement to describe how it is used
3. In case a bioengineered food has a different nutritional profile, the label must reflect that difference
4. Unexpected presence of an allergen in a food based on its name necessitates disclosing of the allergen on the label

It is clear that these requirements aim at avoiding all misleading, unclear, or ambiguous statements included in food labels.

The FDA guidance stresses avoidance of food misbranding. A food would be misbranded if its label includes statements which are false or misleading in any way. If the label does not disclose all facts related to the product, for example, consequences that may result from its use, it will be considered misleading.

Some food manufacturers may want to include informative statements on the food labels of bioengineered foods or foods that contain bioengineered ingredients. FDA recognizes this and allows inclusion of such statements on the condition that they are clear and not misleading the consumer or reader of the statement. In order to illustrate this, the FDA gave examples with specific wording and included their comments on each of them. The following examples have been cited by the FDA (USFDA 2014):

GENETICALLY MODIFIED FOODS

- "Genetically engineered" or "this product contains cornmeal that was produced using biotechnology." FDA comment: "The information that the food was bioengineered is optional and this kind of simple statement is not likely to be misleading."
- "This product contains high oleic acid soybean oil from soybeans developed using biotechnology to decrease the amount of saturated fat." FDA comment: "This example includes both required and optional information."
- "These tomatoes were genetically engineered to improve texture." FDA comment: "If the change in the tomatoes was intended to facilitate processing but did not make a noticeable difference in the processed consumer product, a phrase like 'to improve texture for processing' rather than 'to improve texture' should be used to ensure that the consumer is not misled. The statement that the tomatoes were genetically engineered is optional."

The FDA stresses that terms like "GMO-free" (Figure 5.1) or "not genetically modified" are not technically accurate unless they are clearly used in a context that refers to bioengineering technology.

The use of the term "GMO-free" on food labels (Figure 5.1) may be misleading on some foods because these foods do not actually contain entire organisms. On the other hand, use of the term "free" in a food label to claim the absence of bioengineering would probably be inaccurate. This is because it gives consumers the feeling that it is "zero" bioengineered. The FDA believes that the term "free" can only be guaranteed if there is a definition or a threshold above which the term could not be used. No tests are available at present to achieve this. The agency, therefore, suggests that the term "free" not to be used in bioengineered statements or that a clear indication that a zero level of bioengineered material is not implied.

Substantiation of food label statements is something that is recommended by the FDA. A food manufacturer who wants to claim that a food

Figure 5.1   GM-free label.

or any of its ingredients are not bioengineered should be able to substantiate that the claim is truthful and not misleading. The manufacturer can achieve that through validated testing, which is considered as the most reliable way to identify bioengineered foods or food ingredients.

Many GM foods produced and/or consumed in the United States do not require special regulations and they are not subject to segregation from non-GM foods. With regard to food crops such as soybean or corn, which are exported to Europe, this may result into some problems. This is due to the fact that such foods require labeling under the EU novel food regulation. With regard to imports into Europe, this problem has been addressed for GM soya and corn by assuming that GM components will be present, unless it is certified that the source of the crop is free from GM material. This can be applied as a short-term solution as more GM material is expected to be developed in future.

Although at federal level there are no mandatory labeling regulations for GM foods, some states have moved forward with passing labeling bills. A number of states passed regulations on labeling GM foods. The state of Connecticut approved a GMO labeling bill in May 2013. The bill will not be effective until four other states enact similar legislation (Reilly 2013). On January 9, 2014, Maine's governor signed a bill requiring labeling for foods made with GMOs, with a similar triggering process as Connecticut bill (Herling et al. 2014). The House of Representatives of Vermont state approved a labeling bill on April 23, 2014.

Another effort at state level on mandatory labeling of GM foods is what is known as "California Proposition 37." This is an initiative statute for mandatory labeling of GE foods. The proposition requires labeling of food sold to consumers made from plants or animals with genetic material changed in specified ways and prohibits marketing such food, or other processed food as "natural." The proposition was rejected in California at a state-wide election on November 6, 2012 (Finz 2012).

Efforts to approve and apply mandatory labeling for GM foods in the United States continue. Consumer protection organizations are very active in this respect. In June 2006, the Center for Food Safety filed a lawsuit against the FDA claiming that regulation of GM foods is inadequate.

The United States has constantly criticized the EU for its unscientific GMO regulations which it says amounts to trade protectionism (Davison 2010). After realizing that other countries are no longer relying on GMOs developed in the United States and that these countries are producing and cultivating their own GMOs, United States is now proposing to set up its own system of GMO regulations.

5.4 CONSUMER PERSPECTIVES ON GM FOOD LABELING

Different views have been expressed with regard to whether GM food labeling is needed and whether it should be mandatory or not. Fears that bioengineered foods pose a safety threat to consumers, as well as a "right to know" what is being consumed and to be afforded the choice to avoid bioengineered foods, are the basis for arguments that bioengineered foods should be labeled as such (Degnan 1997). Consumers' views on GE food labeling have been reflected in the results of a number of surveys conducted around the world. In surveys asking whether consumers are satisfied with U.S. food labeling policies, only 18% report that information is missing; among this group, only 3% report that information about bioengineering should be included in the label (Council on Science and Public Health 2012). On the other hand, when consumers were directly asked whether they support mandatory or voluntary GE food labeling policies, an overwhelming majority indicated that they favor mandatory labeling policies (Taylor 2000; Loueiro and Hine 2004; The Mellman Group 2012).

A number of consumer groups have continuously expressed their support of a mandatory labeling policy for GE foods (Consumers Union 2014; Center for Food Safety 2014a,b). A petition calling for mandatory labeling was submitted to the FDA by the Center for Food Safety in the fall of 2011 and more than 400 organizations have expressed their support for the "Just Label It" campaign (Center for Food Safety 2011; Just Label It Campaign 2014). The response of the FDA, which came in the spring of 2012, stated that it had not yet made a decision on the petition and that it would continue to consider it. Other groups have criticized the FDA's approval and labeling policies as inadequate in the face of advancing plant and animal transgenic technologies and have called for reform (Butler 2009; Pelletier 2005). In addition to this, more than a dozen states and the U.S. House and Senate have considered legislation focused on mandatory labeling of bioengineered plants and animals. Only the State of Alaska has passed a law, requiring that bioengineered salmon be labeled.

Surveys of U.S. consumers reveal that while some are willing to pay a premium for foods that do not contain bioengineered ingredients, the majority of consumers are not willing to pay for increases commensurate with the costs of mandatory labeling policies (Chern et al. 2003; Loueiro and Hine 2004; Fernandez-Conejo and Caswell 2006).

With regard to the issue of "consumers' right to know," the court ruling was not in favor of consumers or consumer associations. Consumers' curiosity alone was not enough to require special GE food labeling, the court ruled. The justification of such court ruling was that: (1) special labeling would be a financial burden on the food industries; (2) labeling might mislead consumers into thinking that GE foods are less safer than their conventional counterparts; (3) it would place further burden on the FDA by diverting efforts and resources away from the more important safety-based food labeling; and (4) it would open more, and never-ending, venues for consumers to request manufacturers to disclose information about their products (Council on Science and Public Health 2012).

It is important that consumers clearly understand the meaning of terms used in GM (or non-GM) food labeling. Food manufacturers should make honest efforts to include terms which are understandable and not misleading to consumers. Terms such as "GM-free" or "does not contain GMOs" are not considered accurate terms and may be misleading to consumers. Consumers have the responsibility of educating themselves on the meaning of various terms used in connection with GM and GE technologies as well as terms used in food labeling. This would help them to make an informed decision when purchasing or consuming GM foods.

5.5 INTELLECTUAL PROPERTY (IP) AND PATENTS

One of the most controversial and debated issues related to GMOs, GM crops, and GM foods is the introduction and use of intellectual property (IP) protection and patenting of GMOs and materials produced from them by corporations (Conko 2004; Freedeman 2009). IP rights are the legal rights recognized and offered to creators or inventors. The IP law grants specified exclusive rights to the owner of the discovery or invention. IP rights encompass different types, the most common of which are: patents, copyright, trademarks, industrial design rights, and trade secrets. A patent gives the inventor the right to exclude others from making, using, selling, offering, to sell, and importing an invention for a limited period of time, in exchange for the public disclosure of the invention (WIPO 2008).

Patents are considered a temporary measure of protection of legal rights to use and benefit from an invention. They do not remain forever, but have a specified period of time after which they expire. After elapse of this period, patented products and processes would go in the public domain and anyone would be allowed to use them freely and without any legal implications.

One of the main requirements for patenting an invention or a product is that the inventor must give a comprehensive, written description of the invention with details of the process applied to make it. This is required as it facilitates the reproduction of the technology by any professional in the field once the patent expires and becomes available in the public domain.

The "International Treaty on Plant Genetic Resources for Food and Agriculture" also known as "seed treaty" is an agreement on how to implement the access and benefit-sharing rules of the "Convention on Biological Diversity" (CBD) in the field of food and agriculture (GRAIN 2005). This treaty was adopted by the UN, FAO member states in 2001 and was implemented in 2004. The treaty has been viewed as giving more rights to breeders than to farmers and as granting new privileges to the industry. It has been criticized as giving seed companies free access to most of the world's public "genebanks" without any commitment to share their own material in return (GRAIN 2005). Although the treaty covers a limited number of crops, it includes most of the major food crops. At the same time, it excludes a number of minor food and forage crops of economic importance in the tropics.

The relationships between the IP rights and the right to food has been an issue of concern that lead the United Nations System to assign a special Rapporteur (Professor Olivier De Schutter) to evaluate the situation and present a report on that issue. Part of his report stated that "the current intellectual property rights regime is suboptimal to ensure global food security today. It is unfit to promote the kind of innovation we need to cope with climate change," adding that the United Nations General Assembly needs to develop seed policies that encourage innovation, promote food security, and enhance agro-biodiversity at the same time (Worldhunger.org 2014). He also recommended that there is need to ensure that seed policies respect, protect, and fulfill the right to food of the most vulnerable groups.

Seed saving to be used for replanting in the next harvest and seed sharing have been the normal practice among farmers for thousands of years back and before the introduction of the patent system which does not allow these two practices. Large businesses, such as Monsanto, now require purchasing farmers to sign contracts that prevent the saving of genetically modified seed. The "technology protection system" known as "terminator technology" has been patented in 1998 (Stein 2005). This technology involves genetic modification of seeds in a way that the next generation of seeds self-destructs and is unable to reproduce. Signing a contract by a farmer to use GM seeds produced by the terminator technology will not allow him to use these seeds for replanting during the next growing season. This arrangement allowed the seed industry using this technology to monopolize the seed market and dictate their terms on farmers.

An agreement on trade-related aspects of the intellectual property rights, known as TRIPS, was adopted in 1994. The efforts of biotechnology companies outside the United States, and the help they received from the U.S. government to recognize and enforce their IP rights in patenting their seeds and plants, facilitated the development and adoption of (TRIPS). This agreement controls the patent policy, including plant and seed patents. Any nation wishing to participate in the World Trade Organization (WTO) must adhere to TRIPS (Stein 2005). TRIPS with other international agreements is used to monitor the way local policies of participating nations meet the required international standards. TRIPS has been criticized as being favoring developed nations and disadvantaging developing nations. This is because it encourages strong IP protection, an issue that benefits developed nations since the majority of patents are produced in these countries.

In many countries, political and business motives have pronounced effects on policy decisions made by governments. The United States' policy on IP rights for the life sciences grants broad patent rights to private industry (Stein 2005). This has led to a stringent patent regime for genetically modified seeds in both developed nations (the source for and consumers of IP-protected seeds) and developing nations (the purchasers of IP-protected seeds). The IP protection on seed development has given the chance to private seed industry, which is controlled by a few large corporations, to make substantial commercial gains.

Patents were introduced in the United States as part of the constitution and were included in Article 1, Section 8 of the constitution. The aim of establishing patents was to protect inventions so that the inventor could make profit and at the same time to support innovation and encourage inventors to keep inventing (Anon 2014). At that time, agricultural products, for example, plants and seeds did not fit into the general framework of patents and no patents for these products were published till then. In the year 1930, plants and seeds were considered patentable material. If seeds are patented, farmers would not be allowed to save these seeds according to a technology agreement that farmers sign. By signing this agreement, farmers agree that they will only use the patented seeds for a one time planting, not to supply or share seeds with anyone else for planting, not to save any crop produced from the seed for future planting, and not to use the seeds or provide them to any individual for plant breeding, research, or seed propagation (Anon 2014). This agreement is a legally binding contract, the violation of which can lead to a court case. In general, living things were only allowed to be patented if they have useful and/or novel characteristics.

Patenting of biotechnology in the United States has been encouraged by government statutes, in which the patent law facilitates strong protection for genetically modified plants and seeds. Further, the U.S. Congress is empowered by the constitution to "promote the Progress of Science and useful Arts, by securing for limited Times to Authors and Inventors the exclusive Right to their respective Writings and Discoveries" (Strauss 2009).

The large biotechnology companies in the United States have great influence on the government in shaping the law to favor their property rights in this new technology. The government encourages patenting the genetic modifications of plant material. A large number of patents in plant biotechnology have been awarded to U.S. firms. The U.S. government also owns a significant number of patents, mostly in joint ventures with private industry (Strauss 2009). This lead to existence of huge numbers of patents, mainly concentrated in a small number of companies.

The majority of the world's natural genetic resources are found in the tropics, where most developing nations are geographically located. At the same time, most patent originators and holders of GM plants and seeds are in the developed nations. Large agricultural and seed companies in developed countries take these genetic resources, modify them to develop patentable products, and then sell the patented products to customers including those in developing nations. This led to creation of a continuous one-way flow of genetic resources from developing countries (generally referred to as "the South," since most developing countries fall in the Southern hemisphere) to developed countries (referred to as "the North"), where they are patented and subsequently removed from the public domain. The use of such genetic resources will not be allowed back in the developing countries, where they have been obtained, without permission and payment. If such genetic resources are used without the required permission and payment, this is regarded as "biopiracy," which is regarded as an illegal practice that results in court proceedings.

Biopiracy has caused some concern and attracted wide attention in recent years. The Convention on Biological Diversity had some efforts to address these concerns through adoption of the "principle of state sovereignty over biological resources and the genetic information therein." According to Brody (2010), this approach is considered inadequately justified and that there are alternative solutions to the issue of biopiracy based on different theories of justice. Two alternative solutions suggested by Brody (2010) include the common heritage of mankind principle and the global commons principle.

Access to patented technologies under "liberal humanitarian use exemptions" allows resource-poor farmers around the globe to benefit from new developments in agriculture without the need to use patented resources consequently avoiding any legal implications resulting from unauthorized use of patented material. This public service is made possible and facilitated by the work of some public sector laboratories, which create alternative products specifically targeted to poor farmers throughout the developing world (Conko 2004).

Policy-makers should use a broader perspective to examine the critical implications for the international community and reshape this application of IP in line with the long-term public interest (Strauss 2009).

REFERENCES

American Medical Association (AMA). 2012. Report 2 of the Council on Science and Public Health: Labeling of bioengineered foods. http://factsaboutgmos.org/sites/default/files/AMA%20Report.pdf (accessed May 2, 2014).

Anon. 2014. Biotechnology Intellectual Property Law. http://sites.psu.edu/gmoliteracyproject/current-legislation/biotechnology-intellectual-property-law/ (accessed May 23, 2014).

Auer, C.A. 2003. Tracking genes from seed to supermarket: Techniques and trends. *Trends in Plant Science*, 8:591–7.

Australia New Zealand Food Standards Code. 2014. Standard 1.5.2—Food produced using gene technology. http://www.comlaw.gov.au/Details/F2014C00036 (accessed May 12, 2014).

Better Health Channel. 2014. Regulation of GM foods and ingredients in Australia. http://www.betterhealth.vic.gov.au/bhcv2/bhcarticles.nsf/pages/Food_-_genetically_modified_(GM) (accessed May 3, 2014).

Botha, G.M. and Viljoen, C.D. 2009. South Africa: A case study for voluntary GM labeling. *Food Chemistry*, 112(4):1060–4.

Brody, B.A. 2010. Intellectual property, state sovereignty, and biotechnology. *Kennedy Institute of Ethics Journal*, 20(1):51–73. http://www.researchgate.net/publication/44634331_Intellectual_property_state_sovereignty_and_biotechnology (accessed May 23, 2014).

Butler, J.E.F. 2009. Cloned animal products in the human food chain: The FDA should protect American consumers. *Food and Drug Law Journal*, 64:473–501.

Byrne, P., Pendell, D., and Graf, G. 2010. Labeling of genetically engineered foods. http://www.ext.colostate.edu/pubs/foodnut/09371.html (accessed May 3, 2014).

Canadian Food Inspection Agency. 2014. Biotechnology and the environment. http://www.inspection.gc.ca/plants/plants-with-novel-traits/general-public/eng/1337380923340/1337384231869 (accessed May 7, 2014).

CCHU9005 Group 3. 2014. GM food and labeling: GM food—Add a label to it? https://sites.google.com/site/gmfood11/gm-food-labelling (accessed May 3, 2014).

Center for Food Safety. 2011. Citizen petition before the U.S. Food and Drug Administration. http://justlabelit.org/wp-content/uploads/2011/09/gelabelingpetition.pdf (accessed June 23, 2014).

Center for Food Safety. 2014a. Label genetically engineered foods. http://www.centerforfoodsafety.org/issues/976/ge-food-labeling/ (accessed June 23, 2014).

Center for Food Safety. 2014b. International labeling laws. http://www.centerforfoodsafety.org/issues/976/ge-food-labeling/international-labeling-laws (accessed May 3, 2014).

Charles, B., Okuro, O.J., and Groote, H.D. 2010. Perspectives of gatekeepers in the Kenyan food industry towards genetically modified food. *Food Policy*, 35(4):332–40.

Chern, W.S., Rickertsen, K., Tsuboi, N., and Fun, T.T. 2003. Consumer acceptance and willingness to pay for genetically modified vegetable oil and salmon: A multi-country assessment. *AgBioForum*, 5:105–12.

Codex Alimentarius. International Food Standards. 2014. http://www.codexalimentarius.org/ (accessed May 15, 2014).

Codex Alimentarius Commission. 2003. Principles and guidelines on foods derived from biotechnology.

Codex Alimentarius Commission. 2004. Food import and export.

Conko, G. 2004. GMO patent nonsense. Competitive Enterprise Institute (CEI). http://cei.org/op-eds-and-articles/gmo-patent-nonsense (accessed May 23, 2014).

Consumers Union. 2014. Genetically engineered foods. http://consumersunion.org/wp-content/uploads/2013/05/GE.labeling.bills_.VT_.2.7.13.pdf (accessed June 23, 2014).

Convention on Biological Diversity. 2014. The Cartagena Protocol on biosafety. http://bch.cbd.int/protocol/ (accessed May 15, 2014).

Council for Biotechnology Information. 2001. Substantial equivalence in food safety assessment. http://foodsafety.ksu.edu/en/article-details.php?a=3&c=17&sc=131&id=497 (accessed May 2, 2014).

Council on Science and Public Health. 2012. CSAPH Report 2-A-12 subject: Labeling of bioengineered foods (resolutions 508 and 509-A-11). Reference Committee E.

Daniells, S. 2014. The cost of GMO labeling? Less than a penny a day, says Consumer Union analysis. http://www.foodnavigator-usa.com/Markets/The-cost-of-GMO-labeling-Less-than-a-penny-a-day-says-Consumers-Union-analysis (accessed October 28, 2014).

Davison, J. 2010. GM plants: Science, politics and EC regulations. *Plant Science*, 178(2):94–8.

Degnan, F.H. 1997. The food label and the right-to-know. *Food and Drug Law Journal*, 52:49–60.
European Commission (EC). 2014. Health and consumers. Plants. Existing rules on GM food & animal feed. Regulations 1829/2003 and 1830/2003. http://ec.europa.eu/food/plant/gmo/legislation/gm_food_animal_feed_en.htm (accessed May 4, 2014).
European Food Safety Authority (EFSA). 2012. Scientific opinion addressing the safety assessment of plants developed through cisgenesis and intragenesis. *EFSA Journal 2012*, 10(2):2561. http://www.efsa.europa.eu/en/efsajournal/doc/2561.pdf (accessed May 11, 2014).
FAO/WHO. 1996. Joint FAO/WHO expert consultation on biotechnology and food safety. Rome, Italy.
Fernandez-Conejo, J. and Caswell, M. 2006. The first decade of genetically engineered crops in the United States. http://www.ers.usda.gov/publications/eib11/eib11.pdf (accessed June 23, 2014).
Finz, S. 2012. Prop.37: Genetic food labels loses. Sfgate.com. San Francisco Chronicle. http://www.sfgate.com/news/article/Prop-37-Genetic-food-labels-trailing-4014669.php (accessed May 6, 2014).
Food Standards Agency. 2013. GM labeling. http://www.food.gov.uk/policy-advice/gm/gm_labelling#.U0sVJtKsi1J (accessed May 2, 2014).
Food Standards Australia New Zealand. 2007. Safety assessment of genetically modified foods: Guidance document.
Food Standards Australia New Zealand. 2014. http://www.foodstandards.gov.au/Pages/default.aspx (accessed May 10, 2014).
Freedeman, J. 2009. *Genetically Modified Food: How Biotechnology is Changing What we Eat*. First Edition. The Rosen Publishing Group Inc., New York.
Ghosh, D.K. and Williams, P. 2010. Global food biotechnology regulations and urgency for harmonization, Chapter 29, in Bagchi, D., Francis, C., Lau, F.C., and Ghosh, D.K. (Eds.), *Biotechnology in Functional Foods and Nutraceuticals*. CRC Press, Boca Raton, FL, pp. 531–41.
GMO Compass. 2014. Genetic engineering, plants, and food. The European Regulatory System. http://www.gmo-compass.org/eng/regulation/regulatory_process/156.european_regulatory_system_genetic_engineering.html (accessed May 5, 2014).
GRAIN. 2005. The FAO seed treaty: From farmers' rights to breeders' privileges. *Seedling*. http://www.grain.org/article/entries/585-the-fao-seed-treaty-from-farmers-rights-to-breeders-privileges (accessed May 24, 2014).
Gruère, G.P. and Rao, S.R. 2007. A review of international labeling policies of genetically modified food to evaluate India's propose rule. *AgBioForum*, 10(1):51–64. http://www.agbioforum.org/v10n1/index.htm (accessed May 7, 2014).
Hansen, M. 2001. Genetically engineered food: Make sure it's safe and label it, in Nelson, G.C. (Ed.), *Genetically Modified Organisms in Agriculture*. Academic Press, San Diego, pp. 239–55.

Health Canada. 2014. The regulation of genetically modified food. http://www.hc-sc.gc.ca/sr-sr/biotech/index-eng.php (accessed May 4, 2014).

Helmuth, L. 2000. Biotechnology. Both sides claim victory in trade pact. *Science*, 287(5454):782–3.

Herling, D.J. 2014. As Maine goes, so goes the nation? Labeling for foods made with genetically modified organisms (GMOs). The National Law Review. http://www.natlawreview.com/article/maine-goes-so-goes-nation-labeling-foods-made-genetically-modified-organisms-gmos (accessed May 6, 2014).

IFST. 2008. Genetic modification and food factors affecting EU approach to regulating GM. Information Statement. www.ifst.org (accessed May 3, 2014).

ISAAA. 2014. Global status of commercialized biotech/GM crops. ISAAA Brief 46-2013. http://www.isaaa.org/resources/publications/briefs/46/executivesummary/default.asp (accessed March 18, 2014).

Just Label It Campaign. 2014. Partners. http://justlabelit.org/partners (accessed June 23, 2014).

Key, S., Ma, J.K., and Drake, P.M. 2008. Genetically modified plants and human health. *Journal of the Royal Society of Medicine*, 101(6):290–8.

Loueiro, M.L. and Hine, S. 2004. Preferences and willingness to pay for GM labeling policies. *Food Policy*, 29:467–83.

Organisation for Economic Co-Operation and Development (OECD). 1993. *Safety Evaluation of Foods Derived by Modern Biotechnology: Concepts and Principles*. OECD. pp. 10–13.

Pelletier, D. 2005. Science, law, and politics in FDA's genetically engineered foods policy: Scientific concerns and uncertainties. *Nutrition Reviews*, 63:210–23.

Phillips, P.W.B. and McNeill, H. 2000. A survey of national labeling policies for GM foods. *AgBioForum*, 3(4):219–24. http://www.agbioforum.org/v3n4/v3n4a07-phillipsmcneill.htm (accessed May 7, 2014).

Pinholster, G. 2012. AAAS Board of Directors: Legally mandating food labels could "mislead and falsely alarm consumers". American Association for the Advancement of Science. http://www.aaas.org/news/aaas-board-directors-legally-mandating-gm-food-labels-could-%E2%80%9Cmislead-and-falsely-alarm (accessed May 4, 2014).

Raab, C. and Grobe, D. 2003. Labeling genetically engineered food: The consumer's right to know? *AgBioForum*, 6(4):155–61. http://www.agbioforum.org/v6n4a02-raab.htm (accessed May 7, 2014).

Record of GM Product Dealings. 2014. Food produced using gene technology and approved for sale under the Food Standards Australia New Zealand Act 1991. http://www.ogtr.gov.au/internet/ogtr/publishing.nsf/Content/gmprod-1/$FILE/gmfoodprod4.pdf (accessed May 12, 2014).

Reilly, G. 2013. Malloy signs state GMO labeling law in Fairfield. *ctpost.com*. http://www.ctpost.com/news/article/Malloy-signs-state-GMO-labeling-law-in-Fairfield-5056120.php (accessed May 6, 2014).

Ronald, P. 2011. Plant genetics, sustainable agriculture and global food security. *Genetics*, 188(1):11–20.

Simon, R.R., Nestmann, E.R., Musa-Veloso, K., and Munro, I.C. 2010. Regulations of biotechnology generally recognized as safe (GRAS) and health claims, Chapter 28, in Bagchi, D., Francis, C., Lau, F.C., and Ghosh, D.K. (Eds.), *Biotechnology in Functional Foods and Nutraceuticals*. CRC Press, Boca Raton, FL, pp. 507–30.

Stein, H. 2005. Intellectual property and genetically modified seeds: The United States, trade, and the developing world. *Northwestern Journal of Technology and Intellectual Property*, 3(2):160–78.

Strauss, D.M. 2009. Agreement on trade-related aspects of intellectual property rights. The application of TRIPS to GMOs: International intellectual property rights and biotechnology. *Stanford Journal of International Law*, 45(2):287–320. http://www.thefreelibrary.com/The+application+of+TRIPS+to+GMOs%3a+international+intellectual+property...-a0216486735 (accessed May 26, 2014).

Sundstrom, F.J., Williams, J., VanDeynze, A., and Bradford, K.J. 2002. Identity preservation of agricultural commodities. *Agricultural Biotechnology in California Series* Pub 8077. http://anrcatalog.ucdavis.edu/pdf/8077.pdf (accessed May 3, 2014).

Taylor, H. 2000. Harris Poll #33. Genetically modified foods: An issue waiting to explode? http://www.harrisinteractive.com/vault/Harris-Interactive-Poll-Research-GENETICALLY-MODIFIED-FOODS-AN-ISSUE-WAITING-TO-EXPLODE-2000-06.pdf (accessed June 25, 2014).

The Mellman Group. 2012. Voters overwhelmingly support a labeling requirement for GE foods. http://justlabelit.org/wp-content/uploads/2012/01/Mellman-Survey-Results.pdf (accessed June 25, 2014).

The NCBE Guide. 2014. The novel foods regulations. http://www.ncbe.reading.ac.uk/NCBE/GMFOOD/menu.html (accessed May 3, 2014).

The Organic and Non-GMO Report. 2014. Japan legislation on labeling of genetically engineered foods. http://www.non-gmoreport.com/articles/millenium/japanlegislationlabelinggmfoods.php (accessed May 13, 2014).

USFDA. 1992. Statement of policy—Foods derived from new plant varieties. Guidance to industry for foods derived from new plant varieties. Food for human consumption and animal drugs, feeds, and related products: Foods derived from new plant varieties; policy statement, 22984. FDA Federal Register Volume 57, No. 104. Docket No. 92N-0139. http://www.fda.gov/Food/GuidanceRegulation/GuidanceDocumentsRegulatoryInformation/Biotechnology/ucm096095.htm (accessed October 28, 2014).

USFDA. 2014. Guidance for industry voluntary labeling indicating whether foods have or have not been developed using bioengineering draft guidance. http://www.fda.gov/Food/GuidanceRegulation/GuidanceDocumentsRegulatoryInformation/ucm059098.htm (accessed November 6, 2014).

Weirich, P. (Ed.). 2007. *Labeling Genetically Modified Food: The Philosophical and Legal Debate*. Oxford University Press, Oxford, UK.

Western Australian Agriculture Authority. 2011. An investigation of labelling genetically modified food. Department of Agriculture and Food, Government of Western Australia. http://www.agric.wa.gov.au (accessed May 14, 2014).

Whitman, D.B. 2000. Genetically modified foods: Harmful or helpful? CSA Discovery Guides. http://www.csa.com/discoveryguides/gmfood/overview.php (accessed May 3, 2014).

Worldhunger.org. 2014. Current intellectual property rights, especially those for GMO seeds, threaten poor farmers, food security and the right to food. United Nations. http://www.worldhunger.org/articles/09/global/united_nations.htm (accessed May 27, 2014).

World Intellectual Property Organization (WIPO). 2008. *WIPO Intellectual Property Handbook: Policy, Law and Use*. Chapter 2. Fields of Intellectual Property Protection. Geneva, Switzerland. http://www.wipo.int/edocs/pubdocs/en/intproperty/489/wipo_pub_489.pdf (accessed June 5, 2014).

SUGGESTED REFERENCES

Conover, R. 2004. Biotech labeling still unresolved in codex. *Food Technology*, 58:208.

Graham, V. 2000. *The EU and GM Foods: Current Regulations and Future Trends*. Chandos Publishing, Oxford, UK.

McHughen, A. 2008. Labeling genetically modified (GM) foods. http://www/agribiotech.info/details/McHugen-Labeling%20sent%20to%20web%2002.pdf (accessed June 10, 2014).

Nelson, L. 2004. Labeling laws for transgenic food come into effect. *Nature*, 428:788.

Smyth, S., Phillips, P.W.B., Kerr, W.A., and Khachatourians, G.G. 2004. *Regulating the Liabilities of Agricultural Biotechnology*. CABI Publishing, Wallingford, UK; Cambridge, MA.

Zarrilli, S. 2005. *International Trade in GMOs and GM Products: National and Multilateral Legal Frameworks*. United Nations Conference on Trade and Development, United Nations, New York.

6

GM Foods or Not? The Controversy

6.1 INTRODUCTION

In recent history, there are few issues that have received more attention and wider debate than genetically modified (GM) foods. The world population may be viewed as belonging to one of three groups with regard to production and consumption of GM foods: proponents, opponents, and the confused. The first two groups are a minority, whereas the majority are those in the third category. This same trend is expected to continue for some time in the future. There are huge volumes of information on various issues of GM foods. This makes it a challenging task to find out answers for questions asked by members of the general public and for concerns raised about the effects of GM foods, particularly on human health and the environment. It has been realized that there are different advantages and disadvantages of GM foods. The extent to which they can help or harm humans and the environment is a debatable aspect of this technology. Arguments exist on both sides for and against the growing trend by many of the world's nations as to why they should or should not alter the genetic make-up of plants and animals. The arguments about transgenic crops are about values, which are neither absolute nor universal, and the controversies surrounding transgenic crops have polarized society into proponents and opponents, with once seemingly trustworthy and ethically sound scientists being viewed with suspicion by many (Robinson 1999).

Different terms have been used to describe the two sides of the controversy around GM foods. Some examples include: proponents versus opponents, pros versus cons, supporters versus critics, for versus against, benefits versus harm, lovers versus haters (Squidoo 2014), and advantages versus disadvantages. Whichever two opposing terms are used to describe the controversy, there are always two sides for the debate.

The debate on various issues related to GM foods, for example, production, labeling, consumption, safety, ethics, and socioeconomic, has mounted to a seemingly level of a war. It seems to be a war of words and an exchange of views rather than exchange of facts and credible information. This debate is raging in both developed and developing regions of the world. It is regarded as one of the most controversial and debated issues of recent times. Cook (2005) states that "those promoting GM have mounted an intense campaign, characterizing their opponents as terrorists and Luddites, governed by ignorance, irrationality and hysteria."

Genetic modification is changing, and has the potential for more change of the nature of life as we know it. This statement can be viewed as a positive achievement or as a negative consequence, depending on whether you are supporting or opposing genetic modification. Proponents argue that genetic modification has positive outcomes that outweigh its shortcomings. They focus more on the advantages and pay less, or no, attention to the negatives. On the other side, opponents focus more on the negative consequences, not accepting the positive claims. In reality, both supporters and critics split into different camps. Some critics appreciate some of the benefits of gene technology, whereas some supporters recognize potential limitations and threats. The truth of the matter is that a rational strategy requires an approach that respects and embraces both sets of arguments. The bottom line is that no human activity, including food, is risk-free. However, this is not a reason to trust unquestioningly in genetic modification. Risks must be identified, evaluated, and minimized. Both sides of the debate use science as part of the rationale in their arguments and question the independence or veracity of the other side's research. The question is "who should consumers trust?" Science cannot be an answer to all questions and will not be able to offer people what they really want, namely, total certainty. GM technology should be viewed as being similar to other technologies developed in the sense that it is expected to have some pros and cons. The Sustainable Future (2008) suggests that to obtain a better picture of what both sides of the argument are saying and to develop a more-informed opinion, people need to do more research and to look for credible information on various issues surrounding the technology. Some of the topics, articles, and websites listed in this reference include the following:

- GM crops: the arguments pro and con
- The risks and benefits of GM crops
- GE: the controversy
- GM foods: pros and cons
- GM products: benefits and controversies
- Weighing pros and cons of GM crops in Africa

The dispute surrounding the controversy of GM foods involves different parties including biotechnology companies and institutions, governmental and nongovernmental organizations (NGOs), United Nations organizations, scientists, the media, and the consumers. Supporters of the technology include its developers and users, some governments and governmental institutions, for example, regulators, UN organizations, for example, Food and Agriculture Organization (FAO), World Health Organization (WHO), and some consumers. On the opposing side, there are a number of international groups and institutions which are against the production of GM foods. Some examples include the Greenpeace, Friends of the Earth, the Christian Aid, the Institute for Food and Development Policy, and some consumers.

The main areas of the debate encompass the actual and potential effects of GM crops and GM foods on human and animal health and on the environment, the credibility and objectivity of conducted and published research on gene technology and GM foods, labeling of GM foods, the impact of gene technology on farmers, the conflict of interest among biotechnology companies, researchers, and government regulators, and the possible role of GM foods to help reduce food insecurity worldwide. The website eSSORTMENT (2014) highlights some of the arguments presented by both the pro- and anti-GM foods sides. It asks the question "should we put on the brakes, go ahead full steam, or find the middle ground?" In reality, this question does not have a simple or straight answer. The pro-GM food campaign is expected to suggest the "go ahead," whereas the anti-GM group would be expected to support the "put on brakes." The more reasonable scenario is to "find a middle ground," where some kinds of mutual understanding are maintained and disputed issues are ironed out.

Arguments presented by supporters are that genetic modification could be used as a tool to improve characteristics of plants and animals, it could help reduce the extent of hunger and malnutrition worldwide, it could also reduce cost of production of plants and animals and hence help food producers, and that there are no proven examples or cases of GM products adding health or environmental risks. The concerns raised against production of GM products are that consumption of GM foods may be a health hazard and that these foods may be unsafe for humans. Another concern raised is that GM may lead to a potential damage to the environment and undesirable changes and disruption to the entire ecosystem. The arguments presented to support these concerns are as follows: The technology is unnatural and it may lead to unpredictable results and unexpected and unintended side effects; GM products are hazardous, and their long-term use can produce ill effects on health; it degrades the environment and erodes biodiversity by spreading harmful genes, leading to what is known as genetic pollution; it may cause allergies; it may lead to antibiotic resistance; there is a possibility of emergence of "superweeds"; and there are ethical and moral objections and that people are "messing with nature," raising the question "Do human beings have the right to view nature as their commodity, and alter the genetic structure of living things to correct what some perceive as their deficiencies?"

Some of the terms used by GM food supporters, for example, miracle crops, or by critics of GM foods, for example, Frankenfoods, do not help the debate. It rather makes the waters more muddy and the picture foggy, particularly the main stakeholders—the consumers. The group supporting the technology and its application believes that this technology will be the solution to the world hunger. In contrast, those opposing the technology believe that it will bring environmental and social catastrophes of incalculable consequences (Brac de la PerriFre and Seuret 2000; GRAIN 2004). These extreme and contrasting positions of both groups need to be revisited and addressed in a more sensible, balanced, and objective manner.

The effect that GM foods may have on developing nations is debated. There is also a debate over whether the use of GM crops increases or decreases yield. The market dynamics has also been an issue of debate in relation to the domination and monopoly of the seed industry by a few seed and biotechnology firms. These firms have been engaged in vertical integration, causing structural changes in the seed industry (Hayenga 1998; ETC Group 2008). A large number of social science and public surveys have been conducted in different parts of the globe, and various public and individual views have been expressed on the GM technology and GM foods. Protests against the GM technology and its different applications, particularly in food, have been, and continue to be, held around the world.

Religious concerns about consumption of GM foods have also been raised among different followers of religions, for example, Muslims and Jews. However, no GM foods have been regarded as unacceptable by these groups (Vogt and Parish 2001).

Some environmental groups, for example, Friends of the Earth, express environmental concerns about growing GM crops and that they will negatively affect the environment.

The concerns raised by anti-genetically modified organism (GMO) activists about human health related to GM food consumption lead to a widespread perception among consumers that eating GM foods is harmful. The results of some surveys among the public indicated that consumers are concerned about the possible risks of GM technology and that they need more information about the risks themselves and a desire for choice in being exposed to risks (Lazarus 1991; Hunt 2004).

A more reasonable and prudent approach to deal with GMOs and GM foods would be to address the relevant facts and then conduct a robust and complete analysis of the appropriateness of a specific technology in relation to its intended target country or region. It is not objective to generalize that genetically engineered (GE) crops and GM foods are unequivocally either Frankenfoods or miracle crops. A rational approach would require judging GE crops on a case-by-case basis while considering all the costs, benefits, and risks estimated through robust assessments (Falck-Zepeda et al. 2013). This would lead to objective results that can help resolve many of the debated issues. Many of the attitudes toward the use of GMOs in agriculture and food involve concerns about trust and perceived risks. Public perception of the use of GM in food production is very emotionally charged, and it is therefore essential that the risks and benefits are considered carefully. The polarity and passion of the debate make it essential to weigh the risks and benefits very carefully (Kariyawasam 2010) and draw conclusions that would be helpful to consumers.

There is no doubt that the GM food supply should be closely monitored and regulated, but that does not mean that it should all be banned. GE of plants and animals has much to offer, bearing in mind the potential benefits and possible risks. That is true even for traditional methods of farming, animal husbandry, and medicine (Lei 2005). Due to the concerns from both advocates and opponents, both sides of the debate have come together over the past few years in forums and summits all over the world to discuss the issue (Guidetti 2009). This trend helped to create some common language of understanding between the two opposing parties, together with building some degree of appreciation of both benefits and risks by the supporters and opponents. This can be regarded as a positive step toward coming up with a more reasonable and objective position on GM foods, which is acceptable to both proponents and opponents and which would help consumers to have a better and more clear picture of GM foods and, consequently, make informed decisions about purchase and consumption of GM foods.

In the next sections, the issues of controversy, the arguments presented by each side of the debate, and how each side reacts to those points will be discussed.

6.2 ISSUES OF CONCERN AND CONTROVERSY

Both sides of the debate agree that there are a number of issues and concerns about GM technology and its application to develop GM foods, which need to be considered. The main issues are centered on food safety and human health, potential damage to the environment, disruption of the ecosystem, ethical and moral aspects, and socioeconomic considerations. There is a general agreement between proponents and opponents of GM technology that all of these concerns are valid and must be taken into account seriously, with the attention they deserve. In contrast, each of the two sides views these issues or concerns differently, presents arguments, and draws conclusions that support their stance as "for or against."

6.2.1 Concerns about Food Safety and Human Health

Genetic modification and its application to produce GM foods have the potential to create unpredictable dangers to human health that may be short term or long term, direct or indirect. Specific potential food safety and human health problems that have been expressed are centered on the following points:

- GM foods might be more risky than traditional foods. The health effects of consuming GM foods are not yet fully understood. The most pertinent general question asked by many consumers, which needs a satisfying answer, is: "is it safe to consume GM foods?"
- Potential risks of GM foods related to health have not yet been adequately investigated. There is a need for improved and long-term food safety tests, in addition to reliable technologies and protocols to better identify and manage potential risks (United

States Institute of Medicine and National Research Council 2004; Freedman 2013).
- There is a concern about the use of marker genes in GM. Marker genes are sequences of DNA used in GM to help researchers find out which organisms have taken up the introduced genes. The concern is that do the marker genes allow their recipient organism to produce new proteins? Would these new proteins be present in the GM food? Could the marker gene be transferred to the human gut? Does it have the potential to cause harm to consumers?
- A concern that consumers might be inadvertently exposed to allergens has been expressed. The potential of a GM food to produce an allergic reaction is one of the safety considerations that need careful attention. If a person is allergic to a particular food and a gene from it is transferred to another food, that person could also become allergic to the second food. Transgenic products may adversely affect people suffering from allergies. Testing for allergens in GM foods is part of the research and development of GMOs intended for food. It is essential that tests for allergenicity are improved and that such potential allergens present in food are labeled. On the positive side, the potential that GM can be used to eliminate allergens from foods needs to be investigated.
- Many concerns have been expressed about the use of antibiotic-resistant marker genes, which could reduce the efficacy of antibiotics in treating human and animal diseases. Current transgenic crops may contain antibiotic-resistant marker genes. Such GM crops might contribute to increasing dissemination of antibiotic resistance as an issue with environmental and health implications. If animals are fed antibiotic-resistant plants, the antibiotic-resistant gene might be transferred to bacteria in the animal's gut and subsequently transmit antibiotic-resistant organisms to humans through the food chain.
- A serious concern has been expressed by some consumers about toxicity of the protein products of transgenes themselves or the potential that these proteins might induce unintended effects on plant metabolism, leading to production of toxins.
- The possible carcinogenic effect of GM foods is another concern that has been voiced. Would there be a potential that consuming GM foods may cause cancer in individuals who eat those foods?

A joint FAO/WHO (2000) expert consultation on the safety aspects of GM foods of plant origin studied the GM food safety issue. In its report, the consultation agreed that the safety assessment of GM foods requires an integrated, stepwise, case-by-case approach, which can be aided by a structured series of questions. Similarities and differences between GM foods and their conventional counterparts would help in the identification of potential safety and nutritional issues and are considered the most appropriate strategy for the safety and nutritional assessment of GM foods. The use of the concept of "substantial equivalence" has been recommended by the consultation, provided that refining of some aspects of the steps in the safety assessment process is taken into consideration. In addition, the WHO, together with FAO, has convened several expert consultations on the evaluation of GM foods and provided technical advice for the Codex Alimentarius Commission, which was fed into the Codex Guidelines on safety assessment of GM foods. The WHO promises to pay due attention to the safety of GM foods from the view of public health protection, in close collaboration with FAO and other international bodies.

Frequently asked questions, and answers, have been prepared by WHO (2014) in response to questions and concerns from WHO Member State Governments with regard to the nature and safety of GM foods. The questions and answers cover the main issues of public concern, such as the nature of GM foods, why are they produced, safety of GM foods to human health, their possible effects on the environment, how they are regulated, and the future of GM foods. The WHO emphasizes the fact that GM foods are safe for human consumption because they passed all safety tests in different parts of the world. The WHO also indicates that it has been taking an active role in relation to GM foods based on the benefits and potential of biotechnology to improve public health and at the same time evaluating the need to examine the possible negative effects on human health.

6.2.2 Environmental Concerns

With respect to the environment, the following issues and concerns have been raised:

- A concern has been raised about the uncertain and unpredictable behavior of GM microorganisms in the environment. Microorganisms are known to generate and mutate quickly, exchange genetic information easily among them, and are not

easy to detect in the environment. There is the danger that these GM microorganisms persist in the soil, be absorbed by plants, or contaminate the water system.
- One concern which has been raised with regard to GM crops is that releasing GM organisms into the environment represents a kind of "genetic pollution." There are long held concerns about the transfer from GE crop plants to wild related species to create uncontrollable generation of weeds, referred to as "superweeds." This group of weeds could outcompete and disrupt the natural biodiversity of an area. It would also need stronger types of herbicides to control them.
- There is also a concern that herbicide-tolerant genes will be transferred to neighboring organic crops, non-GM crops, or to other GM crops. This transfer of genes, regarded as a type of contamination, would lead to crossbreeding, which could result in the formation of herbicide-tolerant plants. If a GM crop contaminated a non-GM crop, this would affect farmers' ability to be assured of growing non-GM foods.
- One concern raised has been the possibility of what is referred to as horizontal gene transfer (HGT). HGT is the process through which the genetic material is moved from one organism to another, other than its offspring. This transfer is followed by integration and expression of the genetic material. HGT can take place from plants used as feed to animals that are used for food or from plants used as food to humans.
- Another widespread concern is that herbicide-resistant crops may lead to "sterility" of agricultural fields, leaving them empty of plants used by birds and insects for feeding, thus disrupting the natural ecosystem.
- A concern about insect resistance has been raised. Crops that have been genetically modified to be insect-resistant by adding a toxin could promote a wider problem of resistance to toxins, limiting the use of these toxins by other farmers.
- The use of an antibiotic-resistant gene as a genetic marker in GM technology to develop transgenic crops raises another concern. This gene can potentially be transferred to harmful bacteria, possibly creating superbugs that are resistant to various antibiotics.
- The development of GM crops would result in fewer cultivars being used overall. They might also have an indirect effect on the diversity of other living organisms. This would lead to an

appreciable decrease in the genetic diversity of crops. A serious concern has been brought up in relation to this effect.
- There are also growing concerns that the use of GM for crops aimed at tolerating agrochemicals might lead to the increased use of chemicals, which in turn might cause damage to the environment and to biodiversity (biodiversity or biological diversity means the variability among living organisms from all sources at more than one scale; this includes diversity within species, between species, and of the whole ecosystem) (Biodiversity A to Z 2014).

6.2.3 Ethical Concerns

A range of ethical concerns have been expressed in relation to applications of gene technology in general. The most common immediate response to gene technology is centered on the issue of "messing with nature." There are fears that GM technology might create a wide range of ethical challenges about how human beings relate to nature. Theoretically, gene technology has limitless applications: Gene transfer can be performed among all forms of living organisms and genes can be derived from plants, animals, fish, or microorganisms. This has raised a lot of concerns particularly among religious groups, some of which believe that this represents an interference in natural processes controlled by God.

6.2.4 Regulatory and Legal Concerns

A general concern has been expressed about the role of governments in setting or controlling the rules and regulations pertaining to gene technology and its wide applications. A number of issues have been raised related to this concern. These include the type of responsibilities of governments and what should be left to the industry, the appropriate government agencies to regulate gene technology, regulation monitoring, compliance and enforcing, the requirement of new and separate regulations for gene technology, and the expected role of governments to harmonize international obligations and local demands.

6.2.5 Economic Concerns

Monetary interests are believed, by some, to be the main driving factors behind the development and applications of GM technology. The concern raised here is that it might not be appropriate or practical for a technology driven by profit margins to have the potential for use in humanitarian applications.

There is a noticeable and increasing trend toward becoming dependent on biotech firms. This is particularly true with regard to the development and use of terminator technologies, in which seeds must be bought by farmers each cropping season. Seeds from previous season harvest cannot be grown because they are sterile. This market dominance by biotech companies, which own this technology and genes, is viewed as an ethical issue evolving for economic reasons.

Another economic issue of concern is the search and transfer of genetic resources from developing nations (referred to as the South) to developed nations (referred to as the North). These genetic resources are usually developed and patented by a few biotech companies in developed countries. Economic resources and conditions in developing countries do not allow them to establish genes originating in their respective countries as native resources. This trend of transferring genetic resources from the South to the North harms the economies of developing nations.

There is a concern about the possibility of corporate takeover of the food chain. This results from the observed monopoly of the agricultural biotech business by a few companies, which leads to control of food production. The more the GM technology advances, the more GM foods are produced and the more this monopoly continues.

6.2.6 Animal Health and Welfare

A growing concern has been raised about animal health and welfare. It is believed that classical animal breeding proved to be successful in improving the productivity and well-being of farm animals. There is fear that applying gene technology to animals might compromise their welfare and health. This is due to the fact that it is not easy to predict which genes should be modified to improve animal productivity or health as the majority of traits in livestock are controlled by multiple genes.

6.2.7 Consumer Choice

Consumer choice does not seem to be given the attention it deserves by the GM food producing companies. There is a concern that consumer welfare is not a priority of the food industries that use GM ingredients in their products and that making profit is more important to these industries.

6.2.8 Concerns about Bias in Scientific Publishing

There is a serious concern about scientific publishing related to gene technology and GM foods. Scientific publishing is expected to be honest, credible, unbiased, and objective. This is because the general public has a lot of trust in what scientists reflect on the results and interpretations of their research. Consumers will be confused as who to believe or who to trust and which interpretations are correct. Scientific publishing has been used (or misused) by both sides of the debate. Each side uses it in a tailored manner that suits their purpose of highlighting their views. The results of a specific experiment may be the same but interpretations may be different, depending on which side the author is. Some scientists highlight only the results that support their purpose and hide or disregard results which contradict their assumptions or theories, how insignificant they are. Honest scientific publishing needs to be the norm of both supporters and critics of GM technology, or any kind of research to that matter. Scientists should present all the results and outcomes of their experiments and derive an unbiased interpretation. This way a true and balanced picture of gene technology would be reflected, which is expected to help consumers understand and have the right idea about the technology and its pros and cons. True scientific research cannot arrive at the conclusion that gene technology and GM foods have pros only without any cons or vice versa.

6.3 PROPONENTS

6.3.1 Who Are the Proponents?

The proponents and supporters of the GM technology in general and GM food production advocates fall into different categories, which include the biotech companies and industries, the pro-GM lobby groups, the global network of pro-corporate activists, international bodies and governments, the grocery manufacturers' association, researchers and scientists, and some consumers. In the next section, details of each of those categories and who belongs to them will be discussed.

6.3.1.1 International Bodies and Governments
6.3.1.1.1 UN and Other International Bodies
- FAO (http://www.fao.org)
- WHO (http://www.who.int)

- Consultative Group on International Agricultural Research (http://www.cgiar.org)
- World Trade Organization (http://www.wto.org)
- World Food Programme (http://www.wfp.org)
- World Bank and International Monetary Fund (http://www.imf.org)
- Organization for Economic Cooperation and Development (http://www.oecd.org)

6.3.1.1.2 U.S. Government Departments
- Food and Drug Administration (FDA) (http://www.fda.gov)
- U.S. Department of Agriculture (USDA) (http://www.usda.gov)
- U.S. Agency for International Development (http://www.usaid.gov)

6.3.1.1.3 U.K. Bodies
- Department of Environment, Food, and Rural Affairs (http://www.defra.gov.uk)
- Food Standards Agency (FSA) (http://www.food.gov.uk)
- Advisory Committee on Releases to the Environment (https://www.gov.uk/government/organisations/advisory-committee-on-releases-to-the-environment)
- Advisory Committee on Novel Foods and Processes (ACNFP) (http://acnfp.food.gov.uk/)
- Biotechnology and Biological Sciences Research Council (http://www.bbsrc.ac.uk)
- Department for International Development (https://www.gov.uk/government/organisations/department-for-international-development)
- National Farmers' Union (http://www.nfuonline.com/home/)
- John Innes Centre (http://www.jic.ac.uk/)
- The Royal Society (http://royalsociety.org/)

6.3.1.2 Biotech, Agrochemical, and Associated Companies
Some of the leading companies which fall into this category include Monsanto, Syngenta, Bayer, DuPont, DuPont/Pioneer, Advanta US, Dow Seeds, Dow AgroScienes, and Cargill.

The major GM firms mentioned here started their business as chemical industries and have operated for a long time producing various types of chemical materials. They then moved into the biotechnology business while retaining the chemical industry mainly producing agrochemicals. At present, these companies are considered as the world's biggest manufacturers of agrochemicals, and together they control about 75% of the global pesticide market (GMWatch 2014).

6.3.1.2.1 *Monsanto*
Monsanto Company (http://www.monsanto.com/pages/default.aspx) is an American, multinational corporation specialized in producing chemical substances and agricultural biotechnology products. It is the biggest seed company worldwide and the world's fifth largest agrochemical company. It has its headquarters in Creve Coeur, Missouri, USA. It is one of the leading producers of GE seeds; hence, it has a high interest in supporting GM foods. The company is also specialized in producing the herbicide glyphosate under the brand name of Roundup. The use of this herbicide for food crops is also contested by critics of GM foods. As the case now with the GM products of Monsanto being controversial, some of their earlier chemical products such as dichlorodiphenyltrichloroethane (DDT), polychlorinated biphenyl (PCB), and Agent Orange were also criticized and regarded as unsafe for use. In the field of biotechnology, Monsanto is considered one of the pioneers in genetically modifying plant cells and in applying biotechnology in agriculture. The current controversy surrounding Monsanto and its products emanates from its chemical and biotechnology products, which are controversial, from its lobbying of government agencies and research institutions and from its history as a chemical company (GMWatch 2014).

6.3.1.2.2 *Syngenta*
Syngenta AG (http://www.syngenta.com/) is a Swiss chemical company of worldwide spread specialized in producing seeds and pesticides. It is considered as the world's second largest agrochemical manufacturer and the world's third largest seed company, as reflected by their total sales in 2009 in the commercial market (Shand 2012; GMWatch 2014).

6.3.1.2.3 *Bayer*
Bayer AG (http://www.bayer.com) is a German chemical and pharmaceutical company with its headquarters in Leverkusen, North Rhine-Westphalia. It has worldwide presence and is well known for its original

brand of aspirin. The core business of Bayer at present falls within the areas of health care, nutrition, and high-tech materials.

6.3.1.2.4 DuPont/Pioneer Hi-Bred
DuPont (http://www.dupont.com/) is an American chemical company founded in 1802 by Eleuthere Irenee du Pont as a gunpowder mill. Its business now revolves around five main categories, one of which is agriculture and nutrition. The agriculture division known as "DuPont Pioneer" produces and markets hybrid as well as GM seeds. A number of these seeds are used in the manufacture of some foods, which are consequently regarded as GM foods.

DuPont Pioneer (http://www.pioneer.com/landing), which was formerly known as Pioneer Hi-Bred, is the largest producer of hybrid seeds for agricultural crops in the United States. It has its head office in Johnston, Iowa, USA, with suboffices throughout the globe. Some of the improved agricultural seeds produced and marketed by Pioneer include cereal grains, for example, corn, sorghum, wheat, rice, and oil seeds, for example, soybean, sunflower, canola, in addition to forage and grain additives.

6.3.1.2.5 Cargill
Cargill Inc. (http://www.cargill.com/) is an American multinational corporation with its headquarters in Minnetonka, Minnesota, USA. Cargill conducts a variety of businesses, one of which is trading in agricultural and food commodities. The company is a major distributor of grains and other agricultural products, for example, palm oil. It also operates in the production of some food ingredients such as starch, glucose syrup, and vegetable oils and fats.

6.3.1.2.6 Advanta
Advanta (http://advantaus.com/) is one of the international leaders in the research, development, production, processing, marketing, and sale of high-performance agricultural seeds. It represents a global seed business that combines the most advanced techniques in conventional plant breeding with biotechnology to deliver world-class seeds.

6.3.1.2.7 Dow
Dow Seeds (http://www.dowseeds.eu/) and Dow AgroSciences (http://www.dowagro.com/) are also involved in producing agricultural products using biotechnology, which enter in the production of GM foods.

6.3.1.3 Pro-GM Lobby Groups

A number of organizations and bodies around the world, which are not directly involved in the biotechnology industry, have shown interests in the recent biotechnological developments and products. They are very much involved in supporting, and lobbying for, biotechnological advances and the products resulting from them, including GM foods. This support is reflected in the different forums, which address biotechnology, GMOs, and GM foods. The GMWatch has included details of some of these organizations and the work they do in its website (http://www.gmwatch.org/details). The following are a few examples of these organizations.

6.3.1.3.1 International Service for the Acquisition of Agri-Biotech Applications

International Service for the Acquisition of Agri-Biotech Applications (ISAAA) (http://www.isaaa.org/) is a not-for-profit international organization that shares the benefits of crop biotechnology to various stakeholders, particularly resource-poor farmers in developing countries, through knowledge-sharing initiatives and the transfer and delivery of proprietary biotechnology applications. The organization provides science-based information and appropriate technology to those who need to make informed decisions about their acceptance and use. ISAAA has three centers in South East Asia (ISAAA *SEAsia*Center, hosted by the International Rice Research Institute in Los Baños, Laguna, The Philippines), Africa (ISAAA *Afri*Center, hosted by the International Livestock Research Institute in Nairobi, Kenya), and North America (ISAAA *Ameri*Center, hosted by Cornell University, Ithaca, New York under the International Programs—College of Agriculture and Life Sciences).

6.3.1.3.2 International Food Information Council Foundation

Founded in 1985, the International Food Information Council (IFIC) (http://www.foodinsight.org/) is a nonprofit organization whose mission is to effectively communicate science-based information on food safety and nutrition to health and nutrition professionals, educators, government officials, journalists, and others providing information to consumers, primarily in the United States. The IFIC does not represent any product or company, nor does it lobby for legislative or regulatory action. Its main purpose is to gather and disseminate scientific information on food safety, nutrition, and health. It operates in collaboration with an extensive group of science experts to help translate research results into understandable and useful information for decision-makers as well as for consumers.

6.3.1.3.3 Biotechnology Industry Organization

The Biotechnology Industry Organization (BIO) (http://www.bio.org/) is the largest trade body that represents and serves the biotechnology industry in the United States and around the globe (BioSpace 2014). BIO is the voice of different companies involved in the development of products and services based on biotechnology in the areas of agriculture, food, medicine, pharmaceutical drugs, biofuels, and industrial enzymes. The organization website features news, issues and policies, reports, and other organizational information. A comprehensive BIO profile is given by BioSpace in its website http://www.biospace.com/company_profile.aspx?CompanyId=1311.

6.3.1.3.4 The American Association for the Advancement of Science

The American Association for the Advancement of Science (AAAS) (http://www.aaas.org) is an international nonprofit organization with the stated goals of promoting cooperation among scientists, defending scientific freedom, and supporting scientific education and science outreach for the betterment of all humanity. It is the world's largest and most prestigious general scientific society and is the publisher of the well-known scientific journal *Science*. AAAS serves some 261 affiliated societies and academies of science. AAAS states that "the science is quite clear: crop improvement by modern techniques of biotechnology is safe."

6.3.1.3.5 American Soybean Association

The American Soybean Association (ASA) (http://soygrowers.com/#) is a U.S.-based association which aims at helping soybean farmers in the United States and around the world. The primary focus of the ASA is policy development and implementation, which involves the participation of soybean farmers in addition to other members and voting delegates. The various venues through which the ASA works include the U.S. Congress, the lobbying groups, the administration, and the media.

6.3.1.3.6 National Corn Growers Association

The National Corn Growers Association (NCGA) (http://www.ncga.com/home) is a U.S.-based association. As stated in its mission, the NCGA aims to create and increase opportunities for corn growers. The NCGA anticipates that by the year 2020, they will be able to "maximize opportunities for the corn industry to meet growing domestic and global market demand while increasing corn farmers' environmental and economic sustainability." The association advocates for policies that help risk management, encourage science-based approach, create and sustain a consumer positive image of agricultural production, and build further opportunities for production and utilization of corn.

6.3.1.3.7 European Association for Bioindustries

The European Association for Bioindustries (EuropaBio) (http://www.europabio.org/) is the largest and most influential biotechnology industry group in Europe. It was created in 1996 with the aims of representing the interest of the biotechnology industry in Europe, creating an innovative and sustainable biotech-based industry, and influencing legislation to serve biotechnology industries in Europe. The association's membership comprises 55 corporate members, 15 associate members and bio regions, and 17 national biotechnology associations. Monsanto and Bayer are two of the association's members. Some of the stated association's activities include human and animal health care, diagnostics, bioinformatics, chemicals, crop protection, agriculture, food, and environmental products and services. EuropaBio is engaged in increasing the awareness about the benefits of biotechnology at all levels of the society. This is expected to create a positive image about biotechnology and its applications, which would help the biotechnology industry to flourish.

6.3.1.3.8 AfricaBio

AfricaBio (http://africabio.com/) is an independent, nonprofit biotechnology stakeholders' association established in South Africa, with the aim of creating awareness about biotechnology and its applications and safety. This could be achieved through dissemination of accurate information and knowledge within South Africa and the whole African continent. AfricaBio is targeting local, national, regional, and international levels. It applies different means of knowledge dissemination such as workshops, exhibitions, websites, technology demonstration, and training. It also helps in capacity building in various aspects of biotechnology and in providing services and support to biotechnology stakeholders in the African region.

6.3.1.3.9 AusBiotech Ltd

AusBiotech Ltd (http://www.ausbiotech.org/) is a not-for-profit limited guarantee Australian company that provides representation and services to promote the global growth of Australian biotechnology. It has a wide network among different members of the life sciences, which includes agriculture, food technology, environment, therapeutics, medical technology, and the industrial sectors. The strategic plan of the company builds on the areas of sustainability and growth, outreach and access, and representation and support. Its stated mission is "to cultivate a supportive environment to enable companies to grow and advance their commercial interests and assist them to become global, thereby positioning Australia's biotechnology industry as a significant market for attracting international interest and investment."

6.3.1.3.10 Agricultural Biotechnology Council
The Agricultural Biotechnology Council (ABC) (http://www.abcinformation.org/) comprises six major global biotechnology companies, namely, BASF, Bayer, Dow AgroSciences, Monsanto, Pioneer (DuPont), and Syngenta. The main goal of the council is to "provide factual information and education about the agricultural use of GM technology in the UK, based on respect for public interest, opinions and concerns." ABC deals with different stages of the food chain in relation to the use of biotechnology. It also supports agricultural research applying both traditional and modern biotechnologies, including genetic modification. ABC is a member of EuropaBio and Supply Chain Initiative on Modified Agricultural Crops—http://www.scimac.org.uk/index.html.

6.3.1.3.11 CropGen
CropGen (http://cropgen.org/) is a consumer and media initiative, the mission of which is "to make the case for GM crops and foods by helping to achieve a greater measure of realism and better balance in the UK's public discussions on agriculture and food." CropGen believes that the actual and potential benefits offered by crop biotechnology are not clearly articulated or even intentionally marginalized during public debates on genetic modification. In order to address this issue, CropGen convenes and participates in various media outlets to clarify different issues surrounding GM production and utilization in agriculture and food.

6.3.1.3.12 Crop Protection Association
The Crop Protection Association (CPA) (http://www.cropprotection.org.uk/) is a U.K.-based body acting as the main voice of plant science industry in the United Kingdom. The CPA believes in, and encourages, the positive roles played by modern plant science throughout the food supply chain. All the association team members are actively involved in developing appropriate plant technologies targeting food crops, gardens, woodlands,

infrastructure, and public places. Specific areas covered include biopesticides, seed and plant breeding, agricultural biotechnologies, and bee keeping. The CPA is committed to clarifying the roles and benefits of crop protection industry through involvement in dialogs and debates at the community level. The motto designed by the CPA is "Safeguarding our food supply and quality of life."

6.3.1.3.13 Sense About Science
Sense About Science (SAS) (http://www.senseaboutscience.org/) is a British charity body that works in collaboration with scientists, journalists, and other stakeholders with interest in science, with the aim of promoting the general public understanding of science. It also tries to clarify and correct any unscientific misinformation about new scientific and technological developments encountered in the public or private media or during relevant debates. Issues of public concern are discussed by experts in science and technology in an organized and understandable manner for the benefit of the general public. Open debates on controversial scientific issues are encouraged by SAS.

6.3.1.3.14 The Grocery Manufacturers Association
The Grocery Manufacturers Association (http://www.gmaonline.org) is a trade association of the food industry, representing more than 300 of the largest food, beverage, and other consumer products. It was founded in Washington, DC, USA in 1908, and it helps in disseminating information about the industry affairs through news, publications, and other media resources.

6.3.1.3.15 Food and Drink Federation
The Food and Drink Federation (FDF) (http://www.fdf.org.uk/keyissues.aspx?issue=494) is the voice of the U.K. food and drink industry, the largest manufacturing sector in the country. The FDF believes that modern technologies, including GM, offer considerable potential to improve the quality and quantity of the food supply and could contribute to sustainability by helping to produce more food using fewer resources and with less impact on the environment. The FDF recognizes that the impact of this technology must be objectively assessed through scientific investigation and that robust controls are necessary to protect the consumer and the environment. Consumer understanding through education and information is fundamental to public acceptance of GM in food production.

6.3.1.3.16 The Alliance for Better Foods

The Alliance for Better Foods (http://betterfoods.org/) supports biotechnology as a safe way to provide more abundant, nutritious, and higher quality food supply. It encourages fact-based discussion about development in plant biotechnology. The Alliance membership represents diverse agricultural and food-related groups, which support the use of bioengineered foods and oppose the labeling of GM products. Its membership includes farmers, processors, distributors, retailers, scientists, food technologists, and professionals in other fields dedicated to improving nutrition, protecting the environment, and fighting world hunger. Some of the members are Agricultural Retailers Association, ASA, American Feed Industry Association, American Frozen Food Institute, U.S. Chamber of Commerce, International Dairy Foods Association, International Food Additives Council, and International Pharmaceutical Excipients.

6.3.1.3.17 Other Pro-GM Lobby Groups

In addition to the groups described earlier, the following bodies support biotechnology and genetic modification:

- Institute of Ideas (http://www.instituteofideas.com/index.html)
- Science Media Centre (http://www.sciencemediacentre.org/)
- Scientific Alliance (http://scientific-alliance.org/)
- International Policy Network

6.3.1.4 Global Network of Pro-Corporate Activists

A global network of very vocal pro-corporate activists forms various lobby groups advocating for GM. Their activities are visible through different forums around the world. Some of these groups as listed by Rees (2006) include the following:

- Right-wing think tanks, research bodies, and PR companies
- U.K. anti-environmentalists
- Anti-environmental authors
- Prakash, AgBioView, and AgBioWorld

6.3.2 Arguments in Favor of GM Foods

Supporters of GM technology and its applications believe that there is great potential for GM crops and products and GM foods to give many benefits for farmers, food producers, and manufacturers as well as for consumers. The following arguments are presented by proponents of GM technology and GM foods in support of their case.

6.3.2.1 Benefits for Farmers

Farmers grow various crops including food crops such as cereal crops, legume crops, and oilseed crops. Some farmers also rear animals for milk or meat. Poultry production is another activity that some farmers are involved in. Biotechnology is believed to benefit all farmers whether they are plant or animal producers. Because food mainly comes from agricultural produce, production of GM foods subsequently has many benefits to farmers. The following arguments have been presented in support of food production through the use of GE:

- In general, GM will increase income for farmers. As a result of adopting various biotechnological techniques, farmers have the chance to get more income, which can be used to improve the life and well-being of their families, for example, spending on education of their children or the health of all family members. Increase in farmers' income comes from the following:
   - *Higher crop yields.* One of the aspects achieved through GM is the increase in the quantity of crop yield either per individual plant or per area cropped. This increase in crop yield means that the farmer is getting more produce from the same resources used before using GM. Increase in yield also applies to products obtained from animals genetically modified in one way or the other. Biotechnology has already been shown to increase yields by reducing crop loss to pests through the use of herbicide-tolerant and insect-protected crops (Godfray et al. 2010).
   - *Improved quality attributes of plants and animals.* Desired quality characteristics of plants or animals can be achieved through GM. Examples of improved quality attributes include nutritional, storage, prolonged freshness, and better processing qualities.
- Improved characteristics of plants and animals depend on factors such as improved pest and disease resistance, selective herbicide and insecticide tolerance, ability to withstand weather fluctuations and extreme weather conditions, tolerance of water, and temperature and saline extreme conditions.
- *Better resistance of crops to stress.* The possibility of crop failure due to biotic and nonbiotic factors would be reduced through the

use of GM crops. This comes as a result of developing GM crops which are pest-resistant and which have tolerance to extreme weather conditions such as frost, drought, and high temperatures.
- Less herbicide and pesticide need to be applied by the farmer, which consequently means that the farming expense goes down and revenue increases. Herbicide-tolerant crops allow farmers to control weeds better, which allows crops to thrive (Brookes and Barfoot 2010). Through planting insect-protected crops, farmers will be able to get healthier and damage-free crops (Park et al. 2011). Farmers would be more protected from possible poisoning resulting from insecticide spraying on crops. This is due to the fact that they can spray insecticide less often on *Bacillus thuringiensis* crops, which are developed through GE (Pray et al. 2002; Bill and Melinda Gates Foundation 2013).
- Enabling farmers to market their farm produce in other countries where GM is acceptable.
- *Improved animal welfare*. GM animals are expected to be healthier and have the ability to withstand some diseases or pests if they are modified in a way to improve these characteristics.
- *More productive farm animals*. Farm animals such as cattle, sheep, and poultry can be genetically engineered to increase their products, for example, milk, meat, and eggs.

6.3.2.2 Benefits for Consumers

Consumers are the ultimate and targeted users of the products of GM, including GM foods. It is therefore imperative that clear and convincing details of benefits of GM foods be presented so that consumers have a full picture and decide for themselves whether or not to consume GM foods.

Supporters of GM foods production and consumption state that modern biotechnology provides benefits to consumers at present as well as in the future. The use of biotechnology to produce GM foods is expected to help in reducing hunger and to improve the global food security. Currently, the estimated number of undernourished individuals worldwide is more than one billion. This number is foreseen to be halved by the year 2015, as depicted by the first Millennium Development Goal. The present trend of food availability shows that this is not going to be achieved. On the contrary, both the total number and the proportion of undernourished have increased (Qaim 2011). The application of agricultural biotechnology, in general, and of GM crops in particular, has been presented as one of the best interventions that can help addressing the world hunger and food insecurity problems. It can increase production and at the same time reduce its cost. The food supply would be increased through GM, consequently leading to increased food availability at reduced cost, resulting in improved accessibility by the food insecure. This can be achieved without further damage to the environment, with noticeable decrease in nonrenewable inputs such as fertilizers and pesticides. GM crops are able to produce yields higher than did normal crops. One example of such crops is the "Super Rice" produced under the project "Green Super Rice for the Resources-Poor of Africa and Asia." This project is a Bill and Melinda Gates Foundation-funded rice research project with the aim of developing "Green Super Rice (GSR) varieties that produce high and stable yields under low inputs" and transferring corresponding crop management technology to the resource-poor rice farmers in 15 countries across Africa and four in Asia, including China. The main mission of this project is to help achieve self-sustainable rice production and food security in the target countries and regions (http://www.thegsr.org). Rice production in sub-Saharan Africa and Asia is constantly under the pressure of various kinds of stress—drought, limited nutrient input, poor soils, and pest infections. Recent scientific advances achieved by Chinese rice scientists have accelerated the development of new varieties that can withstand drought, flooding, cold weather, and toxic minerals such as salt and high iron. The GSR project aims at breeding at least 15 elite cultivars suitable for growth in the target countries in Africa and Asia, through using biological technologies and devising new ways of crop management. In addition to that, the GSR project includes building a highly efficient genotyping platform for large-scale molecular breeding activities in the target countries and for the international rice research community.

The food quality can also be improved through GM to meet consumers' needs and preferences. Different quality attributes and parameters, for example, nutritional, sensory, and storage characteristics, can be improved through GM. Production of Golden Rice is one example of GM crops modified to improve nutritional value of foods. Golden Rice is a GM type of rice engineered to contain beta-carotene and to supply vitamin A in the diet. Rice, which is the staple food in many developing countries, is naturally deficient in vitamin A. The lack of vitamin A in rice led to the death of more than one million children and to 350,000 going blind in the past. Consumption of Golden Rice used as one intervention helped to address this malnutrition problem and led to the reduction of malnutrition cases.

Food crops with improved nutritional profiles can be genetically engineered. The natural level of micro- and macronutrients can be increased through GM to address certain nutritional problems or to target particular groups in the society with special nutritional needs. Examples of these crops include wheat with increased levels of folic acid to prevent spina bifida and with increased content of fiber to reduce the risk of colon cancer, tomatoes with higher vitamin content, and peanuts with nonallergenic factors. GM can also be used to modify or improve the processing qualities of food products, for example, production of potatoes with higher starch content which will reduce the amount of oil absorbed during cooking or frying. Shelf life of food crops can also be improved through GM, for example, modifying tomatoes that can ripen on the vine to achieve better and desirable taste and having a longer shelf life. Sensory properties, for example, flavor, texture, and color, can be modified to meet consumers' preferences and tastes. GM has also been used to modify the functional properties of proteins to be used in the manufacture of ice cream to prevent ice crystal formation and to maintain its smooth texture preferred by consumers. The level of naturally occurring harmful and toxic substances present in some food crops can be reduced or eliminated through GM, rendering the foods produced from these crops safer for human consumption.

More nutritious and healthy foods have been produced through GE. Examples of such foods include canola, soybean, and sunflower oils, which do not produce trans fats (Tarrago-Trani et al. 2006; Damude and Kinney 2008; DiRienzo et al. 2008; Mermelstein 2010; Crawford et al. 2011) and which contain higher levels of omega-3 fatty acids (Lichtenstein et al. 2006; Damude and Kinney 2008; DiRienzo et al. 2008; Mermelstein 2010). The levels of omega-3 fats in the meat of cows and pigs have also been improved through GE (Lai et al. 2006; Wu et al. 2012).

GM foods, which are currently available in the market and which have been consumed by people for a number of decades, are reported to be safe for human consumption, with no evidence of harm shown anywhere in the world. This has been documented in, and supported by, a number of studies conducted over the past three decades (WHO 2005; Massengale 2010; American Medical Association 2012; Center for Science in the Public Interest 2012; EPA 2012; FDA 2012; USDA, APHIS 2012). GM foods are reported to be safe for consumption by children and pregnant and nursing women. For those with food allergies, the use of biotechnology itself will not increase the potential for a food to cause an allergic reaction or a new food allergy (International Food Information Council Foundation 2013).

International specialized and professional organizations, such as the WHO, the FAO, the British Medical Association (BMA), and the American

Medical Association, have evaluated evidence regarding the safety and benefits of food biotechnology, and they each support the responsible use of biotechnology for its current and future positive impacts on addressing food insecurity, malnutrition, and sustainability (International Food Information Council Foundation 2013). Results of investigations carried out by the FDA (2008, 2010) indicate that animal products such as meat, eggs, and milk, which are obtained from cloned animals, are the same as those produced from other animals. It is also reported that animal feed containing biotech crops has the same nutritional value as feed obtained from conventionally grown crops (FDA 2012). The BMA believes that there is no substantial evidence to prove that GM foods are unsafe. However, the organization considers that with adequate risk assessment procedures, independent and rigorous testing of novel foods, adequate post marketing surveillance, and proper regulation, GM foods have potential benefits for both the developed and the developing regions of the world in the long term (CropBiotech.net 2004).

GM can be used to improve the safety of some foods, which contain natural toxins or harmful substances or allergens. Some cited examples include the development of potatoes that produce less acrylamide on heating or cooking (Rommens et al. 2008) and production of low-lactose milk using biotechnology-derived enzymes (International Food Information Council Foundation 2011). In the future, it might be possible for scientists to get rid of proteins that lead to allergic reactions to foods such as soy, milk, and peanuts, making the food supply safer for allergic individuals (Lehrer and Bannon 2005; Newell-McGloughlin 2008).

Consumers can benefit from the availability and affordability of sustained foods resulting from a more reliable crop harvest, which is provided by the improved crop disease protection through biotechnology (Gianessi et al. 2003; Brookes 2008; Giddings and Chassy 2009; Brookes and Barfoot 2010; Conservation Technology Information Center 2010).

6.3.2.3 Benefits for the Environment

The GM supporters argue that agricultural biotechnology used in the production of GM foods has a number of positive effects and benefits on the environment when compared with traditional methods of food production. The environmental sustainability of agriculture can be enhanced through the use of modern biotechnology. This may be achieved through improving safe and more effective applications of pesticides, reducing the level of insecticides used on crops, lowering the amount of greenhouse gas emissions, maintaining and improving soil quality, and reducing pre- and postharvest crop losses (Fawcett and Towery 2002; Council for Agricultural Science and Technology 2009; Brookes and Barfoot 2010; Conservation Technology Information Center 2010; National Research Council 2010; Park et al. 2011; Osteen et al. 2012).

The development of more environmentally friendly herbicides and pesticides has been achieved through the application of modern biotechnological techniques (Osteen et al. 2012). The adoption and use of herbicide-tolerant crops, which have been developed through GE, proved to have positive effects both for farmers and on the environment. Farmers can now have more choices in sustainable weed management by using herbicides that disintegrate more rapidly, consequently imposing less negative impact on the environment when compared with older herbicides (Brookes and Barfoot 2010). The development of pest-resistant crops through biotechnology resulted in more precise and more targeted pest control, consequently leading to better crop protection. *B. thuringiensis* crops developed to target only the insects that consume those crops rather than useful insects, for example, honey bees or natural predators of crop pests, proved to be good for the ecosystem (National Research Council 2010).

The following arguments are presented by the supporters of GM techniques as used in agriculture for the production of GM foods:

- Production of GM foods helps to reduce environmental pollution. Traditional methods of agricultural production rely on the use of various chemicals, for example, pesticides and insecticides to control pests and diseases and to protect crops. Farmers have been using these chemicals for centuries with negative effects on the environment. GM reduces the dependence on chemicals through the development of crops, which tend to be more pest-, insect-, and disease-resistant than traditional crops, as have been done with a number of food crops, for example, corn, soybean, and potatoes. The National Center of Food and Agricultural Policy (NCFAP 2013) reports that the use of pesticide application by U.S. GMO farmers has been reduced by 46 million pounds annually since 1996. Reduction of pesticide application also lowers the risk for possible soil or water contamination, which has a positive effect on the environment (BenefitOf.net 2013). The FAO has also stated that GMOs might reduce the environmental impact of food production and industrial processes through reducing the amount of chemicals needed for crop protection (FAO 2003). In contrast, bioherbicides and bioinsecticides have been produced

through GE techniques. These products of GM, which may be applied instead of the traditional chemicals, are considered environmentally friendly.
- *Reduction of soil erosion.* Growing GM crops has been found to decrease soil erosion, which is considered as a positive factor for protection of the environment. The Council for Agricultural Science and Technology in Ames, Iowa, USA, has produced a report which states that planting biotech soy, corn, and cotton has decreased soil erosion by 90%, preserving 37 million tons of topsoil. It also states that biotech crops provide a 70% reduction of herbicide runoff and an 85% reduction in greenhouse gas emissions (Commonground 2014).
- *Reduction of manual labor.* Farmers planting GM crops are expected to spend less time and use less resources and efforts in controlling weeds and pests and in plowing. Less plowing means less soil erosion, runoff, machinery fuel use, and greenhouse gas emissions. All these effects help to conserve the environment through conserving soil and energy.
- *Water conservation.* GM crops can be modified in such a way that they would require less amounts of water to grow, that is, to become drought-tolerant. This means that more water will be conserved and available for use in other crops or in alternative applications; consequently, GM will conserve water, which is a scarce component of the environment.
- *More food from less land.* Agricultural land used for growing food crops is becoming scarce due to the mounting needs for food by the increasing world population. Some farmers have traditionally resolved to use marginal land which is less fertile and less productive. GM of crops has made it possible to produce higher yields from the agricultural land suitable for cropping; consequently, farmers would no longer need to bring marginal land into cultivation, as depicted by the FAO (2003).
- *Rehabilitation of damaged or less fertile land.* The FAO reports indicate that large areas of crop land in developing countries have become saline as a result of unsustainable irrigation practices followed by farmers in these countries. Through GM, salt-tolerant and drought-resistant crop varieties and trees could be produced. It is also possible that GM could be used to select or breed crops and trees suitable for rehabilitation of degraded agricultural land.

- *Bioremediation*. Intensive cropping of agricultural land can lead to damage in the soil structure and/or fertility. GM has made it possible to rehabilitate damaged land through modifying and breeding organisms capable of restoring plant nutrients in the soil as well as the soil structure.
- *Less deforestation*. Higher yields have been obtained from GM food crops. This would help to close the gap in food deficit worldwide and to feed more populations. This is expected to result in less deforestation required for feeding the world's growing population. Less deforestation implies conservation of the environment.

6.3.2.4 Benefits for the Economy

Proponents of GM technology and GM food production argue that this technology has some economic benefits. An accepted fact is that growing GM crops is costly at the beginning, but is cheaper and more economically rewarding in the long run. Initially, extra inputs in the form of research, laboratory testing, field trials, and related inputs need to be used. This comes with additional costs, costs which are higher than those needed to grow non-GM crops. As a result of these costs associated with taking this agricultural technology to the market, companies which are adopting GM technology have initially focused on developing crops which have quick economic returns and benefits. Average agricultural production costs were reduced due to reduced use of herbicides and insecticides. There was an increase in revenues resulting from improved and higher crop yields, in addition to the improvement in produce quality. Growing GM crops is also regarded as more economically friendly because pesticides do not mix with or contaminate air, soil, or water. Production hazards of these chemicals to the environment also diminish, implying improved economic returns.

Economic benefits of growing GM crops are not only confined to developed countries, but also to developing regions of the world. Bennett et al. (2006) report that results of a large-scale survey of resource-poor farmers in South Africa show that adopters of *B. thuringiensis* cotton have benefited in terms of higher yields, lower pesticide use, less labor for pesticide application, and higher gross margins per hectare. They concluded that these benefits observed were clearly related to the GM technology used and not to preferential adoption by farmers. Their results also indicate that the smallest producers benefited from adoption of the technology as much as, if not more than, larger producers. An additional positive effect cited is that hospital records suggested a link between declining pesticide poisonings and adoption of the *B. thuringiensis* variety.

GENETICALLY MODIFIED FOODS

Reports also show that planting GM sweet potatoes has been predicted to raise farmers' income by up to 30% for virus-resistant potatoes and up to 40% for weevil-resistant potatoes (Luttrell 2014). Farmers are able to save money on both the cost of pesticides and the labor required to apply them. Further increase in farmers' income could be derived from higher crop output.

6.3.2.5 Other Points in Favor of GM Foods

Before the advent of GM, foods have been traditionally produced through conventional methods of plant and animal production. Conventional methods of breeding are considered slow when compared with GM methods. It takes a number of years to come up with a new or modified plant or animal product through conventional production processes, whereas the desired GM organisms can be bred in one generation. In that sense, it would be faster and hence more economical to come up with GM foods. In contrast, conventional breeding techniques involve unpredictable processes which might lead to unpredictable results. GM has precise, faster, and more efficient processes that lead to desired outcomes. GM supporters believe that this new technology is merely an extension of the processes which have been applied for centuries to mix and match genes through natural evolution, selection, and conventional crossbreeding.

GM foods are considered safe for human consumption. This has been verified by various tests conducted for them and which are legally required to have them approved as safe. The safety of GM foods currently in the market has been assessed, and the results show that no health problems or hazards have been identified or documented. The basic principle outlining approval of foods as safe is that "foods are considered safe until proven unsafe." Proponents of GM foods indicate that there is yet to be any proof that GM foods are inherently more dangerous than foods produced through conventional methods. There are no recorded proven examples of GM products or foods with additional risks. On the contrary, it is believed that the GM technology can reduce risks that may be encountered.

The conclusions drawn from the work of many scientists on the safety of GM foods and the broad scientific consensus (e.g., FAO 2004; European Commission 2010; Ronald 2011; American Association for the Advancement of Science (AAAS), Board of Directors 2012; American Medical Association 2012) give enough evidence that the currently available transgenic crops and foods derived from them pose no greater risks to human health than foods developed through conventional methods. GM foods have been judged safe to eat, and the methods used to test their safety have been deemed appropriate (FAO 2004). GM crops are also reported to provide a number of ecological benefits (United Nations Industrial Development Organization [UNIDO] 2007). To date, no verifiable untoward toxic or nutritionally deleterious effects resulting from the consumption of foods derived from GM crops have been discovered anywhere in the world (FAO 2004).

Supporters of the GM technology and GM food production accept the fact that this modern technology, similar to any other technology used to produce foods, has some problems and possible drawbacks. They believe that it is important to weigh the benefits against the problems and to make the final judgment based on which outweigh the other. They believe that the benefits of the GM technology used to produce GM foods by far outweigh the problems.

6.4 OPPONENTS

6.4.1 Introduction

Arguments against GM technologies in general and their applications in agriculture and food production come from many venues. These include all the institutions and organizations referred to in Section 6.4.2, activist groups, and individual writers and authors who publish books or articles condemning GM technology and its applications. Their arguments mainly revolve around two main issues: human and animal safety and threats to health and the impact of GM technology on the environment and the ecosystem. Rallies against applications of GM technology and production of GM foods also reflect part of the stance against GMOs and GM foods. Opponents of GM technology and GM foods have presented large volumes of scientific evidence that GM technology has unpredictable results and can lead to unintended side effects on human health and on the environment. A number of publications, including books and articles, with strong arguments against GM technology and GM foods have been published in recent years. A sample of these books and reports includes Batalion (2000), Engdahl (2007), Fagan et al. (2014), Ho (2007), Navdanya et al. (2011), Rees (2006), and Shiva et al. (2011).

Rees (2006) strongly criticizes GE and GM foods and warns about the potential risk on health, the environment, and agronomic calamities. He believes that a handful of corporations want to state their voracious appetite for profit by patenting the seeds of food people eat, thus controlling the global food chain. Big business has control of politicians, legislators, scientists, journalists, government advisory committees, international agencies, science-funding bodies, and scientific journals, states Rees. In an effort to counter the alleged "campaign of misinformation," Rees tries to expose the so-called "wild claims made by the biotech lobby" through listing 14 statements, referred to as "lies."

Lanphier (2014) believes that what is known about GMOs is very little compared with what is unknown and that governments use consumers as guinea pigs for testing GM foods. She also blames the government in failing to require reasonable testing, regulating, and labeling of GMO foods, something that might cause negative health consequences.

The authors of the report "GMO myths and truths" (Fagan et al. 2014) believe that GM crops and foods are neither safe nor necessary to feed the world. The three authors, who are genetic engineers, discuss the different issues related to GM crops and GM foods and try to differentiate between myths and truths surrounding GM foods. They state that the GMO debate is far from over and that the evidence of risk and actual harm from GM foods and crops to health and the environment has grown in the last few years. This report represents a very strong opposition, criticism, and condemnation on GM crops and GM foods.

Engdahl (2009) reports that the American Academy of Environmental Medicine (AAEM), which is an international association of physicians and other professionals dedicated to addressing the clinical aspects of environmental health, has issued a call for an immediate moratorium on genetically manipulated foods. The AAEM states that "GM foods pose a serious health risk in the areas of toxicology, allergy and immune function, reproductive health, and metabolic, physiologic, and genetic health."

6.4.2 Who Are the Opponents?

6.4.2.1 United States
- *Californians for GE-Free Agriculture.* Californians for GE-Free Agriculture (http://www.farmplate.com/local-food/environmental-organization/californians-ge-free-agriculture-occidental-ca/) is a state-wide coalition of farmers and environmental and consumer groups committed to ecologically responsible and economically viable agriculture.
- *Campaign to Label GE Food.* A campaign (http://www.lightparty.com/Health/NoToGMO.html) that calls for the mandatory labeling of GE foods.

- *Center for Food Safety (CFS).* The CFS (http://www.centerforfoodsafety.org/) is a U.S. nonprofit organization, based in Washington, DC, which also maintains an office in San Francisco, California. The center is led by the attorney Andrew Kimbrell, and its listed goal is to protect human health and the environment, focusing on artificial food production technologies such as GM plants and organisms, which according to the CFS have been scientifically proven to be harmful to the environment and to human health. CFS was established by the International Center for Technology Assessment in 1997 and works to achieve its goals through grassroots campaigns, public education, media outreach, and litigation.
- *Council for Responsible Genetics (CRG).* CRG (http://www.councilforresponsiblegenetics.org/) is a nonprofit NGO with a focus on biotechnology. It was founded in 1983 in Cambridge, Massachusetts. An early voice concerned about the social and ethical implications of modern genetic technologies, CRG organized a 1985 Congressional Briefing and a 1986 panel of the American Association for the Advancement of Science, both focusing on the potential dangers of GE biological weapons. Francis Boyle was asked to draft legislation setting limits on the use of genetic engineering, leading to the Biological Weapons Anti-Terrorism Act of 1989.
- *Earth Liberation Front (ELF).* The ELF (http://www.earth-liberation-front.com/), also known as "Elves" or "The Elves," is the collective name for autonomous individuals or covert cells who, according to the ELF Press Office, use "economic sabotage and guerrilla warfare to stop the exploitation and destruction of the environment."
- *Farmer to Farmer Campaign on Genetic Engineering.* The Farmer to Farmer Campaign on Genetic Engineering (Farmer to Farmer) (http://www.nffc.net/Issues/Farmer%20to%20Farmer/page-farmertofarmer.htm) is a network of 34 farm organizations from throughout the United States that endorsed the "Farmer Declaration on Genetic Engineering" released in December 1999. Farmer to Farmer seeks to build a farmer-driven campaign focused on the risks of genetic engineering in agriculture and to provide national forum for farmer action on agricultural biotechnology issues.
- *Friends of the Earth (FoE).* FoE U.S. (http://www.foe.org/) is a part of FoE International, the world's largest grassroots environmental

network. Current campaigns focus on clean energy and solutions to global warming, protecting people from toxic and new, potentially harmful technologies, and promoting smarter, low-pollution transportation alternatives. It is the largest international network of environmental groups and organizations in the world, represented in 74 countries. FoE also focuses on human rights.
- *GE Food Alert.* It (http://urlm.co/www.gefoodalert.org) supplies source of information on GE foods.
- *Genetic Engineering Action Network (GEAN).* GEAN (http://www.geaction.org/) is a network of almost 100 organizations from across the United States working to resist genetic engineering in agriculture.
- *Greenpeace.* Greenpeace (http://www.greenpeace.org/usa/en/) is a nongovernmental environmental organization with offices in over 40 countries and with an international coordinating body in Amsterdam, the Netherlands. Greenpeace states its goal as "to ensure the ability of the Earth to nurture life in all its diversity" and focuses its campaigning on worldwide issues such as global warming, deforestation, overfishing, commercial whaling, genetic engineering, and anti-nuclear issues. "No genetic engineering of nature" is the chosen credo for their campaign. It uses direct action, lobbying, and research to achieve its goals.
- *Food First (Institute for Food and Development Policy).* Food First (http://foodfirst.org/), also known as the Institute for Food and Development Policy, is a nonprofit organization based in Oakland, California, USA. Founded in 1975 by Frances Moore Lappe and Joseph Collins, it describes itself as a "people's think tank and education-for-action center." Its mission is "to eliminate the injustices that cause hunger." According to the Food First website, its main goal is to forge food sovereignty for human rights and sustainable livelihoods, and to do so, it has three programs of development: building local agri-foods systems, farmers forming food sovereignty, and democratizing development.
- *Institute for Responsible Technologies.* The Institute for Responsible Technology (http://www.responsibletechnology.org/) is a world leader in educating policy-makers and the public about GM foods and crops. It investigates and reports the risks and impacts of GM crops and foods on health, environment, the economy, and agriculture, as well as the problems associated with current research, regulation, corporate practices, and reporting.

- *Keep Nature Natural.* It (http://deliciousliving.com/keep-nature-natural-campaign-takes-gmos) is a campaign for labeling and establishment of safety standards for genetically altered foods.
- *Northeast Resistance Against Genetic Engineering* (http://www.upstartsorganicseedlings.com/genetic_page_9.htm).
- *Northwest Resistance Against Genetic Engineering* (*NW RAGE*). NW RAGE (http://www.nwrage.org/content/about-us) is a nonviolent, grassroots organization dedicated to promoting the responsible, sustainable, and just use of agriculture and science. It is working toward a ban on GE and patents on life with a focus on education, community building, advocacy, and action. NW RAGE works with neighbors, teachers, farmers, and friends to stop the reckless splicing and dicing of the world's genetic material. It claims that it is not anti-science, but a pro-precautionary principle. NW RAGE believes in a future that strives to do no harm to humans and to the ecosystems we all depend on. It believes in proceeding with new technologies only if a well-informed civil society decides to do so after proper testing has been conducted.
- *Organic Consumers Association* (*OCA*). The OCA (http://organicconsumers.org/) is an online and grassroots nonprofit 501(c)3 public interest organization campaigning for health, justice, and sustainability. The OCA deals with crucial issues of food safety, industrial agriculture, genetic engineering, children's health, corporate accountability, fair trade, environmental sustainability, and other key topics.
- *Rural Advancement Foundation International* (*RAFI*). The RAFI (http://rafiusa.org/) cultivates markets, policies, and communities that support thriving, socially just, and environmentally sound family farms.
- *Sierra Club.* The Sierra Club (http://www.sierraclub.org/) is one of the oldest, largest, and most influential grassroots environmental organizations in the United States. It was founded on May 28, 1892, in San Francisco, California, by the Scottish-born American conservationist and preservationist John Muir, who became its first president.
- *True Food Network* (http://www.truefood.org.au/).
- *The Union of Concerned Scientists.* The mission of the organization (http://www.ucsusa.org/) in the words of Jane Rissler, Senior Staff Scientist is as follows: "We are campaigning for food policies that will protect people from potentially untreatable and

life-threatening diseases while also promoting sustainable animal agriculture."
- *The AAEM.* The AAEM (http://www.aaemonline.org/) was founded in 1965 and is an international association of physicians and other professionals interested in the clinical aspects of humans and their environment. The AAEM is interested in expanding the knowledge of interactions between human individuals and their environment, as these may be demonstrated to be reflected in their total health. The AAEM provides research and education in the recognition, treatment, and prevention of illnesses induced by exposures to biological and chemical agents encountered in air, food, and water.

6.4.2.2 Europe
- ActionAid (http://www.actionaid.org/).
- Blueridge Institute, Switzerland (http://www.blueridgeinstitute.net/).
- Christian Aid, UK (http://www.christianaid.org.uk/). Christian Aid has a vision—an end to poverty—and its members believe that vision can become a reality.
- Confederation Paysanne, Via Campesina, France (http://www.eurovia.org/spip.php?article1000&lang=en).
- Five Year Freeze (http://www.fiveyearfreeze.org/).
- Foundation for Organic and Biodynamic Research (FIRAB), Italy (http://www.firab.it/site/).
- Friends of the Earth, Europe (http://www.foeeurope.org/).
- Friends of the Earth, UK (http://www.foe.org.uk/).
- Genetics Forum (http://www.topix.com/forum/science/genetics).
- GeneWatch (http://www.genewatch.org/).
- Greenpeace, EU (http://www.greenpeace.org/eu-unit/en/).
- Greenpeace, UK (http://www.greenpeace.org.uk/).
- HDRA, UK (http://www.gardenorganic.org.uk/).
- Institute of Science in Society (ISIS) (http://www.i-sis.org/). ISIS is an interest group that campaigns against what it sees as unethical uses of biotechnology. The group published about climate change, GMOs, homeopathy, traditional Chinese medicine, and water memory.
- International Coalition to Protect the Polish Countryside, Poland (http://icppc.pl/pl/gmo/eng_index.php).

- Italian Association for Organic Agriculture (AIAB), Italy (http://www.aiab.it/).
- National Trust, UK (http://www.nationaltrust.org.uk/).
- Navdanya International (http://www.navdanya.org/).
- NGIN/GM Watch (http://ngin.tripod.com/).
- Oxfam (http://www.oxfam.org/).
- Pesticide Action Network (http://www.panna.org/).
- Save our Seeds (SOS), Germany (http://www.saveourseeds.org/en.html).
- Scientists for Global Responsibility, UK (http://www.sgr.org.uk/).
- Soil Association, UK (http://www.soilassociation.org/). The Soil Association is a charity based in the United Kingdom. Founded in 1946, it has over 27,000 members today. Its activities include campaign work on issues including opposition to intensive farming, support for local purchasing, public education on nutrition, and the certification of organic food. It developed the world's first organic certification system in 1967—standards which have since widened to encompass agriculture, aquaculture, ethical trade, food processing, forestry, health and beauty, horticulture, and textiles. Today, it certifies over 80% of organic produce in the United Kingdom.
- The Green Alliance, UK (http://www.greenalliance.org.uk/).
- The Green Party, UK (http://www.greenparty.org.uk/).
- VAS—Verdi Amviente e Societa Onlus—Italia Green Environment and Society Italy (http://vasonlus.it/).
- Vegetarian Society, UK (https://www.vegsoc.org/). The Vegetarian Society is a British registered charity established on September 30, 1847 to "support, represent, and increase the number of vegetarians in the UK." In the nineteenth century, a number of groups in Britain actively promoted and followed meat-free diets. Key groups involved in the formation of the Vegetarian Society were members of the Bible Christian Church, supporters of the Concordium, and readers of the Truth-Tester journal.
- Wildlife Trust (http://www.wildlifetrusts.org/).
- Women's Environmental Network (http://www.wencal.org/).
- Women's Institute (http://www.womens-institute.co.uk/campaigns/gmfoods-c.shtml).
- World Development Movement (WDM), UK (http://www.wdm.org.uk/). The WDM is a membership organization in the United Kingdom which campaigns on issues of global justice and

development in the Global South. The key to WDM's mission is to promote democratic alternatives, enliven public debate, and attract more members of the public to issues.
- World Peace Culture Fund Genetic Resources Conservation, Russia.

6.4.2.3 Canada, Australia, and New Zealand
- *Earthsave Canada.* Earthsave Canada (http://www.earthsave.ca/) is a registered charity helping people choose foods that benefit our health, the environment, and the lives of all animals. It is a Vancouver-based registered charity since 1992. It advocates the move toward a more whole-food, plant-based diet for environmental sustainability, better health, and compassion toward animals.
- *Greenpeace Australia* (http://www.greenpeace.org/australia/en/).
- *Greenpeace New Zealand* (http://www.greenpeace.org/new-zealand/en/).

6.4.2.4 Africa
- *African Biodiversity Network (ABN).* ABN (http://africanbiodiversity.org/) is a regional network of individuals and organizations seeking African solutions to the ecological and socioeconomic challenges that face the continent. The ABN was first conceived in 1996 in response to growing concern in the region over threats to biodiversity in Africa and the need to develop strong African positions and legal instruments at the national, regional, and international levels. Currently, ABN has 36 partners drawn from 12 African countries: Benin, Botswana, Ethiopia, Ghana, Kenya, Mozambique, South Africa, Tanzania, Togo, Uganda, Zambia, and Zimbabwe. Together, the ABN is finding innovative and pioneering pathways and solutions to the challenges which face the continent.
- *African Centre for Biosafety.* The African Centre for Biosafety (ACB) (http://www.urbansprout.co.za/african_centre_for_biosafety) is a nonprofit organization, based in Johannesburg, South Africa. It provides authoritative, credible, relevant, and current information, research and policy analysis on issues pertaining to genetic engineering, biosafety, and biopiracy in Africa. The ACB is active in playing an effective role in protecting Africa's biodiversity, traditional knowledge, food production systems, and culture

and diversity, from the threats posed by genetic engineering and biopiracy.
- *Biowatch South Africa.* This (http://www.biowatch.org.za/) is established in 1997 to publicize, monitor, and research issues of genetic modification and to promote biological diversity and sustainable livelihoods. Biowatch's head office is in Durban, KwaZulu-Natal. A rural office in Mtubatuba works with small-scale farmers on sustainable agriculture, food and seed security, and farmers' rights.
- *Earth Life Africa.* Earthlife Africa's Johannesburg branch (http://earthlife.org.za/) was founded in 1988 to mobilize civil society around environmental issues in relation to people.
- *Friends of the Earth, Africa.* Friends of the Earth, South Africa/groundWork. (http://www.foei.org/member-groups/africa/south-africa/) groundWork is a nonprofit, environmental justice NGO working in Southern Africa, founded in 1999. The organization's vision is to see communities affected by industrial pollution better equipped to defend and promote their environmental rights at the local, national, and international levels. groundWork seeks to improve the quality of life of vulnerable people in South Africa, and increasingly in Southern Africa, through assisting civil society to have a greater impact on environmental governance. It places particular emphasis on assisting vulnerable and previously disadvantaged people who are most affected by environmental injustices.
- *Gaia Foundation.* The Gaia Foundation (http://www.gaiafoundation.org/) is passionate about regenerating cultural and biological diversity and restoring a respectful relationship with the Earth. Together with long-term partners in Africa, South America, Asia, and Europe, the foundation works with local communities to secure land, seed, food, and water sovereignty. By reviving indigenous knowledge and protecting sacred natural sites, local self-governance is strengthened. This enables communities to become more resilient to climate change and the industrial processes which have caused the many crises we now face. Gaia makes a long-term commitment with other partners to address the root causes of today's most pressing ecological, social, and economic injustices. The foundation represents a small, flexible team, which enables its members to read and respond quickly to emerging issues and opportunities for change.

6.4.2.5 Central and South America
- Amigos della Tiera Uruguay (REDES) (http://www.redes.org.uy/).
- Center of Studies for Rural Change in Mexico (CECCAM), Mexico (http://ceccam.org/).
- Conservation Land Trust, Patagonia (http://theconservationlandtrust.org/).
- Network for a GE-Free Latin America (RALLT) (http://www.rallt.org/).
- Sociedad Peruana de Derecho Ambiental (SPDA), Peru (http://www.spda.org.pe/). The Peruvian Environmental Law Society is an organization dedicated to integrating the environmental component into development policies, in order to achieve sustainable and equitable society and to promote under principles of ethics and respect for nature and responsible citizenship.
- Uruguay Sustentable, Uruguay (http://uruguaysustentable.org/).

6.4.2.6 Asia Pacific
- Gene Ethics, Australia (http://www.geneethics.org/)
- Madge, Australia (Mothers are Demystifying Genetic Engineering) (http://www.madge.org.au/)
- Navdanya, India (http://www.gaiafoundation.org/partner/navdanya)
- No! GMO Campaign, Japan (http://www.worc.org/Japan-no-gm/)

6.4.3 Popular Protests against GMOs and GM Foods

Some consumer group activists in different parts of the world arrange for protests against GM and GM foods. Part of these protests focusses on human and animal health and safety, whereas others are more concerned about the environment and agriculture. In May 2012, a group called "Take the Flour Back" led by Gerald Miles protested against plans by a group from Rothamsted Experimental Station, based in Harpenden, Hertfordshire, England, to stage an experimental trial to use GM wheat to repel aphids (Gaskell et al. 2010).

On May 25, 2013, the "March Against Monsanto" movement held rallies in protest against companies that produce GM seeds (Kopicki 2013). Sleenhoff and Osseweijer (2013) reported that more than two million

GM FOODS OR NOT? THE CONTROVERSY

Figure 6.1 Anti-GM protesters rallying against Monsanto in California, USA.

people in 50 countries worldwide marched against Monsanto. The protest was ignited because of the adoption of what is now called the "Monsanto Protection Act." According to its critics, the act allows for the planting of unapproved GM crops in the United States, overruling previous court orders designed to protect the environment, people's health, and well-being. The rallies took place in different parts of the world, for example, Buenos Aires and other Argentinean cities, Portland, Oregon, Los Angeles, USA (Figure 6.1), and Kitchener, Ontario, Canada. The total number of protesters who took part and reported by different sources varies from hundreds of thousands (ENSSER 2013) to two million (Fusaro 2013). According to organizers, protesters from 436 cities and 52 countries took part in these protests (BBC News 2012; Driver 2012; Sense About Science 2012).

6.4.4 Vandalism and Threats by Anti-GMO Activists

The rage against GMOs and GM foods did not stop at peaceful rallies. Vandalism and threats were reported in some cases. A few activist groups such as ELF and Greenpeace have vandalized GMO research in different parts of the world (CTV News Kitchener 2013; Harmon 2013; Quick 2013; Yahoo News. AP 2013). Some European countries including the United Kingdom witnessed the destruction of crop trials, belonging to academic or governmental research institutions, by protesters. Extreme cases of threats and violence against people and property were also reported (Pollack 2013). In the United States, anti-GMO groups burned the biotechnology laboratory of Michigan State University, causing the destruction of property worth hundreds of thousands of dollars and the loss of research results (Experts 2014). Other incidents of violence and property damage against GMOs have also been reported in other parts of the globe, for example, in Australia in 2011 (CTV News Kitchener 2013; von Mogel 2013) and in the Philippines in 2013 (Australian Broadcasting Corporation [ABC] News 2013; Science Daily 2013).

6.4.5 Arguments against GM Foods

The following sections discuss some of the arguments presented by GM technology and by opponents and critics of GM foods.

GM technology and GM food opponents try to present their case in a logical, organized, and evidence-supported manner. Rees (2006) started by asking the simple question: "does anyone want GM?" The brief answer given was "No," and what followed that was a large volume of evidence justifying the answer "No." The Union of Concerned Scientists (2012) believes that "None of the GM food crops currently on the market offer benefits to consumers in terms of price, nutrition, or new products." The following figures, presented by critics, reflect some of the consumers' perceptions and attitudes toward GM food.

> Eighty percent of consumers in the EU, 86% in the UK, 55% in the USA, 56% in China and 68% in New Zealand don't want GM food at all. With regard to GM food labeling 94% in the EU, 94% in the UK, 92% in the USA, and 87% in China want GM food labeled. Eighty four percent of Australians indicate their worry about eating GM foods and 92% of Canadians were concerned about GM food risks (Rees 2006, p. 9).

The two general themes around which opposition to GM technology and GM foods revolves can be summarized as follows:

1. There are better alternatives to the GM technology with regard to agriculture and food, such as traditional breeding and organic farming.
2. GM technology has many drawbacks, hazards, risks, and threats to humans, animals, agriculture, and the environment.

6.4.5.1 Better Alternatives to GM Technology

The critics of GM technology believe that there are other technologies which prove to be effective, with less drawbacks than GM. Some of these technologies have been successfully used, and they can be improved and developed to give better outcomes. The following technologies have been cited as alternatives to GM.

6.4.5.1.1 Traditional Breeding

This technique has been in existence and successful use for hundreds of years. The case of rice varieties developed through traditional breeding, and which exhibit tolerance to extreme climatic and soil conditions such as drought and flood tolerance, and soil salinity, has been cited as an example of success of traditional breeding. A number of indigenous rice varieties grown in West Bengal, India, which have been developed through traditional breeding, can tolerate drought, flooding, and high soil salinity. No GM was involved or needed to develop such rice varieties and cultivars.

6.4.5.1.2 Non-GM Biotechnology

Traditional breeding can be enhanced and developed through non-GM biotechnological techniques such as marker-assisted selection (MAS). MAS involves the use of a marker for indirect selection of a genetic determinant or determinants of desirable trait or traits such as disease resistance and tolerance to adverse climatic conditions (Guimarães et al. 2007). Application of MAS in plant and animal breeding has been practiced with reasonable success in different parts of the world, for example, in China and India. Over the decades, scientists have discovered that crops are full of dormant characteristics. Rather than using GM, it is possible to turn on a plant's innate ability for improvement (Rees 2006).

6.4.5.1.3 The Use of Organic Agriculture

Organic agriculture does not involve the use of any agrochemicals such as herbicides, pesticides, and artificial fertilizers. No GM products or ingredients are used in organic agriculture. In addition to crop production, organic agriculture is used for livestock. Farm animals do not receive antibiotics, growth promoters, or other drugs to improve production. In this sense, organic agriculture would be more environmentally friendly than GMOs.

6.4.5.2 Hazards and Risks of GM

Critics and opponents of GM technology in general, and its application to produce GM foods in particular, have presented arguments that support their stance against this technology. The following sections cover some of the points which reflect the point of view that GM technology is hazardous and has many risks.

6.4.5.2.1 It Is Unnatural, Has Unpredictable Results, and Can Lead to Unexpected and Unintended Side Effects

Friends of the Earth (FoE 2014) are worried that GE creates whole new life forms. The organisms resulting from this technology are alive, can mutate, multiply, breed with other living things, and continue breeding for generations to come. This trend has been observed and recorded all around the world. FoE cites the following words of Dr Michael Antoniou (Senior Lecturer in Molecular Biology, London) to support their arguments: "This is an imperfect technology with inherent dangers …. It is the unpredictability of the outcomes that is most worrying." The food produced using such unnatural technique is expected to be unnatural too. People fear eating foods that are not natural.

6.4.5.2.2 Genetic Uncertainties and the Disturbance of Nature's Boundaries

Natural boundaries are violated—crossing animals with plants, strawberries with fish, grains, nuts, seeds, and legumes with bacteria, viruses, and fungi, or like human genes with swine (Batalion 2000). This leads to a "control" of living nature and imposing a "non-living" model on it. Critics believe that the term "bioengineering" itself is a contradiction. "Bio" refers to "life"—something that is not mechanically predictable or controllable—and "engineering" implies developing the basics for machines that are predictable, not alive but "dead," so the contradiction.

6.4.5.2.3 Unpredictability and the Unknown

Insertion of the genetic material in a host would lead to uncontrolled and erratic behavior in the host. This is because DNA is complex and there is the potential of complex interactions. These interactions can cause gene suppression or over-expression, causing unpredictable and uncommon changes. The potential hazards are difficult to predict with any certainty. The technique of using the "gunshot" to blast DNA fragments through cell membranes is cited as an example, leading to unpredictable consequences

(Batalion 2000). The foreign genetic material is shot in a random, unpredictable way, possibly resulting in unknown products.

6.4.5.2.4 Hazards from GM due to the Interconnected Genetic Network

Because no gene ever functions in isolation, there will almost always be unexpected and unintended "side effects" from the gene or genes transferred into an organism. One major concern over transgenic foods is their potential to be allergenic (Greenpeace 2007). The case of the transgenic yeast engineered for increased rate of fermentation with multiple copies of its own genes, and which resulted in the accumulation of the metabolite methylglyoxal at toxic mutagenic levels, as reported by Inose and Murata (1995), is cited by Greenpeace as an example for such hazards.

6.4.5.2.5 Instability of Transgenic Lines

GM used in plant tissue culture introduces exotic genes into organisms, and transgenic plants are regenerated from cells after selection in culture. This procedure inherently generates increased genetic instability in the resulting transgenic line (Greenpeace 2010).

6.4.5.2.6 Hazards to Human and Animal Health and Safety

The critics of GM technology argue that GM products, including foods, are hazardous, and their long-term use can produce ill effects on human and animal health. Signing of "Monsanto Protection Act" prevents government restrictions on GMO foods found to cause health risks (NON-GMO Project 2014). Smith (2009) states that "There is more than a casual association between GM foods and adverse health effects. There is causation, 'as defined by recognized scientific criteria.' The strength of association and consistency between GM foods and disease is confirmed in several animal studies."

The following points are presented as evidence of hazards to human and animal health:

1. *GM food may cause allergies.* Perhaps, the number one health concern over GM technology is its capacity to create new allergens in our food supply (Weiss 2013). Proteins are the main components that bring about the observed allergic reactions. It is very common that transfer of genetic material from one living organism to another leads to the creation of a novel protein. GE affects the presence of allergens in different ways. It can increase the levels

of a naturally occurring allergen already available in a food or create new allergenic properties into a food that did not originally contain them. GE also has the potential to introduce brand new allergens man has never known before.

Some plants naturally produce toxins as a defense mechanism to drive away pests. These toxins either are present in very small amounts that would not harm humans or animals when consumed or are inactivated by processing. GE can lower or remove these toxins or allergens, but at the same time, it may increase their levels. GMOs make it possible for allergens in one food type to emerge in a completely different species (Union of Concerned Scientists 2012). Each new food item produced contains many new potentially allergenic proteins (Batalion 2000). Novel proteins produced through GE could raise new allergy outbreaks in humans. One such example is the lectin, which is a protein used in beans to prevent aphids from attacking potato crops. Some people are allergic to lectin.

2. *GMOs are toxic.* The AAEM has come to the conclusion that GMOs and GM foods are toxic. The academy cited results of some scientific and clinical studies that indicate the association of GM foods and specific disease processes (Organic Consumer Association 2014). According to the AAEM, animal studies showed immune dysregulation, altered structure and function of the liver, kidney, pancreas, and spleen, and cellular changes that could lead to accelerated aging. Studies also show intestinal damage in animals fed GM foods, including proliferative cell growth and disruption of the intestinal immune system. The AAEM concluded that GMO foods pose a serious health risk in the areas of toxicology, allergy and immune function, reproductive health, and metabolic, physiological, and genetic health.

3. *Interior toxins in GM foods.* The terms "interior toxins" and "pesticidal foods" have been used by Batalion (2000) to describe the toxic pesticides engineered inside the plant and the foods produced from such plants, respectively. The food is engineered to produce its own built-in pesticide in every cell to kill pests. The potential long-term health impacts of consuming such foods are little known, and people continue to ingest the interior plant toxin from GM foods. Pesticide-related illnesses in farm workers and the contamination of food and drinking water continue to rise due to the increased use of pesticides, with pesticide-resistant transgenic crops.

4. *Vectors can infect mammalian cells and resist breakdown in the gut.* Two extremely important factors in the safety of transgenic organisms used as food are the extent to which DNA, in particular, vector DNA, can resist breakdown in the gut and the extent to which it can infect the cells of higher organisms (Greenpeace 2010). Some studies have shown that bacterial plasmids carrying a mammalian virus are able to infect cultured mammalian cells, which then continue to synthesize the virus. Bacterial viruses can also be taken up by mammalian cells (Heitman and Lopes-Pila 1993).

    Greenpeace advises that people should be extremely cautious about ingesting transgenic foods, as foreign DNA can resist digestion and may be taken up by gut bacteria or absorbed through the gut wall into the bloodstream. DNA uptake into cells could then lead to the regeneration of viruses, or if the DNA integrates into the cell's genome, a range of harmful effects may result, including cancer. Moreover, one cannot assume, without adequate data, that DNA is automatically degraded in processed transgenic foods. GMOs are found to transfer genetically altered DNA into the DNA of bacteria living in the human stomach, reproducing indefinitely (Benson 2013).

5. *Direct and indirect cancer and degenerative disease links.* Rees (2006) and Batalion (2000) warned about possible direct cancer and other degenerative diseases as a result of consuming GM foods. The example of the GM growth hormone rBGH (recombinant bovine growth hormone), used to stimulate more milk production in cows and the possibility that potent chemical substances present in such milk to cause cancer in humans, is cited by Batalion (2000). Conclusions drawn by a number of scientists who support this claim have also been reported. The possibility that the recent increase in the cancer rates might be attributed to the consumption of GM foods has also been cited by Batalion (2000).

6. *Resurgence of infectious diseases.* A number of studies reported by Batalion (2000) indicate that gene technology may be implicated in the resurgence of infectious diseases. The growing resistance to antibiotics used in bioengineering, the formation of new and unknown viral strains, and the lowering of immunity due to consumption of processed and GM foods have been cited as ways by which infectious disease rate increases.

7. *Antibiotic threat via milk.* According to the Center of Food Safety, as reported by Batalion, cows injected with rBGH have a 25%

increase in the frequency of udder infections. Farmers would consequently use more antibiotics, which may eventually end up in the different dairy products people consume. High levels of antibiotic residues in milk may lead to some allergic reactions and could reduce the effects of other antibiotics.

8. *Antibiotic resistance through plants.* When applied to crops, GM technology uses markers to help tracking where the gene goes into the plant cell. The most common marker used is a gene for antibiotic resistance. This gene can be transferred to the food chain. A large number of GM food products contain this gene. The resistant qualities of GM bacteria in food can be transferred to other bacteria in the environment and throughout the human body (Batalion 2000).

9. *GM can lead to creation of superviruses.* Viruses are commonly used in GE. The most common virus used in GE is the cauliflower mosaic virus. It is used in Roundup Ready soy of Monsanto, *B. thuringiensis* maise of Novaris, GM cotton, and canola. Such viruses can mix with genes of other viruses, with the potential of creating more deadly viruses, which can be referred to as "superviruses," according to GM technology opponents.

10. *The safety of GE products is uncertain and is not verified yet.* The consumption of GM foods started in the mid-1990s. There are no long-term data on how GM foods affect human health. Clinical trials carried out on animals fed GM crops were of short duration, and the results and conclusions showed wide variations (Kannall 2014). Short-term studies on the effects of GM foods, for example, for 4 weeks, are not expected to give conclusive results on the safety of these foods for human consumption. Feeding rats GM corn caused negative effects in the functions of the liver, kidney, heart, adrenal gland, and spleen (Kannall 2014). There are no reports of research conducted on the health and safety for people who consumed GM foods. The critics of GM foods argue that it is not logical to say that GM foods are safe simply because people have been eating them for several years with no ill effects. This is because GM foods are not labeled, which makes it difficult to differentiate between GM and non-GM foods. They state that one fact is that in the United States, food-derived illness has doubled since the introduction of GM foods. Another argument is that the safety testing carried out on GM foods is inadequate so far, and at the same time, most of the testing is

conducted by companies producing the food or benefiting from its production. An urgent call for independent, regulated, and long-term studies on the safety of GM foods is made by opponents of GM foods. The current situation of GM food safety as viewed by the geneticist Richard Steinbrecher and reported by Batalion (2000) is summarized as: "to use genetic engineering to manipulate plants, release them into the environment and introduce them into our food chains is scientifically premature, unsafe and irresponsible."

11. *Chemical contamination of GM foods*. GM crops used to produce GM foods are treated with different chemicals, for example, herbicides and pesticides, to promote their growth and production. These chemicals have the potential to cause harm to humans and lead to negative health consequences. One example cited by GM food critics is the soybean which is treated with chemical herbicides, genetically developed to kill all weeds and other plants in the farm field without harming the soy crop. This chemical treatment might result in the food produced from such crops to contain traces of these toxic chemicals. Some pesticides have been associated with some health problems such as nervous system disturbances, visual and skin irritations, endocrine problems, and an increase in some types of cancer.

12. *GM food will not help feed the world, as claimed*. GM food opponents disagree with what they labeled as "myths being promoted by the biotech industry," which states that we need GE to feed a growing world population and to alleviate famine. The anti-GM groups believe that famines occur as a result of a multitude of factors and issues such as economic debt, political instability, and war. GE will not change any of these factors to help achieve the goal of famine eradication. On the contrary, it might even make things worse if farmers in the developing world, where most famines occur, become dependent on GE seeds and the agricultural chemicals related to them. Hunger and poverty go hand in hand. Technological inventions such as GM foods are not going to solve the underlying social, economic, and political problems that cause hunger. Greenpeace (2007) argues that populations in developing countries oppose GM technology and its use to produce GM foods. They present the statement, sent by 24 delegates from 18 African countries to the UN's FAO in 1998, as support to their claims. The statement says: "We strongly object that the

image of the poor and hungry from our countries is being used by giant multinational corporations to push a technology that is neither safe, nor environmentally friendly nor economically beneficial to us. On the contrary, it will thus undermine our capacity to feed ourselves." World hunger has little to do with inadequate production of food or poor quality food. The answers lie rather in redistribution of wealth, equity, and power.

Greenpeace (2010) refutes the argument that GM food is necessary to feed the world because it is based on the assumption that there is not enough food to go around. They believe that the world has one and a half times the amount of food needed to provide nutritious diet and that the solution to ending hunger does not mean simply feeding the world, but enabling the world to feed itself.

6.4.5.3 Impact on the Environment, the Ecosystem, and Farming

Environmentalists and some activist groups are leading the anti-GM group in expressing concerns about GM technology and its potential to cause negative impacts on the environment, the whole ecosystem, and the different farming practices and activities. Critics stress that in addition to their effects on health, GM foods may have negative side effects on the environment and cause hazards to agricultural and natural biodiversity. Environmentalists continue to demand answers to numerous questions pertaining to GM foods.

Arguments against GM foods in relation to the environment, the ecosystem, and agriculture are presented in the following sections.

6.4.5.3.1 Genetic Pollution Caused by Vector-Mediated HGT

Environmentalists are worried, and raise some concerns, about the emergence of what they describe as "genetic pollution." Genetic pollution happens as a result of vector-mediated HGT. These vectors include bacteria and viruses, which are used in GM processes. Pollen of GM crops can be blown into non-GM agricultural area, thus polluting their more traditional relatives. Genes might escape from GM plants into the environment. These modified genes may be transferred to other non-GM plants, thus polluting them with the GM material. HGT to unrelated species via bacteria and viruses has the potential to create new weed species. Transfer of genes between plants is a natural part of the normal evolutionary process and also occurs between unmodified plants. Should this happen repeatedly when genetic modifications have conferred tolerance to herbicides, it has been suggested that a new breed of plants which cannot be controlled with conventional herbicides—superweeds as they have been termed—could emerge (Friends of the Earth 2014). Superweeds can outcompete and disrupt the natural biodiversity of an area. Some GM crops have the ability to crossbreed with other GM or non-GM crops in the same area. If this crossing happens with non-GM or organic crops intended for special markets, it will disrupt the market for these crops and consumers will lose trust in such products. Agricultural products resulting from such crossbreeding may not sell as non-GM or organic produce. This process of crossbreeding is believed to be inevitable and uncontrollable. HGT also has the potential to create new pathogenic bacteria and viruses (Greenpeace 2010).

The vectors carrying the transgene, unlike chemical pollution, can be perpetuated and amplified, given the right environmental conditions. It has the potential to unleash cross-species epizootics and epidemics that will be impossible to control or recall (Colman 1996). Greenpeace (2010) points out that great caution needs to be exercised to avoid releasing of undesirable transgenes and marker genes, through HGT, into the environment.

6.4.5.3.2 *Vector Recombination*
Vectors used in GM can recombine to generate new virulent strains of viruses, particularly in GM plants developed for viral resistance with viral genes. It is also possible that genetic elements recombine with other viruses and bacteria to generate new genetic elements and pathogenic strains of bacteria and viruses, which will also be antibiotic-resistant (Greenpeace 2010). Genes carried by vectors can persist indefinitely in the environment.

6.4.5.3.3 *Possibility of Emergence of "Superweeds"*
Genetic pollution caused by HGT from GE plants to wild relatives can lead to the creation of superweeds, new types of weeds, which are difficult to destroy or eradicate, resulting from crosses between ordinary weeds and GM plants. GM crops are often treated with chemical pesticides and herbicides and are fertilized with chemical fertilizers. These chemicals can contaminate the surrounding environment. Some weeds started to develop a resistance to some of these chemicals, making them more difficult to control. This can also lead to the emergence of superweeds. Examples of these weeds include those that are reported to be resistant to glyphosate (Union of Concerned Scientists 2012; Epoch Times 2013).

6.4.5.3.4 There Is No Need for GM-Developed Methods of Weed Control

Due to the adverse effects of GM, environmentalists prefer, and advocate for, alternative and sustainable methods of weed control such as intercropping, mulching, and use of green manure (FoE). These traditional methods of weed control have been successfully used for centuries; consequently, there is no need to develop other methods, which might have harmful effects, to control weeds.

6.4.5.3.5 GM Causes Loss of Indigenous Biodiversity

The ecosystem in areas where GM is applied to plants will be upset. GM crops which are developed for pest resistance may harm or kill other animals or insects that are useful or which are not harmful.

In addition to being herbicide-resistant, GM crops may also be insect-resistant. By incorporating a modified gene from a soil bacterium, a GM plant can produce throughout its structure, including leaves and fruits, an endotoxin injurious to insects (Friends of the Earth 2014). This particular toxin may contribute to the poisoning of beneficial insects and subsequently their predators. Some studies have shown that one corn GMO type threatened to wipe out Monarch butterfly larvae, which is considered to be a beneficial insect (Batalion 2000; Union of Concerned Scientists 2012).

6.4.5.3.6 Dangers from Creation of Herbicide-Tolerant Plants

Herbicide-tolerant plants have been developed through GM technology to withstand and tolerate herbicides used to kill weeds. Based on this understanding, farmers tend to pour herbicides onto them to kill the invading weeds without damaging their crops. Environmentalists feel that this only serves to promote over dependency on chemical intervention (Friends of the Earth 2014). Herbicides are also known to remain active in the soil for long periods. As GM crop farmers tend to apply herbicides freely and with no fear of crop damage, there is greater potential of polluting groundwater and endangering the ecosystem in general.

6.4.5.3.7 Possible Emergence of "Superpests"

Batalion (2000) reports that laboratory tests indicate that common plant pests such as cotton bollworms can evolve into "superpests," which immune from the *B. thuringiensis* sprays used by organic farmers. Environmentalists are worried that pests which the transgenic cotton was meant to kill, for example, cotton bollworms, pink bollworms, and budworms, were once "secondary pests." Toxic chemicals killed off their predators, unbalanced nature, and thus made them "major- or super-pests."

6.4.5.3.8 GM Might Lead to Elimination of Indigenous Agriculture and the Loss of Natural Species

Non-GM activists and environmentalists believe that the increased use of toxic, nondiscriminating herbicides with herbicide-resistant transgenic plants will lead to large-scale elimination of indigenous agricultural and natural species.

Pollen from GM plants can be transferred by the wind or insects to areas where natural plants are growing. The resulting offspring of the normal plant will now have the GM plants genes as part of its genetic pool. The consequences of this trend are that there will be no more natural plants and as a result kill animals feeding on them.

6.4.5.3.9 GM Technology Can Lead to Pollution and Sterility of Agricultural Soil

Increased use of toxic herbicides would lead to destruction of soil fertility and to reduction of crop yield. It can ultimately result in soil toxicity and sterility and its inability to hold crops. Soil pollution might also result from continued herbicide application. Epoch Times (2013) reports that glyphosate builds up in the soil, leading to its contamination and pollution.

6.4.5.3.10 Possible Extinction of Seed Varieties

The observed monopoly of the seed industry by a few agricultural companies would ultimately result in controlling which seed varieties remain in the market. This might lead to the extinction of a number of known seed varieties from the market. Batalion (2000) believes that "it is now time to start rolling back the monopoly privileges of the seed industry, not to strengthen them further."

6.4.5.3.11 Potential of Destruction of Forest Life

GM technology has plans to develop new types of trees which are nonflowering, herbicide-resistant, and with leaves producing toxic chemicals intended to kill caterpillars and other types of insects in the area. Batalion (2000) reports on such types of trees which are planned to be developed by Monsanto in collaboration with the New Zealand Forest Research Agency. Such trees have been labeled as "terminator trees" and "supertrees." According to Brown (1999), supertrees are being developed, which can be sprayed from the air to kill literally all surrounding life, except the GM trees. This is an attempt to transform international forestry by introducing multiple species of such trees, which are often sterile and flowerless.

6.4.5.3.12 GM Technology Applications May Lead to Antibiotic Resistance

A common practice in the GM process is the use of antibiotic-resistant marker genes to enable researchers to trace transgenes. It has been speculated that the gene sequence controlling antibiotic resistance could be transferred to bacteria, which colonize the human gut and thereby assist in the spread of antibiotic resistance in humans. The spread of resistance to antibiotics is considered a serious threat to human ability to treat diseases. Overuse of antibiotics can potentially cause the development of antibiotic-resistant pathogens. According to the Greenpeace reports, the ACNFP assessed the risk associated with antibiotic-resistant marker genes as extremely low, but nevertheless recommended that researchers should stop using them and develop alternatives.

6.4.5.3.13 Health Hazards Resulting from Pesticide Exposure

The continued application of herbicides such as glyphosate to GM crops which are herbicide-tolerant will increase human and animal exposure to such agricultural chemicals. There is the potential danger of such exposure, which might result in different health problems in humans and farm animals. Glyphosate has been linked to numerous health problems in animal studies, including birth defects, reproductive damage, cancer, and endocrine disruption (Greenpeace 2010).

6.4.5.3.14 GM Technology Leads to Animal "Bio-invasions" and Represents Animal Abuse

Anti-GM activists describe some of the GM processes as "bio-invasions" of animals. Batalion (2000) reports that fish and general marine life have been threatened by the accidental release of GM fish carried out in several parts of the world. The increase in size and in growth rate of various types of fish such as trout, carp, and salmon has been cited as an example of such invasion. Critics of GM technology consider the application of this technology to experiments on animals as a kind of animal abuse. The case of pig number 6706, which was supposed to be a "superpig" through insertion of a particular gene and which became a "supercripple" with arthritis, cross eyes, and mutated figure, is cited as an example of such animal abuse (Batalion 2000).

6.4.5.3.15 Negative Impact on Farming
Anti-GM activists claim that applications of GM technology in agriculture would have negative impacts on small farms and organic agriculture. The following arguments are listed by these groups:

- GM technology leads to a decline in, destruction of, and general economic harm on small self-sufficient family farms. This will have a negative impact on the livelihood and survival of small farmers.
- With regard to organic farming, GMOs can contaminate existing species being grown organically (Benson 2013). This will lead to genetic pollution, loss of crop purity, loss of natural pesticides, and mixing of non-GM crops with GM material. In contrast, the plants that are GM may become more prevalent, choking other plants and causing native plants to die off.
- What is described as "terminator technology" would lead to control of, and dependence on, large agricultural and chemical companies developing and controlling GM products. Owners of small farms may not be able to survive in such an atmosphere.
- With regard to farm production, in general, anti-GM groups argue that GM technology will reduce crop diversity, quality, and quantity. Future agriculture will be more fragile and unstable, if GM technology continues to dominate and expand. It will also cause monopolization of food production, which will have a negative impact on long-term food supply and availability.

6.4.5.4 Ethical and Moral Objections
Anti-GM groups and activists express concerns about ethical and moral issues in relation to GM technology and its applications. The most immediate ethical concern is "messing with nature." This is expressed in the following question asked by GM critics: "Do human beings have the right to view nature as their commodity, and alter the genetic structure of living things to correct what some perceive as their 'deficiencies'?" (Australian Broadcasting Corporation 1999). A number of other questions revolving around related ethical and moral issues are also being raised by the GM technology opponents.

6.4.5.5 GM Technology Creates Socioeconomic and Political Threats
Nearly, all developments and work on GM are carried out in developed regions of the world, sometimes referred to as the "North." At the same

time, most of the genetic resources used in research and development of GMOs are brought from developing countries, also referred to as the "South." This process of obtaining the genetic resources from the South to the North has been going on, and is expected to continue, for a long time. Anti-GM activists view this as an increased drain of genetic resources from the South to the North, and they are very much concerned about it.

Another socioeconomic threat viewed by anti-GM groups is the increased marginalization of small farmers associated with the intellectual property rights and other restrictive practices related to seed certification. Large seed companies develop the seeds, own the right to sell them, and control and monitor their use. This is regarded as a violation of human rights by the seed companies which are only concerned about their profit.

Adoption of GM technology in agriculture is regarded by anti-GM activists as a substitution of traditional technologies and produce. This causes a threat and has the potential to cause harm to individuals or companies who are not willing to follow this trend or do not have the means or resources to use this new technology.

The term "biocolonization" is used to describe the combined control of genetic and agricultural resources by a few multinational firms, which holds a powerful weapon for the invasion of cultures. Biocolonization is regarded as a real threat facing large sectors worldwide, particularly when it affects human food. When people lose food self-sufficiency, they become wholly dependent and subservient (Batalion 2000). In many parts of the world, particularly developing regions, major agricultural resources such as seeds, water, and land can be bought by foreign bodies and subsequently converted to exported cash, instead of using them for local survival crops. This trend is viewed as a threat to self-sufficiency of the affected cultures and populations.

The lack of labeling of GM foods in many parts of the world is considered as a kind of violation of human rights as it does not allow the consumer to know what is in the food they are purchasing. Labeling GM foods would allow the consumer to make an informed decision as to whether to buy and consume the food or to opt for other non-GM or organic foods. Some consumers need to know the food ingredients for health or religious reasons. The lack of labeling of GM foods leaves these groups of consumers in the dark and makes it difficult for them to observe their dietary regimes or customs.

6.5 AND THE DEBATE CONTINUES!

The debate continues. It will and it should. It will continue because some groups of both sides of the debate do not want to view the full picture of the GM technology and GM foods. Some GM food supporters focus only on the benefits that the GM technology offers and on the positive aspects of GM foods, while keeping a blind eye on the possible risks. On the other side, GM technology and GM food critics highlight and try to magnify the potential risks to human health, the environment, and the whole ecosystem, without paying much attention to the positive aspects the technology has. The debate will continue until a common ground of understanding is reached and both sides fully appreciate the benefits and risks GM technology and GM foods have. It will also continue until researchers and scientists present credible results and interpretation of their work, without being affected by external pressures to skew results or to emphasize aspects that fulfill the goals of their research financers. The debate will continue, and issues will remain controversial until consumers find out the facts and then give support or oppose accordingly in an informed manner. However, until further studies show that GM foods and crops do not pose serious threats to human and animal health, the environment, or the world's ecosystems, the debate over their release will continue.

The debate should continue for the benefit of the confused consumers, believed to be a majority. Large sectors of consumers are confused because every day they find conflicting information and reports about GM foods. Each side of the debate tries to convince consumers that their arguments are truthful and that they represent the real picture. The technology supporters try to portray a rosy picture without mentioning the risks, whereas the opponents discuss the risks and threats of the GM technology. Most consumers are confused and do not know who to trust. The debate should continue in a professional and objective manner and should put the welfare of the consumers at the heart of discussion agenda.

Both sides need to realize and express potential benefits and possible risks, hazards, and threats of GM foods. It needs to be emphasized that no human activity is entirely risk-free and that certainly no food, whether produced through GM or traditional techniques, is. This does not mean that we should believe in or fully trust modern technologies, such as GM, without thorough study. Possible risks should be identified, evaluated, and reduced as much as possible, to a safe level. There is an urgent need to emphasize that scientific findings are not expected to give clear-cut answers to pertinent questions on GM technology and

GM foods and should not be taken as the ultimate goal to end the debate surrounding GM foods.

REFERENCES

American Association for the Advancement of Science (AAAS), Board of Directors. 2012. Legally mandating GM food labels could mislead and falsely alarm consumers. http://www.aaas.org/news/aaas-board-directors-legally-mandating-gm-food-labels-could-%E2%80%9Cmislead-and-falsely-alarm (accessed August 20, 2014).

American Medical Association. 2012. Report 2 of the Council on Science and Public Health: Labeling of bioengineered foods. http://factsaboutgmos.org/sites/default/files/AMA%20Report.pdf (accessed August 20, 2014).

Australian Broadcasting Corporation (ABC) News. 2013. Scientists speak out against vandalism of genetically modified rice. http://www.abc.net.au/news/2013-09-20/scientists-speak-out-against-vandalism-of-gm-rice/4970626 (accessed August 26, 2014).

Australian Broadcasting Corporation. 1999. Waiter, there is a gene in my food. The hot issues. http://www.abc.net.au/science/slab/consconf/hot.htm (accessed August 26, 2014).

Batalion, N.B. 2000. *50 Harmful Effects of Genetically Modified Foods*. Americans for Safe Food. Oneonta, NY.

BBC News. 2012. GM wheat trial belongs in a laboratory. http://www.bbc.com/news/science-environment-17928172 (accessed August 26, 2014).

BenefitOf.net. 2013. Environmental benefits of GMOs. http://benefitof.net/environmental-benefits-of-gmos/ (accessed October 10, 2014).

Bennett, R.M., Morse, S., and Ismael, Y. 2006. The economic impact of genetically modified cotton on South African smallholders: Yield, profit and health effects. *Journal of Development Studies*, 42(4):662–77.

Benson, J. 2013. Seven ways GMO toxicity affects animals, plants, and soil. *Natural News*. http://www.naturalnews.com/038792_GMO_toxicity_digestion_cancer.html (accessed September 1, 2014).

Bill and Melinda Gates Foundation. 2013. Agricultural development: Strategy overview. http://www.gatesfoundation.org/What-We-Do/Global-Development/Agricultural-Development (accessed August 10, 2014).

Biodiversity A to Z. 2014. Biodiversity terms: Biodiversity. http://www.biodiversitya-z.org/content/biodiversity (accessed October 10, 2014).

BioSpace. 2014. Biotechnology Industry Organization (BIO). http://www.biospace.com/company_profile.aspx?CompanyId=1311 (accessed September 20, 2014).

Brac de la PerriFre, R.A. and Seuret, F. 2000. *Brave New Seeds: The Threat of GM Crops to Farmers*. Zed Books, London.

Brookes, G. 2008. The impact of using GM insect resistant maize in Europe since 1998. *International Journal of Biotechnology*, 10:148–66.

Brookes, G. and Barfoot, P. 2010. Global impact of biotech crops: Environmental effects, 1996–2010. *GM Crops and Food: Biotechnology in Agriculture and the Food Chain*, 3(2):129–37.

Brown, P. 1999. Forests in danger from GM super-tree says WWF (World Wide Fund for Nature). *The Guardian*. http://www.theguardian.com/environment/1999/nov/10/gmcrops.gm (accessed August 14, 2014).

Center for Science in the Public Interest. 2012. Straight talk on genetically engineered foods: Answers to frequently asked questions. http://cspinet.org/new/pdf/biotech-faq.pdf (accessed August 14, 2014).

Colman, A. 1996. Production of proteins in the milk of transgenic livestock—Problems, solutions and successes. *American Journal of Clinical Nutrition*, 63: S639–45.

Commonground. 2014. GMO foods. http://findourcommonground.com/food-facts/gmo-foods/ (accessed October 10, 2014).

Conservation Technology Information Center (CTIC). 2010. Facilitating conservation farming practices and enhancing environmental sustainability with agricultural biotechnology. CTIC, West Lafayette, IN.

Cook, G. 2005. *Genetically Modified Language: The Discourse of Arguments for GM Crops and Food*. Routledge Publisher (Taylor & Francis Group), London, UK.

Council for Agricultural Science and Technology. 2009. U.S. soybean production sustainability: A comparative analysis. Special Publication 30. https://www.cast-science.org/ (accessed August 14, 2014).

Crawford, A.W., Wang, C., Jenkins, D.J., and Lemke, S.L. 2011. Estimated effect on fatty acid intake of substituting a low-saturated, high-oleic, low linolenic soybean oil for liquid oils. *Nutrition Today*, 46(4):189–96.

CropBiotech.net. 2004. British Medical Association statement on GM foods. K Sheet No. 3. http://www.isaaa.org/kc/Publications/pdfs/ksheets/K%20Sheet%20%28BMA%29.pdf (accessed March 6, 2014).

CTV News Kitchener. 2013. March against Monsanto comes to King Street in Kitchener. CTV Television Network. http://kitchener.ctvnews.ca/march-against-monsanto-comes-to-king-street-in-kitchener-1.1296971 (accessed August 26, 2014).

Damude, H. and Kinney, A. 2008. Enhancing plant seed oils for human nutrition. *Plant Physiology*, 147(3):962–8.

DiRienzo, M.A., Lemke, S.L., Petersen, B.J., and Smith, K.M. 2008. Effect of substitution of high stearic low linolenic acid soybean oil for hydrogenated soybean oil on fatty acid intake. *Lipids*, 43(5):451–6.

Driver, A. 2012. Scientists urge protestors not to trash GM material. *Farmers Guardian*. http://www.farmersguardian.com/home/arable/scientists-urge-protestors-not-to-trash-gm-trials/46673.article (accessed August 26, 2014).

Engdahl, F.W. 2007. *Seeds of Destruction: The Hidden Agenda of Genetic Manipulation*. Global Research Publishers. Montreal, Canada. http://www.global-research.ca/.

Engdahl, F.W. 2009. *A Moratorium on Genetically Manipulated (GMO) Foods*. Global Research. http://globalresearch.ca/a-moratorium-on-genetically-manipulated-gmo-foods/13701 (accessed January 15, 2014).

EPA. 2012. United States Regulatory Agencies Unified Biotechnology. http://usbiotechreg.epa.gov/usbiotechreg/ (accessed February 7, 2014).

Epoch Times. 2013. GMO toxicity affects animals, plants and soil. http://www.theepochtimes.com/n3/4680-gmo-toxicity-affects-animals-plants-and-soil/ (accessed September 1, 2014).

eSSORTMENT—Your Source of Knowledge. 2014. Genetically altered food debate. http://www.essortment.com/genetically-altered-food-debate-40664.html (accessed August 24, 2014).

ETC Group. 2008. Who owns nature? Corporate power and the final frontier in the commodification of life. Issue No. 100, p. 11. http://www.etcgroup.org/sites/www.etcgroup.org/files/publication/707/01/etc_won_report_final_color.pdf (accessed February 10, 2014).

European Commission. 2010. A decade of EU-funded GMO research (2001–2010). Directorate General for Research and Innovation Biotechnologies, Agriculture, Food. EUR 24473 EN. http://ec.europa.eu/research/biosociety/pdf/a_decade_of_eu-funded_gmo_research.pdf (accessed August 20, 2014).

European Network of Scientists for Social and Environmental Responsibility (ENSSER). 2013. Statement: No consensus on GMO safety. http://www.ensser.org/increasing-public-information/no-scientific-consensus-on-gmo-safety/ (accessed August 26, 2014).

Experts. 2014. GMO answers. http://gmoanswers.com/experts (accessed August 26, 2014).

Fagan, J., Antoniou, M., and Robinson, C. 2014. GMO myths and truths. An evidence-based examination of the claims made for the safety and efficacy of genetically modified crops and foods. Second Edition, Version 1.0, Earth Open Source©. www.earthopensource.org. (accessed October 9, 2014).

Falck-Zepeda, J., Gruère, G., and Sithole-Niang, I. (Eds.). 2013. *Genetically Modified Crops in Africa: Economic and Policy Lessons from Countries South of the Sahara.* International Food Policy Research Institute, Washington, DC.

FAO. 2003. Weighing the GMO arguments: For. http://www.fao.org/english/newsroom/focus/2003/gmo7.htm (accessed February 10, 2014).

FAO. 2004. State of food and agriculture 2003–2004. Agricultural biotechnology: Meeting the needs of the poor. http://www.fao.org/docrep/006/Y5160E/y5160e10.htm#P3_1651The (accessed August 20, 2014).

FAO/WHO. 2000. Safety aspects of genetically modified foods of plant origin. Report of a Joint FAO/WHO Expert Consultation on Foods Derived from Biotechnology. World Health Organization.

Fawcett, R. and Towery, D. 2002. Conservation tillage and plant biotechnology: How new technologies can improve the environment by reducing the need to plow. CTIC, West Lafayette, IN.

FDA. 2008. Guidance for industry: Use of animal clones and clone progeny for human. http://www.fda.gov/downloads/AnimalVeterinary/GuidanceComplianceEnforcement/GuidanceforIndustry/UCM052469.pdf (accessed August 7, 2014).

FDA. 2010. Animal cloning. http://www.fda.gov/AnimalVeterinary/Safety Health/AnimalCloning/default.htm (accessed August 7, 2014).
FDA. 2012. Genetically engineered plants for food and feed. http://www/fda.gov/Food/GuidanceRegulation/GuidanceDocumentsRegulatoryInformation/Biotechnology/ucm096126.htm (accessed August 7, 2014).
Freedman, D.H. 2013. The truth about genetically modified food. *Scientific American.* http://www.scientificamerican.com/article/the-truth-about-genetically-modified-food/ (accessed October 12, 2014).
Friends of the Earth. 2014. Genetic engineering. http://www.foe.org/projects/food-and-technology/genetic-engineering (accessed August 26, 2014).
Fusaro, D. 2013. European scientists ask for GMO research. *Food Processing.* http://www.foodprocessing.com/articles/2013/european-scientists-ask-for-gmo-research/ (accessed August 26, 2014).
Gaskell, G., Stares, S., Allansdottir, A. et al. 2010. Europeans and biotechnology in 2010: Winds of change? A report to the European Commission's Directorate-General for Research. http://ec.europa.eu/research/science-society/document_library/pdf_06/europeans-biotechnology-in-2010_en.pdf/ (accessed August 26, 2014).
Gianessi, L., Sankula, S., and Reigner, N. 2003. *Plant Biotechnology: Potential Impact for Improving Pest Management in European Agriculture.* The National Center for Food and Agricultural Policy, Washington, DC.
Giddings, L.V. and Chassy, B.M. 2009. Igniting agricultural innovation: Biotechnology policy prescriptions for a new administration. *Science Progress.* http://scienceprogress.org/2009/07/igniting-agricultural-innovation/ (accessed August 8, 2014).
GMWatch. 2014. GM firms. http://www.gmwatch.org/index.php/articles/gm-firms (accessed February 10, 2014).
Godfray, H., Beddington, J., Crute, I. et al. 2010. Food security: The challenge of feeding 9 billion people. *Science*, 327(5967):812–8.
GRAIN. 2004. Bt cotton at Mali's doorstep: Time to act! www.grain.org/briefings_files/btcotton-synthesis-feb-2004-en.pdf (accessed March 7, 2014).
Greenpeace. 2007. Will GE feed the world? Genetic engineering will not feed the world. http://www.greenpeace.org/new-zealand/en/campaigns/genetic-engineering/understanding-ge/will-ge-feed-the-world/ (accessed September 5, 2014).
Greenpeace. 2010. Greenpeace, GM crops and world hunger. http://www.greenpeace.org.uk/about/greenpeace-gm-crops-and-world-hunger (accessed September 5, 2014).
Guidetti, G. 2009. Pros and cons of genetically modified foods. http://ezinearticles.com/?Pros-and-Cons-of-Genetically-Modified-Foods&id=3415986 (accessed March 12, 2014).
Guimarães, E., Ruane, J., Scherf, B., Sonnino, A., and Dargie, J. (Eds.). 2007. Marker-assisted selection. Current status and future perspectives in crops, livestock, forestry and fish. FAO. https://www.scribd.com/doc/61148662/

Marker-Assisted-Selection-Current-Status-and-Future-Perspectives-in-Crops-Livestock-Forestry-and-Fish-FAO (accessed March 12, 2014).

Harmon, A. 2013. A race to save the orange by altering its DNA. *The New York Times*. http://www.nytimes.com/2013/07/28/science/a-race-to-save-the-orange-by-altering-its-dna.html?pagewanted=all&_r=1& (accessed August 26, 2014).

Hayenga, M. 1998. Structural changes in the biotech seed and chemical industrial complex. *AgBioForum*, 1(2):43–55. http://www.agbioforum.org/v1n2/v1n2a02-hayenga.htm (accessed February 10, 2014).

Heitman, D. and Lopes-Pila, J.M. 1993. Frequency and conditions of spontaneous plasmid transfer from *E. coli* to cultured mammalian cells. *BioSystems*, 29:37–48.

Ho, M. 2007. *GMO Free: Exploring the Hazards of Biotechnology to Ensure the Integrity of Our Food Supply*. Institute of Science in Society and Third World Network, Vital Health Publishing, Ridgefield, CT.

Hunt, L. 2004. Factors determining the public understanding of GM technologies. Review article. *AgBiotechNet*, 6(128):1–8. CAB International Publishing. http://www.ctu.edu.vn/~dvxe/doc/Factors%20determining%20%20understandingGMO.pdf (accessed March 2014).

Inose, T. and Murata, K. 1995. Enhanced accumulation of toxic compounds in yeast cells having high glycolytic activity: A case study on the safety of genetically engineered yeast. *International Journal of Food Science and Technology*, 30:141–6.

International Food Information Council (IFIC) Foundation. 2011. Questions and answers about food biotechnology. http://www.foodinsight.org/Resources/Detail.aspx?topic=Questions_and_Answers_About_Food_Biotechnology.

International Food Information Council Foundation. 2013. Fact sheet: Benefits of food biotechnology. *Food Insight*. http://www.foodinsight.org/Resources/Detail.aspx?topic=Fact_Sheet_Benefits_of_Food_Biotechnology (accessed January 17, 2014).

Kannall, E. 2014. Disadvantages of genetically modified food. LIVINGSTRONG.COM. http://www.livestrong.com/article/345554-disadvantages-of-genetically-modified-food/ (accessed January 20, 2014).

Kariyawasam, K. 2010. Legal liability, intellectual property and genetically modified crops: Their impact on world agriculture. *Pacific Rim Law and Policy Journal Association*, 19(3):459–85.

Kopicki, A. 2013. Strong support for labeling modified foods. *The New York Times*. http://www.nytimes.com/2013/07/28/science/strong-support-for-labeling-modified-foods.html?_r=1& (accessed August 26, 2014).

Lai, L., Kang, J.X., Li, R. et al. 2006. Generation of cloned transgenic pigs rich in omega-3 fatty acids. *Nature Biotechnology*, 24(4):435–6.

Lanphier, L. 2014. More GMO info—What you must know. Exhibit health—A better way to health and wellness. http://www.exhibithealth.com/health-education/more-gmo-info-what-you-must-know-1518/ (accessed September 1, 2014).

Lazarus, R.J. 1991. The tragedy of distrust in the implementation of Federal Environmental Law. *Law and Contemporary Problems*, 54(4):311–74.
Lehrer, S.B. and Bannon, G.A. 2005. Risks of allergic reactions to biotech proteins in foods: Perception and reality. *Allergy*, 60(5):559–64.
Lei, H. 2005. Genetically modified foods: Pros and cons. http://www.blisstree.com/2005/07/26/mental-health-well-being/genetically-modified-food-pros-and-cons/ (accessed March 12, 2014).
Lichtenstein, A.H., Matthan, N.R., Jalbert, S.M. et al. 2006. Novel soybean oils with different fatty acid profiles alter cardiovascular disease risk factors in moderately hyperlipidemic subjects. *American Journal of Clinical Nutrition*, 84(3):497–504.
Luttrell, A. 2014. The benefits of genetically modified food crops. Economic benefits of genetically modified crops. https://suite101.com/a/the-benefits-of-genetically-modified-food-crops-a218670 (accessed February 10, 2014).
Massengale, R.D. 2010. Biotechnology: Going beyond GMOs. *Food Technology*, 64(11):30–5.
Mermelstein, N.H. 2010. Improving soybean oil. *Food Technology*, 64(5):72–6.
National Center of Food and Agricultural Policy (NCFAP). 2013. Report. http://www.ncfap.org/ (accessed October 10, 2014).
National Research Council. 2010. *Impact of Genetically Engineered Crops on Farm Sustainability in the United States*. The National Academies Press, Washington, DC.
Navdanya, Navdanya International, the International Commission on the Future of Food and Agriculture, and The Center for Food Safety (Coordinators). 2011. The GMO Emperor has no clothes. A global citizens report on the state of GMOs—False promises, failed technologies: Synthesis report. Navdanya International.
Newell-McGloughlin, M. 2008. Nutritionally improved agricultural crops. *Plant Physiology*, 147:939–53.
NON-GMO Project. 2014. GMO facts: Frequently asked questions. http://www.nongmoproject.org/learn-more/ (accessed September 1, 2014).
Organic Consumer Association. 2014. The American Association of Environmental Medicine (AAEM) calls for immediate moratorium on genetically modified foods. http://www.organicconsumers.org/articles/article_18080.cfm (accessed September 3, 2014).
Osteen, C., Gottlieb, J., and Vasavada, U. (Eds.). 2012. Agricultural resources and environmental indicators, EIB-98, USDA, Economic Research Service (ERS), August 2012.
Park, J.R., McFarlane, I., Phipps, R.H., and G. Ceddia. 2011. The role of transgenic crops in sustainable development. *Plant Biotechnology Journal*, 9:2–21.
Pollack, A. 2013. Seeking support, biotech food companies pledge transparency. *The New York Times*. http://www.nytimes.com/2013/07/29/business/seeking-support-biotech-food-companies-pledge-transparency.html (accessed August 26, 2014).

Pray, C.E., Huang, J., Hu, R., and Rozelle, S. 2002. Five years of Bt cotton in China—The benefits continue. *The Plant Journal*, 31(4):423–30.

Qaim, M. 2011. Genetically modified crops and global food security, Chapter 2, in C.A. Carter, G.C. Moschini, and I. Sheldon (Eds.), *Frontiers of Economics and Globalization*, Volume 10. Genetically Modified Food and Global Welfare. Emerald Group Publishing Limited, Bingley, UK, pp. 29–54.

Quick, D. 2013. More than 100 participate in Charleston's march against Monsanto, One of 300+ in world on Saturday. *The Post and Courier*. http://www.postandcourier.com/article/20130526/PC16/130529414 (accessed August 26, 2014).

Rees, A. 2006. *Genetically Modified Food: A Short Guide for the Confused*. Organic Consumer Association.

Robinson, J. 1999. Ethics and transgenic crops: A review. *Electronic Journal of Biotechnology*. http://www.ejbiotechnology.info/index.php/ejbiotechnology/article/view/v2n2-3/821 (accessed November 12, 2014).

Rommens, C., Yan, H., Swords, K., Richael, C., and Ye, J. 2008. Low-acrylamide French fries and potato chips. *Plant Biotechnology Journal*, 6(8):843–53.

Ronald, P. 2011. Plant genetics, sustainable agriculture and global food security. *Genetics*, 188(1):11–20. http://www.genetics.org/content/188/1/11.long (accessed August 20, 2014).

Science Daily. 2013. Fighting GM crop vandalism with a government-protected research site. http://www.sciencedaily.com/releases/2013/02/130228124134.htm (accessed August 26, 2014).

Sense About Science. 2012. Don't destroy research Q&A: Plant scientists answer your questions. http://www.senseaboutscience.org/pages/plant-science-qa.html (accessed August 26, 2014).

Shand, H. 2012. The big six: A profile of corporate power in seeds. Agrochemicals and Biotech (PDF). *The Heritage Farm Companion*. http://www.seedsavers.org/site/pdf/HeritageFarmCompanion_BigSix.pdf (accessed July 10, 2014).

Shiva, V., Barker, D., and Lockhart, C. 2011. The GMO Emperor has no clothes: A global citizens report on the state of GMOs—False promises, failed technologies. Synthesis Report, Navdanya International.

Sleenhoff, S. and Osseweijer, P. 2013. Consumer choice: Linking consumer intentions to actual purchase of GM labeled food products. *GM Crops and Food: Biotechnology in Agriculture and the Food Chain*, 4(3):166–71.

Smith, J.M. 2009. Doctors warn: Avoid genetically modified food. Institute for Responsible Technology. http://www.responsibletechnology.org/doctors-warn (accessed January 15, 2014).

Social Issues Research Centre (SIRC). 2014. Genetically modified food. http://www.sirc.org/gate/genetically_modified_food.html (accessed January 15, 2014).

Squidoo. 2014. Genetically modified foods: The debate. http://www.squidoo.com/GM-food (accessed March 12, 2014).

Tarrago-Trani, M.T., Phillips, K.M., Lemar, L.E., and Holden, J.M. 2006. New and existing oils and fats used in products with reduced trans-fatty acid content. *Journal of American Dietetic Association*, 106(6):867–80.

The Sustainable Future. 2008. Genetically modified food: Pros and cons. http://curtrosengren.typepad.com/sustainable/ (accessed March 10, 2014).

Union of Concerned Scientists. 2012. Genetic engineering in agriculture. http://www.ucsusa.org/food_and_agriculture/our-failing-food-system/genetic-engineering/#monarch (accessed September 1, 2014).

United Nations Industrial Development Organization (UNIDO). 2007. Industrial biotechnology and biomass utilization: Prospects and challenges for the developing world.

United States Institute of Medicine and National Research Council. 2004. *Safety of Genetically Engineered Foods: Approaches to Assessing Unintended Health Effects*, pp. R9–10. National Academies Press, Washington, DC.

USDA. Animal and Plant Health Inspection Service (APHIS). 2012. Questions and answers: Okanagan Specialty Fruits' non-browning apple (Events GD743 and GS784). http://www.aphis.usda.gov/publications/biotechnology/2012/faq_okanagan_apple.pdf (accessed August 7, 2014).

Vogt, D.U. and Parish, M. 2001. Food biotechnology in the United States: Science, regulation, and issues. Congressional Research Service, The Library of Congress (CRS Report for Congress, Order Code RL30198), Washington, DC.

von Mogel, K.H. 2013. GMO crops vandalized in Oregon. Biology Fortified. http://www.biofortified.org/2013/06/gmo-crops-vandalized-in-oregon/ (accessed August 26, 2014).

Weiss, T.C. 2013. Genetically modified foods and health risks. http://www.disabled-world.com/fitness/nutrition/foodsecurity/gm-health-risks.php/ (accessed August 26, 2014).

WHO. 2005. Modern biotechnology, human health, and development: An evidence-based study. http://www.who.int/foodsafety/publications/biotech/biotech_en.pdf?ua=1 (accessed August 7, 2014).

WHO. 2014. Food safety. Frequently asked questions on genetically modified foods. http://www.who.int/foodsafety/areas_work/food-technology/faq-genetically-modified-food/en/ (accessed October 16, 2014).

Wu, X., Ouyang, H., Duan, B. et al. 2012. Production of cloned transgenic cow expressing omega-3 fatty acids. *Transgenic Research*, 21(3):537–43.

Yahoo News. AP. 2013. Millions march against Monsanto in over 400 cities. http://news.yahoo.com/millions-march-against-monsanto-over-400-cities-222259976.html (accessed August 26, 2014).

SUGGESTED REFERENCES

Africa Biofortified Sorghum (ABS) Project. 2012. ABS project: Technology development. http://biosorghum.org/abs_tech.php (accessed May 22, 2014).

African Agricultural Technology Foundation. 2012. GMO crops riding high. http://www.aatf-africa.org/ (accessed May 20, 2014).

African Agricultural Technology Foundation. 2012. Water Efficient Maize for Africa (WEMA). http://wema.aatf-africa.org/about-wema-project (accessed May 22, 2014).

Batista, R. and Oliveira, M.M. 2009. Facts and fiction of genetically engineered food. *Trends in Biotechnology*, 2:277–86.

Boyens, I. 1999. *Unnatural Harvest: How Corporate Science is Secretly Altering Our Food*. Doubleday Books, New York.

Capper, J.L., Castañeda-Gutiérrez, E., Cady, R.A., and Bauman, D.E. 2008. The environmental impact of recombinant bovine somatotropin (rbST) use in dairy production. *Proceedings of National Academy of Sciences*, 105(28): 9668–73.

Carter, N. 2012. Petition for determination of nonregulated status: ArcticTM Apple (*Malus domestica*). Events GD743 and GS784. http://www.aphis.usda.gov/brs/aphisdocs/10_16101p.pdf (accessed May 22, 2014).

Charlebois, S., Labrecque, J., and Spiers, E. 2007. Can genetically modified foods be considered as a dominant design? *British Food Journal*, 109:81–91.

Chassy, B., Hlywka, J., Kleter, G. et al. 2008. Nutritional and safety assessments of foods and feeds nutritionally improved through biotechnology. *Comprehensive Reviews in Food Science and Food Safety*, 7:50–113.

CIAT (International Center for Tropical Agriculture) & IFPRI (International Food Policy Research Institute). 2002. Biofortified crops for improved human nutrition. A challenge program proposal. http://www.cgiar.org/pdf/biofortification.pdf (accessed March 10, 2014).

Codex Alimentarius Commission. 2003. Alinorm 03/34: Joint FAO/WHO Food Standard Programme, Codex Alimentarius Commission, 25th Session, Rome, June 30–July 5, 2003. Appendix III, Guideline for the conduct of food safety assessment of foods derived from recombinant-DNA plants and Appendix IV, Annex on the assessment of possible allergenicity.

Conko, G. and Miller, H.I. 2011. The rush to condemn genetically modified crops: Impractical regulations and nuisance lawsuits. Academic journal article from *Policy Review*, No. 165.

Conko, G. and Prakash, C.S. 2004. Can GM crops play a role in developing countries? *AgBioWorld*. http://www.agbioworld.org/biotech-info/articles/agbio-articles/gm-crop-role.html (accessed January 6, 2014).

Conner, A.J., Glare, T.R., and Nap, J.P. 2003. The release of genetically modified crops into the environment: Part II. Overview of ecological risk assessment. *Plant Journal*, 33:19–46.

Conway, G. 2000. Genetically modified crops: Risks and promise. *Conservation Ecology*, 4(1):2. http://www.consecol.org/vol4/iss1/art2 (accessed May 10, 2014).

Dawkins, K. 1997. *Gene Wars: The Politics of Biotechnology*. Open Media Pamphlet Series. Seven Stories Press, New York.

Edgerton, M.D. 2009. Increasing crop productivity to meet global needs for feed, food, and fuel. *Plant Physiology*, 149(1):7–13.

Ellstrand, N. 2000. The elephant that is biotechnology: Comments on "Genetically modified crops: Risks and promise" by Gordon Conway. *Conservation*

Ecology, 4(1):8. http://www.consecol.org/vol4/iss1/art8 (accessed May 10, 2014).

Falk, M.C., Chassy, B.M., Harlander, S.K., Hoban, T.J., IV, McGloughlin, M.N., and Akhlaghi, A.R. 2002. Food biotechnology: Benefits and concerns. *Journal of Nutrition*, 132(6):1384–90.

Falkner, R. Ed. 2007. *The International Politics of Genetically Modified Foods: Diplomacy, Trade and Law*. Palgrave Macmillan, Basingstoke, England, New York.

FAO. 2000. FAO statement on biotechnology. http://www.fao.org/biotech/fao-statement-on-biotechnology/en/ (accessed May 20, 2014).

FAO. 2009a. Feed the world, eradicating hunger. Paper presented at World Summit on Food Security.

FAO. 2009b. G.M. Food safety assessment: Tools for trainers. Food and Agriculture Organization of the United Nations, Rome.

FAO. 2012. Coping with water scarcity: An action framework for agriculture and food safety. FAO, Rome.

FAO/WHO. 2000. Safety aspects of genetically modified foods of plant origin. Report of a Joint FAO/WHO Expert Consultation on Foods Derived from Biotechnology, Geneva, Switzerland, May 29–June 2, 2000. ftp://ftp.fao.org/es/esn/food/gmreport.pdf (accessed March 15, 2014).

FAO/WHO. 2003a. Principles for the risk analysis of foods derived from modern biotechnology. Rome. ftp://ftp.fao.org/es/esn/food/princ_gmfoods_en.pdf (accessed March 16, 2014).

FAO/WHO. 2003b. Guideline for the conduct of food safety assessment of foods derived from recombinant-DNA plants. Rome. ftp://ftp.fao.org/es/esn/food/guide_plants_en.pdf (accessed March 16, 2014).

FDA. 2011. Bovine somatotropin. http://www.fda.gov/AnimalVeterinary/SafetyHealth/ProductSafetyInformation/ucm055435.htm (accessed May 20, 2014).

FDA. 2012a. Genetically engineered animals. http://www.fda.gov/AnimalVeterinary/DevelopmentApprovalProcess/GeneticEngineering/GeneticallyEngineeredAnimals/default.htm (accessed May 20, 2014).

FDA. 2012b. Regulation of genetically engineered animals. http://www.fda.gov/ForConsumers/ConsumerUpdates/ucm048106.htm (accessed May 20, 2014).

Fedoroff, N.V. and Brown, N.M. 2004. *Mendel in the Kitchen: A Scientist's View of Genetically Modified Foods*. Joseph Henry Press. Imprint of National Academies Press, Washington, DC.

Five Year Freeze. 2002. Feeding or fooling the world? Can GM really feed the world? http://www.fiveyearfreeze.org/Feed_Fool_World.pdf (accessed March 16, 2014).

Floros, J.D., Newsome, R., Fisher, W. et al. 2010. Feeding the world today and tomorrow: The importance of food science and technology. An IFT scientific review. *Comprehensive Reviews in Food Science and Food Safety*, 9:572–99.

Food and Drink Federation. 1998. *Genetically Modified Foods. A Flavour of the Current Issues*. Food and Drink Federation, London.

Food and Drink Federation. 1999. *GM Crops and the Environment. Benefits and Risk.* Food and Drink Federation, London.
Gaskell, G. and Bauer, M.W. (Eds.). 2001. *Biotechnology 1996–2000. The Years of Controversy.* Science Museum, London.
Goldstein, D.A., Tinland, B., Gilbertson, L.A. et al. 2005. Human safety and genetically modified plants: A review of antibiotic resistance markers and future transformation selection technologies. *Journal of Applied Microbiology*, 99:7–23.
Gonsalves, D. 2004. Virus-resistant transgenic papaya helps save Hawaiian industry. *California Agriculture*, 58(2):92–3.
Greenpeace. 2010. GM crops and world hunger. http://www.greenpeace.org.uk/about/greenpeace-gm-crops-and-world-hunger (accessed July 20, 2014).
Gruère, G.P. 2011. Global welfare and trade-related regulations of GM food: Biosafety, markets, and politics, Chapter 13, in Carter, C.A., Moschini, G.C., and Sheldon, I. (Eds.), *Frontiers of Economics and Globalization*, Volume 10. Genetically Modified Food and Global Welfare. Emerald Group Publishing Limited, Bingley, UK, pp. 309–336.
Guidetti, G. 2009. Pros and cons of genetically modified foods. http://ezinearticles.com/?Pros-and-Cons-of-Genetically-Modified-Foods&id=3415986 (accessed July 20, 2014).
Hails, R.S. 2000. Genetically modified plants—The debate continues. *Trends in Evolution and Ecology*, 15(1):14–8.
Hammond, B.G. and Jez, J.M. 2011. Impact of food processing on the safety assessment for proteins introduced into biotechnology-derived soybean and corn crops. *Food and Chemical Toxicology*, 49:711–21.
Heinberg, R. 1999. Cloning of the Buddha: The moral implications of biotechnology. *Quest*.
Herrera-Estrella, L. and Alvarez-Morales, A. 2001. Genetically modified crops: Hope for developing countries? The current GM debate widely ignores the specific problems of farmers and consumers in the developing world. *EMBO Reports*, 2(4):256–8.
Ho, M.-W. 1998. *Genetic Engineering Dream or Nightmare? The Brave New World of Science and Business.* Gateway Books, Nevada City, CA.
Ho, M.-W. 2007. *GMO Free: Exploring the Hazards of Biotechnology to Ensure the Integrity of Our Food Supply.* Institute of Science in Society and Third World Network. Vital Health Publishing, Ridgefield, CT.
Hodge, R. 2009. *Genetic Engineering: Manipulating the Mechanisms of Life. Facts on File.* New York.
Hubbard, R. and Ward, E. 1996. *Exploding the Gene Myth.* Beacon Press, Boston.
Huffman, W.E., Rousu, M., Shogren, J.F., and Tegene, A. 2003. Better dead than GM fed? Information and the effects of consumers' resistance to GM foods in high-income countries. Department of Economics, Iowa State University, Iowa.
Hug, K. 2008. Genetically modified organisms: Do the benefits outweigh the risks? *Medicina (Kaunas)*, 44:87–99.

Hutchison, W.D., Burkness, E.C., Mitchell, P.D. et al. 2010. Area wide suppression of European corn borer with Bt maize reaps savings to non-Bt maize growers. *Science*, 330(6001):222–5.

IFIC. 2012. Consumer perceptions of food technology survey. http://www.foodinsight.org/Resources/Detail.aspx?topic=2012ConsumerPerceptionsofTechnologySurvey (accessed May 20, 2014).

International Institute of Tropical Agriculture. 2012. IITA in the News 2012 Archive. http://www.iita.org/ (accessed May 21, 2014).

International Rice Research Institute (IRRI). 2012. Golden Rice Project. http://www.irri.org/goldenrice/ (accessed June 10, 2014).

International Service for the Acquisition of Agri-Biotech Applications (ISAAA). 2004. Pocket K No. 12: Delayed ripening technology. ISAAA, Manila.

Isaac, G. 2002. *Agricultural Biotechnology and Transatlantic Trade: Regulatory Barriers to GM Crops*. CABI Publishing, Wallingford, Oxon, UK, New York.

James, C. 2012. Global status of commercialized biotech/GM crops. ISAAA Brief No. 44. ISAAA, Ithaca, NY.

Jefferson, V. 2006. The ethical dilemma of genetically modified food. *Journal of Environmental Health*, 69(1):33–5.

Kier, L.D. and Petrick, J.S. 2008. Safety assessment considerations for food and feed derived from plants with genetic modifications that modulate endogenous gene expression and pathways. *Food and Chemical Toxicology*, 46:2591–605.

König, A., Cockburn, A., Crevel, R.W. et al. 2004. Assessment of the safety of foods derived from genetically modified (GM) crops. *Food and Chemical Toxicology*, 42:1047–88.

Kotak, P. 2014. Are you aware of dangerous genetically modified foods that you might already be eating? HomeRemedies foryou.com. http://www.home-remedies-for-you.com/articles/585/general-wellness/are-you-aware-of-dangerous-genetically-modified-foods-that-you-might-already-be-eating.html (accessed May 15, 2014).

Kowalski, K.M. 2002. *The Debate over Genetically Engineered Food: Healthy or Harmful?* Enslow Publishers, Berkeley Heights, NJ.

Lambrecht, B. 2002. *Dinner at the New Gene Café: How Genetic Engineering is Changing What we Eat, How we Live, and the Global Politics of Food*. Macmillan, London, UK.

Langhof, O. 2011. Why GM foods won't solve hunger in Africa. *Greenpeace Africa*. http://www.greenpeace.org/africa/en/News/Blog/why-gm-food-wont-solve-hunger-in-africa/blog/36437/ (accessed May 15, 2014).

Lappe, M. and Bailey, B. 2002. *Against the Grain: Biotechnology and the Corporate Takeover of Your Food*. Common Courage Press, Monroe, ME.

Lawrence, G. and Hindmarsh, R. (Eds.). 2004. *Recoding Nature: Critical Perspectives on Genetic Engineering*. UNSW Press, Sydney.

Lei, H. 2007. Genetically modified foods: Pros and cons. http://www.geneticsandhealth.com/2005/07/26/genetically-modified-food-pros-and-cons/ (accessed May 15, 2014).

Lotter, D. 2009a. The genetic engineering of food and the failure of science—Part 1: The development of a flawed enterprise. *International Journal of Sociology of Agriculture and Food*, 16:31–49.

Lotter, D. 2009b. The genetic engineering of food and the failure of science—Part 2: Academic capitalism and the loss of scientific integrity. *International Journal of Sociology of Agriculture and Food*, 16:50–68.

MacKenzie, D. and McLean, M. 2002. Who's afraid of GM feeds? *Feed Mix*, 10(3): 16–19. http://www.agbios.com/docroot/articles/02-232-001.pdf (accessed March 25, 2014).

Mather, R. 2012. The threat from genetically modified foods. *Mother Earth News*, April/May, 42–51.

May, R.M. 1999. Genetically modified foods: Facts, worries, policies and public confidence. Office of Science and Technology, London, UK. http://www.dti.gov.uk/ost/genetic/geni.htm (accessed March 25, 2014).

Mayer, S., Hill, J., Grove-White, R., and Wynne, B. 1996. *Uncertainty, Precaution and Decision Making: The Release of Genetically Modified Organisms into the Environment*. The Green Alliance, Publication BIO3, London.

McNight, C. 2014. List of foods that are GM foods. LIVESTRONG.COM. http://www.livestrong.com/article/419389-list-of-foods-that-are-gm-foods/ (accessed May 25, 2014).

Mendoza, E.M.T., Laurena, A.C., and Botella, J.R. 2008. Recent advances in the development of transgenic papaya technology. *Biotechnology Annual Review*, 14:423–62.

Meron, T.M. 2002. Africa bites the bullet on genetically modified food aid. Worldpress.org. http://www.worldpress.org/Africa/737.cfm (accessed May 25, 2014).

Monsour, M. 2013. Measure M: The pros and cons of banning GMOs. http://www.des.ucdavis.edu/ (accessed May 25, 2014).

Morris, J. 2006. *The Ethics of Biotechnology*. Chelsea House Publishers, Philadelphia.

National Centre for Biotechnology Education. 2014. Genetically-modified food 2014. http://www.ncbe.reading.ac.uk/NCBE/GMFOOD/menu.html (accessed August 24, 2014).

National Center for Food and Agricultural Policy. 2014. News and events. http://www.ncfap.org/ (accessed May 10, 2014).

National Research Council of the National Academies. 2012. *National Summit on Strategies to Manage Herbicide-Resistant Weeds: Proceedings of a Symposium*. The National Academies Press, Washington, DC.

Nestle, M. 2003. *Safe Food: Bacteria, Biotechnology, and Bioterrorism*. University of California Press, Berkeley, CA.

Netherwood, T., Martín-Orúe, S.M., O'Donnell, A.G. et al. 2004. Assessing the survival of transgenic plant DNA in the human gastrointestinal tract. *Nature Biotechnology*, 22:204–9.

Newell-McGloughlin, M. 2012. Transgenic crops, next generation, in R.A. Meyers (Ed.), *Encyclopedia of Sustainability Science and Technology*, Volume 15. Springer Science + Business Media, LLC, New York, pp. 10732–65.

Nottingham, S. 2003. *Eat Your Genes. How Genetically Modified Food is Entering Our Diet*. Zed Books, London, UK.
Nuffield Council on Bioethics. 1999. *Genetically Modified Crops: The Ethical and Social Issues*. London, UK.
Ochman, H., Lawrence, J.G., and Groisman, E.A. 2000. Lateral gene transfer and the nature of bacterial innovation. *Nature*, 405:299–304.
Omar, R. 2009. The genetic engineering of rice and other crops. The genetic engineering industry: Impacts and regulation. http://www.fomca.org.my/v2/images/stories/pdf/071009_GE_Seminar_ppt6.pdf (accessed January 20, 2014).
Organic Food for Everyone. 2014. GM foods—Part 2: Genetically modified foods (GM) and substantial equivalence. http://www.organic-food-for-everyone.com/gm-foods.html (accessed July 10, 2014).
Owens, S. 2001. Salt of the Earth: Genetic engineering may help to reclaim agricultural land lost due to salinisation. *EMBO Reports*, 2(10):877–9.
Paparini, A. and Romano-Spica, V. 2004. Public health issues related with the consumption of food obtained from genetically modified organisms. *Biotechnology Annual Reviews*, 10:85–122.
Parr, D. 1997. *Genetic Engineering. Too Good to go Wrong?* Greenpeace UK, London.
Phillips, P.W.B. and McNeill, H. 2002. Labelling for GM foods: Theory and practice, in Santaniello, V., Evenson, R.E., and Zilberman, D. (Eds.), *Market Development for Genetically Modified Foods*. CABI Publishing, Wallingford, UK, pp. 245–260.
Pimentel, D., Hunter, M.S., Lagro, J.A. et al. 1989. Benefits and risks of genetic engineering in agriculture. *BioScience*, 39(9):606–14.
Pinstrup-Andersen, P. and Schiøler, E. 2000. *Seeds of Contention: World Hunger and the Global Controversy over GM Crops*. International Food Policy Research Institute. The Johns Hopkins University Press, Baltimore.
Pirondini, A. and Marmiroli, N. 2008. Environmental risk assessment in GMO analysis. *Rivista di Biologia*, 101:215–46.
Pringle, P. 2003. *Food Inc.: Mendel to Monsanto—The Promises and Perils of the Biotech Harvest*. Simon and Schuster, New York.
Raeburn, P. 1996. The last harvest: The genetic gamble that threatens to destroy American agriculture. University of Nebraska.
Rajasekaran, K., Cary, J.W., and Cleveland, T.E. 2006. Prevention of preharvest aflatoxin contamination through genetic engineering of crops. *Mycotoxin Research*, 22(2):118–24.
Raney, T. and Matuschke, I. 2011. Current and potential farm-level impacts of genetically modified crops in developing countries, Chapter 3, in Carter, C.A., Moschini, G.C., and Sheldon, I. (Eds.), *Frontiers of Economics and Globalization*, Volume 10. Genetically Modified Food and Global Welfare. Emerald Group Publishing Limited, Bingley, UK, pp. 55–82.
Rifkin, J. 1999. *The Biotech Century: Harnessing the Gene and Remaking the World*. J.P. Tarcher Inc. An imprint of Penguin Group, New York.
Rissler, J. and Mellon, M. 1996. *The Ecological Risks of Engineered Crops*. MIT Press, Cambridge, MA.

Rowell, A. 2004. *Don't Worry (It's Safe to Eat): The True Story of GM Food, BSE and Foot and Mouth*. Reprint. Earthscan Publications Limited, London, UK.

Ruse, M. and Castle, D. (Eds.). 2002. *Genetically Modified Foods: Debating Biotechnology*. Prometheus Books, Amherst.

Sakko, K. 2002. The debate over genetically modified foods. http://www.actionbioscience.org/biotech/sakko.html (accessed June 20, 2014).

Sasson, A. 1998. *Biotechnologies in Developing Countries: Present and Future*. Volume 2: International Cooperation. UNESCO Publishing, Paris, France.

Schonwald, J. 2012. The future of food. *The Futurist*, pp. 24–8. www.wfs.org (accessed January 11, 2014).

Science and Society. 2014. How do economic factors affect GMO development? http://www.scienceandsociety.emory.edu/GMO/Economics.htm (accessed February 10, 2014).

Science Museum. 1992. *Biotechnology in Public. A Review of Recent Research*. Science Museum, London.

Scorza, R. and Ravelonandro, M. 2006. Control of plum pox virus through the use of genetically modified plants. *OEPP/EPPO Bulletin*, 36:337–40.

Shah, A. 2002. Genetically engineered food. Global issues. http://www.globalissues.org/issue/188/genetically-engineered-food (accessed September 12, 2014).

Shiva, V. 1997. *Biopiracy: The Plunder of Nature and Knowledge*. South End Press, Cambridge, MA.

Shiva, V. 1999a. *Monocultures of the Mind: Perspectives on Biodiversity and Biotechnology*. South End Press, Cambridge, MA.

Shiva, V. 1999b. *Stolen Harvest: The Highjacking of the Global Food Supply*. South End Press, Cambridge, MA.

Shiva, V., Barker, D., and Lockhart, C. 2011. The GMO Emperor has no clothes: A global citizens report on the state of GMOs—False promises, failed technologies. Synthesis Report. Navdanya International, Brooklyn, New York.

Smith, J.M. 2003. *Seeds of Deception: Exposing Industry and Government Lies About the Safety of the Genetically Engineered Foods You're Eating*. Yes Books, Portland, ME.

Smith, J.M. 2007. *Genetic Roulette: The Documented Health Risks of Genetically Engineered Foods*. Yes Books, Fairfield, Iowa.

Snell, C., Bernheim, A., Bergé, J.B. et al. 2011. Assessment of the health impact of GM plant diets in long-term and multigenerational animal feeding trials: A literature review. *Food and Chemical Toxicology*, 50:1134–48.

Social Issues Research Centre (SIRC). 2014. Genetically modified food. http://www.sirc.org/gate/genetically_modified_food.html (accessed January 15, 2014).

Stone, G.D. 2002. Both sides now: Fallacies in the genetic modification wars, implications for developing countries, and anthropological perspectives. *Current Anthropology*, 43(4):611–30.

Stonebrook, S. 2013. 4 Potential health risks of eating GMO foods. Care2. http://www.care2.com/greenliving/health-risks-of-eating-gmo-foods.html (accessed January 20, 2014).

Straughan, R. and Reiss, M. 2001. *Improving Nature? The Science and Ethics of Genetic Engineering.* Cambridge University Press, Cambridge.

Teitel, M. and Wilson, K.A. 2001. *Genetically Engineered Food: Changing the Nature of Nature.* Park Street Press: A division of Inner Traditions International, Rochester, VT.

The Green Alliance. 1994. *Why Are Environmental Groups Concerned About Release of Genetically Modified Organisms into the Environment?* Publication BIO1. The Green Alliance, London.

The Green Alliance. 1997. How can biotechnology benefit the environment? Report of the European federation of biotechnology. Green Alliance Workshop. Publication BIO4. London.

Third World Network (TWN). 1999. British Medical Association Board of Science and Education. The impact of genetic modification on agriculture, food and health—An interim statement. http://www.twnside.org.sg/title/genmo-cn.htm (accessed January 6, 2014).

Thomas, K., Herouet-Guicheney, C., Ladics, G. et al. 2007. Evaluating the effect of food processing on the potential human allergenicity of novel proteins: International workshop report. *Food and Chemical Toxicology*, 45:1116–22.

Thompson, P.B. 1997. *Food Biotechnology in Ethical Perspective.* Blackie Academic & Professional, London.

Thomson, J.A. 2007. *Seeds for the Future: The Impact of Genetically Modified Crops on the Environment.* Comstock Publishing Associates, Ithaca, NY.

Ticciati, L. and Ticciati, R. 1998. *Genetically Engineered Foods: Are They Safe? You Decide.* McGraw-Hill, New York.

Tollefson, J. 2011. Brazil cooks up transgenic bean. *Nature*, 478(7368):168.

United Nations Department of Economic and Social Affairs (UNDESA). 2010. Water scarcity. http://www.un.org/waterforlifedecade/scarcity.shtml (accessed January 20, 2014).

United Nations University, Institute of Advanced Studies. 2005. Food and nutrition biotechnology: Achievements, prospects and perceptions.

UNU-IAS Report. United Nations University. Yokohama, Japan. http://www.eldis.org/vfile/upload/1/document/0708/DOC19681.pdf (accessed May 22, 2014).

USDA, Agricultural Research Services (ARS). 2009. HoneySweet plum trees: A transgenic answer to the plum pox problem. http://www.ars.usda.gov/is/br/plumpox/ (accessed May 20, 2014).

USDA, Agricultural Research Services (ARS). 2010. Improving rice, a staple crop worldwide. *Agricultural Research Magazine*, 58(5):4–7.

USDA, APHIS. 2012. Biotechnology. http://www.aphis.usda.gov/biotechnology/ (accessed May 20, 2014).

van den Eede, G., Aarts, H., Buhk, H.J. et al. 2004. The relevance of gene transfer to the safety of food and feed derived from genetically modified (GM) plants. *Food and Chemical Toxicology*, 42:1127–56.

Weasel, L.H. 2008. *Food Fray: Inside the Controversy over Genetically Modified Food.* AMACOM.

WebMD. 2013. Are biotech foods safe to eat? http://www.webmd.com/food-recipes/features/are-biotech-foods-safe-to-eat (accessed July 24, 2014).

WHO. 2009. Global prevalence of vitamin A deficiency in populations at risk 1995–2005: WHO global database on vitamin A deficiency http://www.who.int/vmnis/database/vitamina/x/en/index.html (accessed May 25, 2014).

Wohlers, A.E. 2010. Regulating genetically modified food. Policy trajectories, political culture, and risk perceptions in the U.S., Canada, and EU. *Politics and the Life Sciences*, 29(2):17–39.

7
Consumer Issues

7.1 INTRODUCTION

Consumers' opinion and reaction toward different issues are considered as critical factors and powerful drivers that help to shape governmental policies in different parts of the globe. Recently, consumers' opinion on genetically modified (GM) foods has been in the spotlight and has a great influence on governmental policies related to GM technology in general and to GM foods in particular. It is extremely important for policy-makers, regulators, and food producers to understand consumers' opinion because it very much influences the ultimate acceptance or rejection of GM foods in the marketplace. Governments usually design programs that aim at improving and maintaining citizens' health. Part of these programs encompasses availing and disseminating information related to health, which includes education, public information campaigns, and regulation of advertising and labeling (Aldrich 1999). Information on various issues related to GM foods, as well as other sources, is made available by the government, through its different departments, and made accessible to consumers.

Consumers in different parts of the world are looking at various issues surrounding GM foods, with a lot of caution. This cautious approach resulted from a number of factors. The highly publicized negative image created by anti-GM groups caused a food scare among consumers and developed a negative attitude toward GM foods. Other possible factors include the limited knowledge and awareness of consumers toward GM foods, the lack of trust in government experts and food regulators, and the difficulty to manage food safety due to the increased food trade and globalization (Kim 2012).

Generally, consumers fall into different categories with regard to perceptions and attitudes toward GM foods. Some have expressed their concerns about safety of GM foods and are hesitant to consume such modified foods, whereas others believe that this technology has benefits that outweigh the drawbacks and that it is safe to consume GM foods as there is no reported evidence of health risks. A third group has doubt and insecurity toward GM foods, but at the same time acknowledges the benefits of biotechnology as an effective tool for sustainable food supply and availability. The anticipated potential risks linked to consumption of GM foods, and the negative effects of the technology on the environment as well as the ecosystem, are believed to be the main factors that cause fear, uncertainty, and doubt among some consumers (Phillips and Corkindale 2002). The level of consumers' concern on food safety increases with involuntary exposure to food risks, when food risk is believed to be uncontrollable and lacks scientific proof (Slovic 1999).

A number of factors may affect consumers' perceptions and attitudes toward GM foods. Consumers use cognitive, affective, or behavioral responses to accept or reject GM foods (Frewer 2003). Many researchers (Scholderer et al. 1999; Boccaletti and Moro 2000; Rosati and Saba 2000; Subrahmanyan and Cheng 2000; Moon and Balasubramanian 2004; Lusk and Coble 2005; Curtis and Moeltner 2006; Chen and Li 2007) conclude that acceptance or rejection of GM foods by consumers is likely to be associated with their risk/benefit beliefs about biotechnology. It is, therefore, logical to say that consumers who believe that GM food has benefits will be more willing to buy and consume it, whereas consumers who perceive GM food as a health risk and risky to the environment will be less willing to purchase it (Han and Harrison 2007).

Consumers may show some level of tolerance to risk associated with consumption of GM foods if that is linked to some direct benefit to themselves (Frewer 2003). This can be true on condition that the risk is not too large to be totally intolerable. In that case, the main factor determining consumers' acceptance of a particular technology will be the perception of personal benefits.

Acceptability or rejection of any new technology is very much affected by risks and benefits of its users or the general public perceptions. Many such technologies may take a long time to deliver noticeable benefits to the public. This would create a potential risk if the technology has wide benefits, which are not clear to the public and the risks are more clear. Further exploitation of such technologies would become nonviable. Consumers' decision-making is found to be very much influenced by direct and immediate benefits they get from a product or a technology. This applies to GM foods, but it is the assessment of risks and benefits that affects consumers' perceptions and attitudes. Other factors such as ethical and moral aspects, information and knowledge of the technology, and the level of trust in governmental regulatory bodies play a part. In addition to the innate factors that affect consumers' choice and acceptability to any product, GM foods in this case, other external factors are found to have influence on consumers' decisions. Examples of external factors include wide campaigns against GM applications, which are spearheaded by established and influential lobby groups such as Greenpeace and Friends of the Earth. The media too have played a major role in influencing public attitudes toward GM foods. The tone of the media reports on benefits and risks of GM technology has noticeable impact on the public decisions.

The adoption of GM technology in countries other than the United States met a slow start. This resulted in a widespread resistance from consumers toward the technology and its various applications. Consumers' objection to the technology was, to a large extent, fueled by various news against GM, which is expressed in the different media outlets (Knight et al. 2014). Due to the lack of global agreement on the issue of food labeling, consumers in different parts of the world have different perceptions and attitudes toward GM foods. Although the United States does not require mandatory labeling of GM foods, other countries such as European countries, Japan, Australia, New Zealand, Brazil, China, and South Korea require labeling of products that contain GM ingredients (Huffman et al. 2003; Thorpe and Robinson 2005; O'Fallon et al. 2007).

A societal anxiety is felt among consumers over the GM tools. This effect is caused by multiple factors such as consumer unfamiliarity, lack of reliable information, a steady stream of negative opinion in the news media, opposition by activist groups, growing mistrust of industry, and a general lack of awareness of how the food production system has evolved (Conko and Prakash 2004). Some of the reported factors that affect consumers' attitudes toward GM foods include socioeconomic factors such as level of education, gender, and religious beliefs. Introduction of new information about GM technology would also affect consumers' attitudes and reactions toward GM. It is suggested that the existence of negative new information on GM would increase negative attitudes toward the technology. At the same time, this negative attitude may be reversed with the introduction of new positive information (Heiman and Zilberman 2011) and the existence of evidence of benefits from the use of genetically modified organisms (GMOs).

7.2 CONSUMER RIGHTS

Consumers include everybody in the society. They are considered a major contributor to the economy because they affect, and are affected by, different public and private economic decisions. Consumers have legal rights as well as responsibilities. Steel (2014) lists eight consumer rights, four of which were included in the "Consumer Bill of Rights" enacted by the U.S. Congress in the early 1960s and which was later recognized by the United Nations and its member states. The consumer rights recognized by the United Nations are the right to safety (of products and services); the right to be informed (about products and services); the right to choose (from among a variety of products and services); the right to be heard (on issues affecting their lives); the right to satisfaction (of basic needs); the right to redress; the right to consumer education; and the right to healthy environment. Considering GM foods as a product of the food industry using the GM technology, it is imperative that these eight consumer rights apply to GM products. It would be the responsibility of the various government authorities, as well as the GM food producers and processors, to ensure that these rights are operational. At the same time, consumers have a huge responsibility to know what their rights are and to ensure that they enjoy them. It is important that consumers understand what the food standards, regulations, and labeling entail. Inadequate knowledge and understanding of these aspects may have negative consequences on consumers' choices and consumption that might lead to ill health. For example, consumers need to understand what the PLU Code (Price LookUp Code—Produce Code—Produce Label), shown on the small sticker on fruits and vegetables, means. The code has the following meanings: labels beginning with "9" indicate organic; labels beginning with "4" or "3" indicate conventional; and labels beginning with "8" indicate GM. The International Federation for Produce Standards (IFPS 2012) prepared a document that contains rules for the international PLU numbering scheme. The international PLU system is governed by voluntary cooperation of participating countries that are represented by national or regional representatives on the IFPS board of directors. PLU codes have been used by supermarkets since 1990 to make checkout and inventory control easier, faster, and more accurate. Fresh fruit and vegetable PLU codes are used to identify bulk produce (and related items such as nuts and herbs). Some consumers are usually in a hurry and do not spend time looking at these labels, thus ignoring a right they have.

Direct and indirect involvement of the public through offering the opportunity to provide information and comments on proposed regulations, and having a high level of transparency, would have a positive impact on increasing trust in what food regulators provide as laws and regulations. Consumers need to be involved at different stages of drafting food standards, laws, and regulations. Their views should be given due consideration and should be used in defining the required regulations. This will give the consumers some degree of ownership of the food laws and regulations and increase their trust and confidence in the regulating authorities, the resulting regulations, and the food products. At the same time, consumers' involvement ensures that they are practicing their rights with adequate satisfaction.

Consumers' education about various aspects of GM foods is believed to be a useful strategy to address concerns around potential health risks and adverse effects on the environment and the ecosystem (Han and Harrison 2007). Educating consumers and raising their awareness on different GM food issues are not expected to bear fruits in the short run. Consumers' knowledge and full understanding of GM foods would be a long process. In the meantime, it is expected that there will be an increasing demand for non-GM and organic foods. Nonetheless, effective management of safety verification and continuous public education on GM foods may assure consumers of the safety of GM foods in the long term, eventually leading to more acceptability of GM foods (Kim 2012).

Some of the consumers' rights are the right to choose, the right to availability and access to credible information, and the right to participation in discussions and decisions of issues that affect their lives and well-being. Respect of consumers' rights has been a central issue and a focal area in the debate around GM technology and GM foods in different parts of the globe. The debate about consumers' right to information on GM foods is centered on labeling and the potential for some hazards and risks related to human health. Some of the pertinent questions that require detailed and unbiased answers to inform the public and to give them the chance to make informed decisions and the right choice include the following:

- Should all GMOs in food be labeled to identify them?
- What types of words or descriptions would help consumers and not mislead or confuse them?
- Is the use of negative language in GM food labeling appropriate and should it be allowed?
- If GM food labeling increases the cost of production, who should pay for the extra cost?

- What should be the balance between mandatory and voluntary labeling?

7.3 STUDIES ON CONSUMERS' PERCEPTIONS, ATTITUDES, AND PREFERENCES FOR GM FOODS

Since the time GM foods have been introduced in the market, consumers started to express some concerns and different views about consumption of GM foods. According to Lusk (2011), majority of the surveyed European consumers were unwilling to eat GM foods. In 1997, more than half the population of many European countries were unwilling to eat GM foods, and nearly two-thirds of Swedes, Germans, and Austrians considered GM food to be a serious health hazard. This wide concern about GM foods among consumers prompted many economists to carry out economic analysis on various issues related to GM foods. Other researchers also showed interests in, and conducted many studies on, consumers' perceptions, attitudes, and views on GM foods.

Many consumer surveys have been conducted in different parts of the world, and the findings and interpretation of the results are widely published. Subsequently, the scientific and nonscientific literatures are now flooded with findings from research on consumer issues related to GM foods.

Despite the large volume of work published on consumer research on GM foods, Lusk (2011) believes that it is difficult to distill what is known about consumer preferences for GM foods. The apparent problem is that there is lack of consensus among applied theorists and producer and industry groups in trusting the results of research conducted on consumer issues related to GM foods. This might lead to more confusion among consumers about safety and benefits of GM foods.

Consumers' perceptions, attitudes, and responses toward GM foods are usually influenced by the type of information they receive about GM foods, for example, positive statements, negative statements, both negative and positive, or no statements. These statements can come from pro- or anti-GM groups, the media, or from the food producers and processors as part of the food labels. How consumers react to any of these statements is expected to depend on the strength of the message, that is, how the message providers emphasize the benefits and advantages or the risks and disadvantages of GM foods. The provision of positive, negative, and third-party information has been reported to significantly influence consumers' views on GM foods (Rousu et al. 2007). Consumers usually get information about GM foods, or any other related issue, from various media channels such as newspapers, specialized magazines, periodicals, radio, TV, or the internet. In most cases, views expressed in these media venues represent the author's or writer's views. In scientific and professional publications, interpretation of research results may also be subjective and painted by the author's views and beliefs. All these media outlets present positive, negative, or balanced views and information about GM foods. This information has direct impact on the consumers' views, perceptions, and attitudes toward GM foods. Consumers need to be careful in evaluating the information they receive from media to help them make the right decisions with regard to purchasing and consuming GM foods.

Opposition to GM technology and GM foods has been reported to be stronger among European and Asian nations than in the United States (Gaskell et al. 2003). Different interpretations have been cited to account for the European rejection of GM foods. Laros and Steenkamp (2004) suggest that European rejection is based on fear of the unknown and avoidance of risk, whereas Poortinga and Pidgeon (2005) believe that the absence of tangible benefits to the consumer is a strong reason for the rejection of GM foods than risk version or fear. Some explanations have been presented to account for the noted differences between American consumers and consumers in other parts of the world. One suggestion is that cultural factors can be major determinants in shaping consumers' approval of a specific technology. Cultural factors are usually deep-rooted and therefore would be difficult to change (Bredahl 2001). A consumer's evaluation of GM may be affected both by their world view and by their perception of benefits and risks of the technology (Siegrist 1999).

The International Food Information Council (IFIC) conducted a survey in 2012 about "Understanding consumer perceptions of food technology and sustainability." According to the reported results (IFIC 2012), a majority (69%) of the participating consumers have confidence in the U.S. food supply safety. The results of that survey also showed that the majority (69%) of the consumers would likely buy foods improved through biotechnology to provide better nutrition. Among all the participants, none reported avoiding GM foods, in spite of the fact that around half of the respondents reported avoidance of certain foods or ingredients.

In spite of the fact that nearly all consumers in the United States have consumed GM foods or foods containing GM ingredients, only 26% of the consumers believe that they have eaten a GM food (Pew Initiative on Food and Biotechnology 2006). This might indicate that lack of information or knowledge about GM foods among the average U.S. consumers is the main reason behind this discrepancy. Puduri et al. (2010) conducted a study that aimed to quantify consumers' perceptions in support of GM products in the United States. The study attempted to analyze and compare the effects of consumers' socioeconomic characteristics and their expressed value judgment on their intensity of support toward usefulness of GM foods. Results of the study indicate that public support for GM foods is significant and that people accept the usefulness of these foods for better future. It also established that factors such as gender, education, trust on government and its regulatory bodies, confidence on scientists, income levels, and ethnicity have significant influence on respondents' support for GM foods. The authors believe that an open national debate among various supporting and opposing groups in the society on GM and its products and dealing with issues of concern and the long-term health, environmental, social, and ethical implications would help consumers understand more about GM foods. Consumers feel that they need adequate information to educate and help them make more informed decisions with regard to consumption of GM foods. Consumers' perception about biotech food products may be influenced by producers' general attitude as well as through media reports.

One of the surveys about American consumers' attitudes toward GM foods was carried out in Colorado, USA (Byrne 2010). The survey covered 437 supermarket shoppers representing four different communities. Results of the survey showed that 78% of the participants supported mandatory GM food labeling, with women favoring labeling more than men and younger consumer less likely to support mandatory labeling. The respondents showed their unwillingness to pay a premium for such labeling. Another survey carried out in 2010 by Deloitte Development LLC (Deloitte Development 2010) reported that 34% of the U.S. consumers were very or extremely concerned about GM foods. With regard to gender differences, the survey found that 10% of men were extremely concerned compared with 16% of women and that 16% of women were unconcerned, compared with 27% of men.

Hebden et al. (2005) noticed a contradiction in the way Americans deal with GM foods. On the one hand, America is the chief producer and consumer of GM foods, whereas on the other hand, Americans seem to know less about the presence of GM foods than people living in other nations. A number of published reports (Gaskell et al. 2003; Hallman et al. 2004, 2005; Huang et al. 2004) indicate that American consumers perform better than Europeans and Asians on questions about the principles behind GM foods, but at the same time show some degree of unawareness about agricultural biotechnology (Pew Initiative on Food and Biotechnology 2006). It is also observed that more than half of the Americans do not realize that foods containing GM ingredients are sold in supermarkets and that less than a third believe that they have consumed GM foods (Hallman et al. 2004). Other observations on the knowledge and behavior of Americans in relation to GM foods when compared with consumers in other nations are also cited (Hallman et al. 2004). About a third of the Americans indicated that they have heard about European demonstrations opposing GM foods, and only about 20% say that they have discussed the GM food issue more than once or twice with anyone.

Many researchers published reports on consumer attitudes in Europe toward GM foods (Frewer et al. 1995; Grunert et al. 2000; Bredahl 2001). Negative views about GM foods have been reported in Europe as well as in other countries such as Singapore (Subrahmanyan and Cheng 2000) and New Zealand (Campbell et al. 2000; Gamble and Gunson 2002). Miles and Frewer (2003) found that the main concerns among U.K. consumers were human health, the environment, and animal welfare.

Legge and Durant (2010) studied the public opinion in the European Union (EU) with regard to GM foods. They stated that proponents of biotechnology argue that citizens' opposition to GM foods is rooted in emotionalism, media, and nongovernmental organizations' distortions of good science. Critics believe that this argument is too reductionist, exaggerates scientific capacity, inappropriately privileges scientific values over social and political values, and inaccurately captures how citizens evaluate biotechnology.

European consumer polls carried out in 2009 (GMO Compass 2009) showed a gradual decline in opposition to GMOs in Europe and that around 80% of the respondents did not actively avoid GM products when shopping.

A survey was conducted by Eurobarometer (2010) to find out the views of European consumers on different applications of biotechnology. One of the areas surveyed was GM and GM foods. The key findings of the survey, which dealt with GM and GM foods, indicate that Europeans are divided in their optimism about biotechnology and GE, do not see benefits of GM foods, consider GM foods to be probably unsafe or even harmful, and are not in favor of development of GM foods. With regard to who to trust as the best advisers for issues of biotechnology, survey participants think that medical professionals and university academics are the best advisers. The survey results also revealed that there was a widespread awareness of GM foods among the European participants. A large majority (84%) indicated that they have heard of GM foods. With regard to the attitude of respondents toward GM foods, results show that there was an overall suspicion of GM foods among the European public. There was a general negative attitude toward GM foods as indicated by the following reported results: 70% agree that GM foods are fundamentally unnatural, 61% agree that GM foods make them feel uneasy, 61% disagree that the development of GM foods should be encouraged, 59% disagree that GM foods are safe for their health and that of their families, and 58% disagree that GM foods are safe for future generations. When the survey results were further analyzed at the country level, the following findings were reported:

- More than two-thirds believe that GM food production is not good for the economy.
- Majority of the respondents believes that GM foods are not good for them.
- About 40% feel that GM foods help developing countries' populations.
- A majority agrees that GM food has benefits for some people, but at the same time has risks for others.
- There was a general agreement among participants that GM foods are fundamentally unnatural.
- The majority feels uneasy about GM foods.
- The majority does not consider GM food safe for their health.
- More than 75% believe that GM food production would harm the environment.
- The development of GM foods should not be encouraged.

As can be seen from the overall results and results at the country level, there is a general agreement at the two levels.

A recent survey was conducted by Eurobarometer in 2013 (European Commission 2013). This survey, conducted in the EU member states, was meant to evaluate European citizens' attitudes toward science and innovation in general. Considering that GM food production as very much related to science and innovation, the results of this survey may be reflective of European consumers' attitudes toward GM foods. Results of the survey show that 77% of the Europeans interviewed think that science and technology have a positive influence on society. In contrast, a concern over risks, such as to human health and the environment, emanating from the new technologies, was expressed. More than three quarters (76%) of the respondents believe that research and innovation should be conducted with due attention to ethical principles, have reasonable gender balance (84%), and have public involvement (55%). With regard to public interest in developments in science and technology, 53% of the respondents indicate that they have interest, but feel that they are not well-informed (58%). The sources of information most Europeans rely on to get idea about new developments in science and technology include television (65%), newspapers (33%), websites (32%), and magazines (26%).

The differences in attitudes toward GM observed between U.S. and EU consumers may be attributed to what is referred to as the "invisible flag." Coval and Moskowitz (1999) state that consumers generally prefer products and ideas originating from their own countries, described as "home bias." Many people have the perception that GM technologies are mainly linked to Monsanto, which is an American company; therefore, some of the negative attitudes by European consumers toward GM technology as a whole may be based on the perception that this is a technology that has been imposed on Europeans by an American company.

In a study to understand the differences between Europe and North America in acceptance of GM crops, Zilberman et al. (2013) presented some explanation to the causes of differences. The reasons cited include the lack of support by the European chemical companies to introduce GM technology in Europe, the technology did not seem to provide benefits to consumers, the crops in which the technology was used were not significant in Europe, and that the introduction of the technology came during a time when the green parties in Europe had significant influence.

A large sector of American and European consumers is reported to have a negative prior attitude toward GM foods and that they would be ready to pay a premium to label GM foods and to avoid them (Food Standards Australia New Zealand 2008). The stated willingness-to-pay (WTP), that is, the willingness that respondents report in a survey to avoid GM, is expected to be much higher than the actual WTP, that is, how consumers actually behave when making purchase decisions. The stated WTP may change when additional new information becomes available (Fusaro 2013). Lusk and Coble (2005) state that the presumed risk related to GM foods is considered to be a main reason for avoiding or rejecting them and that the WTP to avoid GM foods relates more to the level of risk aversion of an individual.

A study on consumers' attitude of risks and benefits toward GM foods in South Korea was conducted by Kim (2012). The study explored the importance of South Korean consumers' perception and attitude toward GM foods and assessed how their perception of risks and benefits of GM foods and individual socioeconomic background affect their acceptance of GM foods. As is the case in many countries, a number of consumer groups and nongovernmental organizations lead anti-GM food campaigns. This resulted in highly negative publicity of GM foods in South Korea, making it a great challenge for marketers intending to enter the South Korean market. With regard to local Korean consumers, results of the study showed that consumers' socioeconomic status and their perceived benefits associated with GM foods were found to be strong indicators of consumers' GM food purchase intention. The study concludes that consumers' background and diversity in South Korean demographic may have significant effect on their purchasing intention for GM foods.

Fishman (2014) conducted a study that aimed at determining the role of mass media in shaping consumers' perceptions and knowledge of GM foods. The survey concluded that 58% of the participants do consult mass media, specifically newspapers (44%) for information on GM foods. Television was found to produce a more positive perception of GM foods than radio, whereas radio was found to give a more negative perception of GM foods than newspapers. With regard to supplying knowledge on GM foods, radio and newspapers were found to be better sources than television.

Some studies on consumers' perceptions toward GM foods have also been conducted in developing countries. A study carried out in Botswana (Oladele and Subair 2009) compared university lecturers' perception toward consumption of GM foods in Nigeria and Botswana. The results show that lecturers of Botswana are more favorably disposed to GM foods than those of Nigeria. Many factors such as information, availability and first experience to GM foods, and national debates on GM products, may have led to this difference in perception among this group of educated individuals. More studies of such nature are needed in developing countries in order to help governments to plan for relevant and appropriate strategies to portray a clear picture of GM products and to help consumers make their own decisions.

Consumer purchasing behavior for GM foods is not necessarily similar to, or same as, their attitudes toward the GM technology or GM foods. In spite of the fact that consumers' attitudes toward GM foods may be negative, there is a chance that these attitudes will not directly correlate with a negative purchasing behavior (Gaskell et al. 2003). This idea has been supported by surveys conducted in the United Kingdom and reported by Halford and Shewry (2000).

Consumers' WTP and preferences for GM foods are some of the parameters studied to measure consumers' acceptance for GM foods.

Hoban (1997) states that consumers' willingness to buy biotechnology products will depend on biotechnology's willingness to educate consumers. Consumers' knowledge and awareness about GM foods are expected to help them make informed decisions with regard to buying, or not, GM foods. Consumers' WTP for foods that contain GMO ingredients was studied by Noussair et al. (2004), among a demographically representative sample of French consumers. Their results showed that 35% of the participants were unwilling to purchase products made with GMOs, 23% were indifferent or value the presence of GMOs, and 42% were willing to purchase them, provided that they were sufficiently inexpensive. These findings are not in agreement with most of the surveys conducted in Europe, which indicated overwhelming opposition to GM foods.

A study reported by Lusk et al. (2005) and conducted on U.S., European, Chinese, Japanese, and Taiwanese consumers as well as consumers from other parts of the world showed that the majority (82%) of the surveyed consumers are willing to pay sizeable premiums to avoid GM foods. Consumers who were willing to purchase GM foods were those who were told that GM foods provide direct health or nutrition benefits. The study also reported that people were willing to pay much more in order to avoid GM food if it does not create tangible benefits for the consumer. On the basis of these results, it is logical to expect farmers and food producers to respond to consumers' wants and put more non-GM food in the market, something that did not happen.

The actual purchase behavior of European consumers toward GM foods when compared with their attitude and behavioral intention for buying GM foods was studied by Sleenhoff and Osseweijer (2013). Results of the study indicated that a majority of respondents showed a negative attitude toward GM foods, but despite that more than 50% of the respondents stated that they did not actively avoid purchase of GM foods and 6% actually bought one of the labeled GM foods. Those results may be interpreted to mean that a voiced negative attitude of consumers toward GM foods cannot be used as an indicator or a reliable guide for what consumers actually do in grocery stores.

Generally, consumers are found to be more concerned about cost and perceived quality when making nutritional choices (GMO Compass 2009). Low-income consumers are more likely to buy low-quality products because these are expected to be cheaper in price (Gaskell et al. 2010). Notwithstanding these observations on consumer behavior, a considerable heterogeneity among consumers, with respect to their product choices, has been noticed (Gaskell et al. 2011).

Sometimes, consumers are uniformed or not adequately informed about GM technology but still show concern about its risks and hazards. This attitude might be due to the negative information about GM, which they receive from various sources such as the media or the views of GM opponent lobby groups.

The level of knowledge on GM and non-GM foods among consumers can be improved through adequate and relevant advertising by producing companies. This does not happen due to the fact that most large food companies that produce non-GM foods are the same ones that sell food that contains some GM ingredients. The level of lack of relevant knowledge among U.S. consumers is reported by Pew Initiative on Food and Biotechnology (2006), which stated that about 60% of the consumers, then, have not seen, read, or heard about GM foods available in grocery stores. Although this high level of lack of knowledge about GM foods among U.S. consumers might have changed with time, the general trend among consumers of not troubling themselves to actively search for information about GM foods continues.

Ellahi (1996) conducted a survey that studied awareness of, and policy implications for, the U.K. food retailers and manufacturers. Results of the survey indicated that the U.K. food retailing industry was generally well-informed, whereas food manufacturers showed lack of awareness of possible products and the implications their usage may have on their companies. In contrast, both food retailers and manufacturers believed that biotechnology has positive aspects for food production and that they were aware of consumers' concerns that need to be addressed.

One hypothesis about inadequate knowledge on GM foods among consumers is that it is not worth it (Lusk 2011). The logical question that follows is that if that is the case, why do large sectors of consumers avoid purchasing or eating GM foods? Brossard and Nisbet (2007) postulate that most citizens lack the ability, and/or the motivation, to be fully informed about an issue and rely on easy-to-get information from sources such as media, to help in forming an opinion. In contrast, consumers' trust in government and food regulatory agencies is found to be directly related to their acceptance to GM foods (Kuznesof and Ritson 1996; House et al. 2004). The issue of increasing the level of desire of consumers to acquire knowledge by themselves rather than relying on, or blindly trusting, other resources has been debated. Some authors believe that public-intensive education would help to improve consumers' acceptance of biotechnology (Lusk 2011). This view relies on the observation that people with higher level of education generally accept GM foods.

The Pew Initiative on Food and Biotechnology review (2006) conducted during the period 2001–2006 on U.S. consumers indicated that American's knowledge of GM foods and animals was low. Some consumers were confused about the source of genes contained in some food commodities, for example, Flavr Savr tomato to contain fish genes. This confusion might have come as a result of wide propaganda against GM technology in general and GM foods in particular. A lot of information about GM technology which was going around came from different sources with different views.

GM food labeling, whether mandatory or voluntary, has been an issue of debate and controversy among different countries as well as among consumers and consumerist groups. Some countries such as the United States and Canada do not require labeling for GM foods, whereas others such as EU countries, Japan, and Russia adopted mandatory GM food labeling. The effect of GM labeling regime on market outcomes has been investigated by Golan and Kuchler (2011). They studied the role that mandatory GM labeling versus voluntary labeling has played in the split between those countries with small GM markets and those with large GM markets. Their findings indicate that labeling has negligible effects on consumers' choice or on GM differentiation costs and therefore does not explain the split in GM market outcomes. There seems to be some sort of contradiction with regard to attitudes of consumers toward food labels. On the one hand, food labels are supposed to help consumers to choose foods that suit their taste or health status or meet any demand they have. Labels help consumers to know what is in the food and consequently to make an informed decision and appropriate choice. On the other hand, food labels have been found to be a weak tool to affect food consumption behavior simply because consumers do not pay much attention to them. Aldrich (1999) reported that there is empirical evidence which indicates that consumers often make hasty food choices when shopping for groceries and that they do not pay much attention to food labels. Some possible explanations for consumers' disregard for food labels have been suggested. One possibility is that if the food label includes too much information, consumers would be reluctant to pay attention to all the details. Some research results suggest that a lot of information on a label may reduce the chance that consumers will read or correctly interpret it (Golan and Kuchler 2011). Even if consumers read all the information on a food label, there is little chance that they order the information according to importance. Sometimes, consumers are unaware of the food labeling that indicates the presence of GM ingredients. Noussair et al. (2002) conducted an experiment to study the discrepancy between European public opinion and consumer purchase behavior with regard to GMOs in the food supply. Their results indicated that consumers are typically unaware of the labeling indicating GMO content.

In Australia, where GM food labeling is mandatory, survey results reported by the Food Standards Australia and New Zealand in 2007 (Food Standards Australia New Zealand 2008) showed that 27% of the surveyed Australians checked the food label to find out if it contained GM components, when seeing the product for the first time.

The New York Times conducted a poll in 2013 (Kopicki 2013), the results of which showed that 93% of the Americans indicated that they want GMO labeling. This came at the same time when the World Food Prize was awarded to Robert Farley and Mary-Bell Chilton, who are considered as employees of Monsanto and Syngenta. This prompted a lot of reaction in the form of rallies against the award and against GM foods. At the same time, the European Network of Scientists for Social and Environmental Responsibilities posted a statement, arguing that there is no scientific consensus on the safety of GM foods, which was signed by about 200 scientists.

The already existing consumers' attitudes toward GM foods seem to have more influence than the provision of new information about GM foods. Frewer et al. (2003) conclude that the extent to which people trusted the source of information about GM foods appeared to be determined by existing consumers' attitudes toward GM foods. Two views to explain the reasons for public opposition to GM technology have been presented. One view interprets the public opposition to this technology as resulting from the misinterpretation of the risks associated with it. This interpretation has been recently questioned, and the public opposition to GM technology has been attributed to the absence of perceived benefits to the consumer (Gaskell et al. 2004).

Supporters and critics of GM technology and GM foods try to convince consumers in what they say by presenting the benefits or drawbacks and risks of the technology, depending on what they want to portray about the technology. All these information are expected to be available to consumers to help them decide for themselves on various issues related to GM technology. It is important and more ethical for both supporters and critics of GM foods to give all and honest information, and the points that support their cause, so that consumers get the whole true picture and make informed decisions.

Some researchers felt a need to study the possibility that expressed benefits to health and the environment would represent more acceptable reasons for accepting GM than reduced cost or improved shelf life of the produce (Frewer et al. 1995). A number of studies were conducted to investigate that possibility (Bech-Larsen and Grunert 2000; Boccaletti and Moro 2000; Fortin and Renton 2003). Fortin and Renton (2003) reported that consumers' resistance to purchase or consumption of GM foods was unlikely to be changed by learning that their shelf life will be increased. A study in four Scandinavian countries (Bech-Larsen and Grunert 2000) found that resistance to food products may be lessened if consumers were informed about the benefits of the food and were allowed to sample the product before making a judgment.

A new program known as "Horizon 2020" is planned to be the next EU research and innovation program. This program is expected to run during the period 2014–2020. It has a strong focus and orientation toward addressing societal challenges that have direct effects on people's lives, such as better healthcare, greener transport, and food and energy security (Horizon 2020, 2014). The program will allocate a special budget for "Science with and for society," aiming at integrating scientific and technological aspects into the European society. It is expected that the program addresses some of the issues pertaining to GM foods.

It is evident that many studies, surveys, reviews, and polls have been, and continue to be, conducted among consumers on various issues surrounding GM foods in an effort to arrive at a conclusive, or at least satisfactory, answer to the pertinent question "is it safe to eat GM foods?" These studies have been conducted in different parts of the world, at different times, under different conditions, and with different goals. Generally, it is a healthy sign to have such studies going on, but before that, consumers need to be fully aware of benefits and risks of GM technology in general, as well as the safety and health issues surrounding the consumption of GM foods. Results of such studies, surveys, and polls would carry more weight and be more meaningful if consumers are adequately informed and are knowledgeable on various issues surrounding GM technology and GM foods. Part of the responsibility to gain that knowledge falls on the consumers themselves. Consumers need to search for credible information from unbiased sources and understand it before expressing their views or making decisions on purchase or consumption of GM foods. Government authorities, food producers and processors, as well as various media outlets have great responsibility in raising public awareness on issues related to GM foods. This has to be done in a way that does not mislead the consumers but help them to get the right information that would help them to decide for themselves to purchase or consume GM foods.

7.4 IMPACT OF MOVING FROM INCLUDING GM COMPONENTS TO NON-GM COMPONENTS ON FOOD SALES

Some food manufacturing companies, which used to incorporate GM ingredients in their products, shifted toward replacing those ingredients with non-GM components in an attempt to attract more consumers and to boost sales of their products. The expected improvement in sales did not match manufacturers' anticipation. Boulder Brands and General Mills are two companies which reported such consumer trends. Boulder Brands—which announced a high-profile re-launch of its Smart Balance spreads on a non-GMO platform in March—has admitted that the move has not transformed the brand's flagging fortunes (Watson 2014a). The food manufacturer did not notice improvements in the sales of products not incorporating GM ingredients, which replaced products containing GM components. In spite of the shift from using GMOs to non-GMO ingredients in their food products, the company did not notice expected improvement in the sales of these products. This indicates that consumers' trends in purchasing foods are not affected by replacing GM with non-GM ingredients in the food products, at least in the case of Boulder Brands. A similar trend, that is, of food purchases not being affected by replacing GM ingredients in food products with non-GM ingredients, has also been noticed by General Mills, as reported by Watson (2014b). Although General Mills' decision to rid Original Cheerios of GMOs has generated an enormous amount of publicity, it has not, apparently, translated to the top line (Watson 2014b). Sales of Original Cheerios manufactured using non-GM ingredients did not show improvement when compared with Cheerios produced using GM components. These consumer trends, of not changing food purchase behavior due to replacing GM ingredients with non-GM ingredients, may be explained in a number of ways. One explanation is that consumers did not notice the change in the new product. It can also be attributed to consumers' indifference of consuming GM- or non-GM-based food products. Consumers might also need to be more educated about the difference between GM and non-GM components in food products. This can be part of the general consumer awareness about biotechnology, GM foods, and their benefits and possible risks. The fact that these food products' sales did not improve may be a short-term effect, and sales might improve in the long run. The shift from using GM ingredients to non-GM ones might need to be coupled with efforts from the manufacturer to draw the attention of consumers to the formulation changes. This can be done through more advertisements emphasizing the change.

7.5 ROLES OF MASS MEDIA

Historically, media was designed to reach large audiences or viewers, originally through newspapers and magazines. The meaning of the term and its outreach broadened with the inventions of radio, TV, cinemas, and the internet. Mass media can be used for various purposes. Pushparaj (2014) lists the following aspects, which can be achieved through media: entertainment, news and current affairs, political awareness, education, public announcements, and advertisements. Media has a huge power to achieve these purposes at a very little cost. Achieving all these purposes helps to educate people and to increase their awareness of current issues that affect their lives. It helps to educate the general public, policy-makers, and the different industrialists. The power of media in disseminating messages of different types is widely used by different authorities and agencies to help them reach wider audiences.

Mass media has very strong influence in shaping people's views and perceptions of many things in human life. People tend to believe in what they read in different media sources and become affected by the views expressed in them. This is possibly due to the easiness of access to information from media than from other professional or technical sources or due to the easy-to-understand and comprehend language in which media reports are disseminated. In contrast, media sources are at hand every day or even every minute, whereas technical, scientific, or professional information is usually published less frequently than media reports. If media is misused in a way that helps to spread baseless ideas and news, it would have a negative effect on its audience. Sometimes, the media develops unnecessary sensation and distortion of truth to attract attention (Pushparaj 2014).

The role of mass media on consumers' perception, attitudes, and action in relation to GM technology applications, particularly those related to GM foods, has been studied and reported by many researchers (Pattison et al. 1996; Frewer et al. 2003; Gaskell et al. 2004; Laros and Steenkamp 2004; Knight et al. 2014). Mass media reports on GM foods are widely controversial. Some reports support GM foods whereas others criticize it. At the same time, some balanced reports present both sides of the debate without any bias toward one side. These controversial media reports have widely spread fear of GM foods and GM crops among consumers (Laros and Steenkamp 2004). Citing terms such as "Frankenfoods," "unreliable," "disaster," "environmental risks," "risks of cancer," and "food health fears" in mass media outlets or in some professional publications is expected to spread fear among consumers. Such terms have been commonly used by mass media in Britain, Canada, the Netherlands, and the United States.

Differences in "who to trust," when it comes to getting credible sources of information on GM technology, also exist between American consumers and other consumers around the world. Europeans are reported to place their trust in consumer and environmental groups, whereas Americans tend to trust scientific and academic resources more than consumer and environmental groups (Lang and Hallman 2005). The role of press in educating the public about technology issues has also been found to be different in Europe and the United States. The European press has covered the various biotechnology issues more extensively than press in the United States. This wide press coverage in Europe resulted in more public awareness, making European consumers to be both cognizant and wary of the GM technology (Durant et al. 1999; McInerney et al. 2004).

There are great expectations and hopes that mass media plays a pivotal role in disseminating credible information and gains further consumer trust as a source of reliable reporting on various issues including biotechnology and GM foods.

REFERENCES

Aldrich, L. 1999. Consumer use of information: Implications for food policy. Food and Rural Economics Division, Economic Research Service, U.S. Department of Agriculture, Agricultural Handbook No. 715.

Bech-Larsen, T. and Grunert, K. 2000. Can health benefits break down Nordic consumers' rejection of genetically modified foods? A conjoint study of Danish, Norwegian, Swedish and Finnish consumers' preferences for hard cheese. Australia and New Zealand Marketing Academy, Gold Coast, Australia.

Boccaletti, S. and Moro, D. 2000. Consumer willingness-to-pay for GM food products in Italy. *AgBioForum* 3:259–67.

Bredahl, L. 2001. Determinants of consumer attitudes and purchase intentions with regard to genetically modified foods—results of a cross-national survey. *Journal of Consumer Policy*, 24:23–61.

Brossard, D. and Nisbet, M.C. 2007. Deference to scientific authority among a low information public: Understanding U.S. opinion on agricultural biotechnology. *International Journal of Public Opinion Research*, 19:24–52.

Byrne, P. 2010. Labeling of genetically engineered foods. http://www.ext.colostate.edu/pubs/foodnut/09371.html (accessed May 3, 2014).

Campbell, H., Fitzgerald, R., Saunders, C., and Sivak, L. 2000. *Strategic Issues for GMOs in Primary Production: Key Economic Drivers and Emerging Issues*. Centre for the Study of Agriculture, Food and Environment, University of Otago, Dunedin, New Zealand.

Chen, M.-F. and Li, H.-L. 2007. The consumer's attitude toward genetically modified foods in Taiwan. *Food Quality and Preference*, 18(4):662–74.

Conko, G. and Prakash, C.S. 2004. Can GM crops play a role in developing countries? *AgBioWorld*. http://www.agbioworld.org/biotech-info/articles/agbio-articles/gm-crop-role.html (accessed January 6, 2014).

Coval, J.D. and Moskowitz, T.J. 1999. Home bias at home: Local equity preference in domestic portfolios. *Journal of Finance*, 54:2045–73.

Curtis, K.R. and Moeltner, K. 2006. Genetically modified food market participation and consumer risk perceptions: A cross-country comparison. *Canadian Journal of Agricultural Economics*, 54(2):289–310.

Deloitte Development. 2010. Deloitte 2010 food survey: Genetically modified foods. http://www.deloitte.com/assets/Dcom-UnitedStates/Local%20Assets/Documents/Consumer%20Business/us_cp_2010FoodSurveyFactSheetGeneticallyModifiedFoods_05022010.pdf (accessed September 10, 2014).

Durant, J., Bauer, M., and Gaskell, G. 1999. *Biotechnology in the Public Sphere: A European Source Book*. Science Museum Press, London, UK.

Ellahi, B. 1996. Genetic modification for the production of food: The food industry's response. *British Food Journal*, 98(4/5):53–72.

Eurobarometer. 2010. Biotechnology report. Special Eurobarometer 341/Wave 73.1–TNS Opinion and Social. http://ec.europa.eu/public_opinion/archives/ebs/ebs_341_en.pdf (accessed September 14, 2014).

European Commission. 2013. EU-wide poll shows public support for responsible research and innovation. http://europa.eu/rapid/press-release_IP-13-1075_en.htm (accessed September 16, 2014).

Fishman, K. 2014. Genetically modified foods: What is the mass media's role in shaping consumer perception's and knowledge? http://nature.berkeley.edu/classes/es196/projects/2002final/Fishman.pdf (accessed September 14, 2014).

Food Standards Australia New Zealand. 2008. Consumer attitudes survey 2007. A benchmark survey of consumers' attitudes to food issues. http://www.foodstandards.gov.au/scienceandeducation/publications/consumerattitudes/ (accessed September 10, 2014).

Fortin, D. and Renton, M. 2003. Consumer acceptance of genetically modified foods in New Zealand. *British Food Journal*, 105:42–58.

Frewer, L. 2003. Societal issues and public attitudes towards genetically modified foods. *Trends in Food Science and Technology*, 14:319–32.

Frewer, L.J., Howard, C., and Shepherd, R. 1995. Genetic engineering and food: What determines consumer acceptance? *British Food Journal*, 97:31–6.

Frewer, L.J., Scholderer, J., and Bredahl, L. 2003. Communicating about the risks and benefits of genetically modified foods. The mediating role of trust. *Risk Analysis*, 23:117–33.

Fusaro, D. 2013. European scientists ask for GMO research. *Food Processing*. http://www.foodprocessing.com/articles/2013/european-scientists-ask-for-gmo-research/ (accessed September 12, 2014).

Gamble, J. and Gunson, A. 2002. *The New Zealand Public's Attitudes Regarding Genetically Modified Food*. HortResearch, Mt Albert, New Zealand.

Gaskell, G., Allum, N.C., and Stares, S.R. 2003. *Europeans and Biotechnology in 2002: Eurobarometer 58.0*. European Commission, Brussels.

Gaskell, G., Allum, N., Wagner, W. et al. 2004. GM foods and the misperception of risk perception. *Risk Analysis*, 24:185–94.

Gaskell, G., Stares, S., Allansdottir, A. et al. 2010. Europeans and biotechnology in 2010: Winds of change? A report to the European Commission's Directorate-General for Research. European Commission Directorate-General for Research 2010 Science in Society and Food, Agriculture and Fisheries, and Biotechnology, EUR 24537 EN. http://ec.europa.eu/public_opinion/archives/ebs/ebs_341_winds_en.pdf (accessed September 10, 2014).

Gaskell, G., Allansdottir, A., Allum, N. et al. 2011. The 2010 Eurobarometer on the life sciences. *Nature Biotechnology*, 29(2):113–4.

GMO Compass. 2009. Opposition decreasing or acceptance increasing? An overview of European consumer polls on attitudes to GMOs. http://www.gmo-compass.org/eng/news/stories/415.an_overview_european_consumer_polls_attitudes_gmos.html (accessed September 10, 2014).

Golan, E. and Kuchler, F. 2011. The effect of GM labeling regime on market outcomes, Chapter 11, in Carter, C.A., Moschini, G.C., and Sheldon, I. (Eds.), *Frontiers of Economics and Globalization*, Volume 10. Genetically Modified Foods and Global Welfare. Emerald Group Publishing Limited, Bingley, UK, pp. 263–82.

Grunert, K.G., Lahteenmaki, L., Nielsen, N.A., Poulsen, J.B., Ueland, O., and Astrom, A. 2000. Consumer perception of food products involving genetic modification: Results from a qualitative study in four Nordic countries. Working Paper no 72. Nordic Industrial Fund, Denmark.

Halford, N. and Shewry, P. 2000. Genetically modified crops: Methodology, benefits, regulation and public concerns. *British Medical Bulletin*, 56:62–73.

Hallman, W.K., Hebden, W.C., Cuite, C.L., Aquino, H.L., and Lang, J.T. 2004. Americans and GM food: Knowledge, opinion and interest in 2004 (Food Policy Institute Report No. RR-1104-007). Rutgers University, Food Policy Institute, New Brunswick, NJ.

Hallman, W.K., Jang, H.M., Hebden, C.W., and Shin, H.K. 2005. *Consumer Acceptance of GM Food: A Cross-cultural Comparison of Korea and the United States*. Food Policy Institute, Rutgers University.

Han, J.H. and Harrison, R.W. 2007. Factors influencing urban consumers' acceptance of genetically modified foods. *Review of Agricultural Economics*, 29(4):700–19.

Hebden, W.C., Shin, H.K., and Hallman, W.K. 2005. Consumer responses to GM foods: Why are Americans so different? CHOICES: The Magazine of Food, Farm and Resource Issues. http://www.choicesmagazine.org/2005-4/GMOs/2005-4-06.htm (accessed October 15, 2014).

Heiman, A. and Zilberman, D. 2011. The effects of farming on consumers' choice of GM foods. Agbioforum.org 14, article 9.

Hoban, T.J. 1997. Consumer acceptance of biotechnology: An international perspective. *Nature Biotechnology*, 15:232–4.

Horizon 2020. 2014. http://www.h2020.net/ (accessed October 20, 2014).

House, L., Lusk, J., Jaeger, S. et al. 2004. Objective and subjective knowledge: Impacts on consumer demand for genetically modified foods in the United States and the European Union. *AgBioForum* 7:113–23.

Huang, J., Bai, J., Pray, C., and Tuan, F. 2004. Public awareness, acceptance of and willingness to buy genetically modified foods in China. Unpublished report, Rutgers University.

Huffman, W.E., Shogren, J.F., Rousu, M., and Tengene, A. 2003. Consumer willingness to pay for genetically modified food labels in a market with diverse information: Evidence from experimental auctions. *Journal of Agricultural and Resource Economics*, 28(3):481–502.

IFIC. 2012. Consumer perceptions of food technology survey. http://www.foodinsight.org/Resources/Detail.aspx?topic=2012ConsumerPerceptionsofTechnologySurvey (accessed September 14, 2014).

International Federation for Produce Standards. 2012. Produce PLU Codes: A Users' Guide-2012. http://www.ifpsglobal.com/ (accessed July 10, 2014).

Kim, R.B. 2012. Consumer attitude of risk and benefits toward genetically modified (GM) foods in South Korea: Implications for food policy. *Inzinerine Ekonomika-Engineering Economics*, 23(2):189–99.

Knight, J.G., Mather, D.W., and Holdsworth, D.K. 2014. Consumer benefits and acceptance of genetically modified food. http://otago.ourarchive.ac.nz/bitstream/handle/10523/1605/ConsumerbenefitsGMOs.pdf (accessed September 14, 2014).

Kopicki, A. 2013. Strong support for labeling modified foods. *New York Times*. http://www.nytimes.com/2013/07/28/science/strong-support-for-labeling-modified-foods.html?_r=2& (accessed October 19, 2014).

Kuznesof, S. and Ritson, C. 1996. Consumer acceptability of genetically modified foods with special reference to farmed salmon. *British Food Journal*, 98:39–47.

Lang, J.T. and Hallman, W.K. 2005. Who does the public trust? The case of genetically modified food in the United States. *Risk Analysis*, 25(5):1241–52.

Laros, F. and Steenkamp, J.-B.E.M. 2004. Importance of fear in the case of genetically modified food. *Psychology and Marketing*, 21:889–908.

Legge, J.S. Jr. and Durant, R.F. 2010. Public opinion, risk assessment, and biotechnology: Lessons from attitudes toward genetically modified foods in the European Union. *Review of Policy Research*, 27(1):59–76.

Lusk, J.L. 2011. Consumer preferences for genetically modified food, in Carter, C.A., Moschini, G.C., and Sheldon, I. (Eds.), *Frontiers of Economics and Globalization,*

Volume 10. Genetically Modified Foods and Global Welfare. Emerald Group Publishing Limited, Bingley, UK, pp. 243–62.

Lusk, J.L. and Coble, K.H. 2005. Risk perceptions, risk preference, and acceptance of risky food. *American Journal of Agricultural Economics*, 87(2):393–405.

Lusk, J.L., Jamal, M., Kurlander, L., Roucan, M., and Taulman, L. 2005. A meta analysis of genetically modified food valuation studies. *Journal of Agricultural and Resource Economics*, 30:28–44.

McInerney, C., Bird, N., and Nucci, M. 2004. The flow of scientific knowledge from lab to the lay public: The case of genetically modified food. *Science Communication* 26:75–106.

Miles, S. and Frewer, L. 2003. Public perception of scientific uncertainty in relation to food hazards. *Journal of Risk Research*, 6:267–83.

Moon, W. and Balasubramanian, S.K. 2004. Public attitudes toward agrobiotechnology: The mediating role of risk perceptions on the impact of trust, awareness, and outrage. *Review of Agricultural Economics*, 26(2):186–208.

Noussair, C., Robin, S., and Ruffieux, B. 2002. Do consumers not care about biotech foods or do they just not read labels? *Economics Letters*, 75:47–53.

Noussair, C., Robin, S., and Ruffieux, B. 2004. Do consumers really refuse to buy genetically modified food? *The Economic Journal*, 114(492):102–20.

O'Fallon, M.J., Gursoy, D., and Swanger, N. 2007. To buy or not to buy: Impact of labeling on purchasing intentions of genetically modified foods. *International Journal of Hospitality Management*, 26:117–30.

Oladele, O.I. and Subair, S.K. 2009. Perception of university lecturers towards consumption of genetically modified foods in Nigeria and Botswana. *Agriculturae Conspectus Scientificus*, 74(1):55–9.

Pattison, C., Hedderley, D., and Frewer, L.J. 1996. Modelling public risk perceptions for lifestyle and technological hazards. http://www.riskworld.com/Abstract/1996/sraeurop/ab6ad008.htm (accessed September 12, 2014).

Pew Initiative on Food and Biotechnology. 2006. Review of public opinion research. http://www.pewtrusts.org/en/archived-projects/pew-initiative-on-food-and-biotechnology (accessed September 7, 2014).

Phillips, W.B. and Corkindale, D. 2002. Marketing GM food: The way forward. *AgBioForum*, 5(3):113–21.

Poortinga, W. and Pidgeon, N. 2005. Trust in risk regulation: Cause or consequence of the acceptability of GM food? *Risk Analysis*, 25(1):199–209.

Puduri, V.S., Govindasamy, R., and Nettimi, N. 2010. Consumers' perceptions toward usefulness of genetically modified foods: A study of select consumers in USA. *The IUP Journal of Agricultural Economics*, VII(3):7–17.

Pushparaj, A. 2014. An essay on the role of media. Publish your articles. http://www.publishyourarticles.net/eng/articles/an-essay-on-the-role-of-media.html (accessed October 19, 2014).

Rosati, S. and Saba, A. 2000. Factors influencing the acceptance of food biotechnology. *Italian Journal of Food Science*, 12(4):425–34.

Rousu, M.C., Huffman, W.E., Shogren, J.F., and Tegene, A. 2007. Effects and value of verifiable information in a controversial market: Evidence from lab auctions of genetically modified food. *Economic Inquiry*, 45:409–32.

Scholderer, J., Balderjahn, I., Bredahl, L., and Grunert, K.G. 1999. The perceived risks and benefits of genetically modified food products: Experts versus consumes. *European Advances in Consumer Research*, 4:123–9.

Siegrist, M. 1999. A causal model explaining the perception and acceptance of gene technology. *Journal of Applied Social Psychology*, 29(10):2093–106.

Sleenhoff, S. and Osseweijer, P. 2013. Consumer choice linking consumer intentions to actual purchase of GM labeled food products. *GM Crops and Food: Biotechnology in Agriculture and the Food Chain*, 4(3):166–71. Landes Bioscience http://www.landesbioscience.com/journals/gmcrops/2013GMC0015.pdf (accessed September 10, 2014).

Slovic, P. 1999. Trust, emotion, sex, politics and science: Surveying the risk-assessment battlefield. *Risk Analysis*, 19(4):689–701.

Steel, C. 2014. List of consumer rights and responsibilities. eHow. http://www.ehow.com/list_7686826_list-consumer-rights-responsibilities.html (accessed October 19, 2014).

Subrahmanyan, S. and Cheng, P.S. 2000. Perceptions and attitudes of Singaporeans toward genetically modified food. *Journal of Consumer Affairs*, 34(2):269–90.

Thorpe, A. and Robinson, C. 2005. When Goliaths clash: US and EU differences over the labeling of food products derived from genetically modified organisms. *Agriculture and Human Values*, 21:287–98.

Watson, E. 2014a. Non-GMO reboot has not revitalized sales of Smart Balance spread, says Boulder Brands. http://www.foodnavigator-usa.com/Manufacturers/Boulder-Brands-sees-no-lift-from-non-GMO-Smart-Balance/ (accessed November 7, 2014).

Watson, E. 2014b. Axing GMOs from Original Cheerios has not boosted sales, says General Mills. http://www.foodnavigator-usa.com/Manufacturers/Axing-GMOs-from-Original-Cheerios-has-not-boosted-sales-says-General-Mills (accessed November 7, 2014).

Zilberman, D., Kaplan, S., Kim, E., Hochman, G., and Graff, G. 2013. Continents divided: Understanding differences between Europe and North America in acceptance of GM crops. *GM Crops and Food: Biotechnology in Agriculture and the Food Chain*, 4(3):202–8.

SUGGESTED REFERENCES

Arvanitoyannis, I. and Krystallis, A. 2005. Consumers' beliefs, attitudes and intentions towards genetically modified foods, based on the "perceived safety vs. benefits" perspective. *International Journal of Food Science and Technology*, 40(4):343–60.

Barnard, N.D. 2011. Should you say no to GM foods? Ask the doc. weird science. 26–7. http://www.vegetariantimes.com/article/ask-the-doc-weird-science/ (accessed October 21, 2014).

Bredahl, L., Grunert, K.G., and Frewer, L.J. 1998. Consumer attitudes and decision-making with regard to genetically engineered food products—a review of the literature and a presentation of models for future research. *Journal of Consumer Policy*, 21:251–77.

Brossard, D., Shanahan, J., and Nesbitt, T.C. 2007. *The Media, the Public and Agricultural Biotechnology*. CAB International, Wallingford.

Carlsson, F., Frykblom, P., and Lagervist, C.J. 2007. Consumer benefits of labels and bans on GM foods-choice experiments with Swedish consumers. *American Journal of Agricultural Economics*, 89:152–61.

Costa-Fonta, M., Gila, J.M., and Traill, W.B. 2008. Consumer acceptance, valuation of and attitudes towards genetically modified food: Review and implications for food policy. *Food Policy*, 33:99–111.

Crespi, J.M. and Marette, S. 2003. "Does contain" vs. "does not contain": Does it matter which GMO label is used? *European Journal of Law and Economics*, 16:327–44.

Cummins, R. and Lillisto, B. 2004. *Genetically Engineered Food: A Self-Defense Guide for Consumers*. Marlowe & Company, an Imprint of Avalon Publishing Group Inc, New York.

Dannenberg, A. 2009. The dispersion and development of consumer preferences for genetically modified food—a meta-analysis. *Ecological Economics*, 68:2182–92.

Dibb, S. and Lobstein, T. 1999. *GM Free. A Shopper's Guide to Genetically Modified Food*. Virgin Publishing Ltd, London.

Fortin, D. and Renton, M. 2003. Consumer acceptance of genetically modified foods in New Zealand. *British Food Journal*, 105:42–58.

Frewer, L.J., Howard, C., Hedderley, D., and Shepherd, R. 1996. What determines trust in information about food-related risks? Underlying psychological constructs. *Risk Analysis*, 16:473–86.

Frewer, L.J., Howard, C., and Shepherd, R. 1998. The importance of initial attitudes on responses to communication about genetic engineering in food production. *Agriculture and Human Values*, 15:15–30.

Gaskell, G., Bauer, M.W., Durant, J., and Allum, N.C. 1999. Worlds apart? The reception of genetically modified foods in Europe and the U.S. *Science*, 285:384–7.

Hamstra, A.M. 1993. *Consumer Acceptance of Food Biotechnology—The Relation Between Product Evaluation and Acceptance*. The SWOKA Institute, The Hague.

Hoban, T.J. 1996. Consumers will accept biotech foods. BT Catalyst 10, No. 4.

Hoban, T.J. 1997. Consumer acceptance of biotechnology: An international perspective. *Nature Biotechnology*, 15:232–4.

Hu, W., Veeman, M.M., and Adamowicz, W.L. 2005. Labelling genetically modified food: Heterogeneous consumer preferences and the value of information. *Canadian Journal of Agricultural Economics*, 53:83–102.

Hunter, B.T. 2003. Bioengineered crops. *Consumers' Research Magazine*, 86(11):8–9.

Institute of Grocery Distribution. 2000. *GM Foods—Past, Present, Future?* IGD Business Publications, Institute of Grocery Distribution, Watford.
Kalaitzandonakes, N. and Bijman, J. 2003. Who is driving biotechnology acceptance? *Nature Biotechnology*, 21:366–9.
Kalaitzandonakes, N., Marks, L.A., and Vickner, S.S. 2004. Media coverage of biotech foods and influence on consumer choice. *American Journal of Agricultural Economics*, 86:1238–46.
Lemkow, L. 1993. *Public Attitudes to Genetic Engineering: Some European Perspectives.* Loughlinstown House, Dublin.
Loureiro, M.L. and Hine, S. 2004. Preferences and willingness to pay for GM labeling policies. *Food Policy*, 29(5):467–83.
Lusk, J.L., House, L.O., Valli, C. et al. 2004. Effect of information about benefits of biotechnology on consumer acceptance of genetically modified food: Evidence from experimental auctions in United States, England, and France. *European Review of Agricultural Economics*, 31:179–204.
Lusk, J.L. and Rozan, A. 2008. Public policy and endogenous beliefs: The case of genetically modified food. *Journal of Agricultural and Resource Economics*, 33:270–89.
Lusk, J.L., Traill, W.B., House, L.O. et al. 2006. Comparative advantage in demand: Experimental evidence of preferences for genetically modified food in the United States and European Union. *Journal of Agricultural Economics*, 57:1–21.
Marchant, G.E., Cardineau, G.A., and Redick, T.P. 2010. *Thwarting Consumer Choice: The Case Against Mandatory Labeling for Genetically Modified Foods*. AEI Press, Washington, DC.
McInerney, C., Bird, N., and Nucci, M. 2004. The flow of scientific knowledge from lab to the lay public: The case of genetically modified food. *Science Communication*, 26:75–106.
Moon, W. and Balasubramanian, S. 2003. Is there a market for genetically modified foods in Europe? Contingent valuation of GM and non-GM breakfast cereals in the United Kingdom. *Journal of Agrobiotechnology Management and Economics*, 6: Article 6.
Mucci, A., Hough, G., and Ziliani, C. 2004. Factors that influence purchase intent and perceptions of genetically modified foods among Argentine consumers. *Food Quality and Preference*, 15:559–67.
Pew Initiative on Food and Biotechnology. 2005. U.S. vs. E.U.: An examination of the trade issues surrounding genetically modified food. http://www.pewtrusts.org/~/media/legacy/uploadedfiles/wwwpewtrustsorg/reports/food_and_biotechnology/BiotechUSEU1205pdf.pdf (accessed October 21, 2014).
Robinson, C. 2002. *Genetic Modification Technology and Food. Consumer Health and Safety*. ILSI Press, Brussels.
Rossiter, L. (Ed.). 1997. *Biotechnology Factfile. A Quick Reference Guide.* Institute of Grocery Distribution, Watford.
Rousu, M.C. and Lusk, J.L. 2009. Valuing information on GM foods in a WTA market: What information is most valuable? *AgBioForum*, 12:226–31.

Sadler, M.J. 2000. *GM Foods: Past, Present, Future: Industry's Approach, Consumer Attitudes, Expectations for the Future.* Institute of Grocery Distribution (IGD), Watford.

Santaniello, V. and Evenson, R.E. 2004. *Consumer Acceptance of Genetically Modified Foods.* International Consortium on Agricultural Biotechnology Research, CAB International, Wallingford, UK.

Santaniello, V., Evenson, R.E., and Zilberman, D. (Eds.). 2002. *Market Development for Genetically Modified Foods.* CABI Publishing, Wallingford, UK.

Scully, J. 2003. Genetic engineering and perceived levels of risk. *British Food Journal,* 105:59–77.

Siipi, H. and Uusitalo, S. 2011. Consumer autonomy and availability of genetically modified food. *Journal of Agricultural and Environmental Ethics,* 24:147–63.

Smink, G.C.J. and Hamstra, A.M. 1994. *Impacts of New Biotechnology in Food Production on Consumers.* The SWOKA Institute, The Hague.

Tegene, A., Huffman, W.E., Rousu, M., and Shogren, J.F. 2003. The effects of information on consumer demand for biotech foods: Evidence from experimental auctions. Technical Bulletin No. 1903, USDA Economic Research Service, Washington, DC.

The Consumers' Association. 1997. Gene cuisine—A consumer agenda for genetically modified foods (Policy report). The Consumers' Association, London.

The Consumers' Association. 2002. GM dilemmas (Policy report). The Consumers' Association, London.

Townsend, E. and Campbell, S. 2004. Psychological determinants of willingness to taste and purchase genetically modified food. *Risk Analysis,* 24:1385–93.

Zechendorf, B. 1994. What the public thinks about biotechnology. *Biotechnology,* 12:870–5.

Zechendorf, B. 1998. Agricultural biotechnology: Why do Europeans have difficulty accepting it? *AgBioForum,* 1(1):8–13. http://www.agbioforum.org/v1n1/v1n1a03-zechen.htm (accessed October 21, 2014).

GLOSSARY

A

Abiotic: Absence of living organisms.

Abnormal: Any change from the usual or "correct." It may not necessarily mean harmful or undesirable; it equally can mean atypical, unusual, or uncommon. When used in reference to genes, an abnormal gene can result in a specific disorder.

Abortion: Termination of a pregnancy before birth.

Absorption: The digestibility of a dietary supplement into the bloodstream.

Acclimatization: The adaptation of a living organism such as plant, animal, or microorganism to a changed environment that subjects it to physiological stress.

Acellular: Tissues or organisms that are not made up of separate cells but often have more than one nucleus.

Acquired: Developed in response to the environment, not inherited, such as a character trait (acquired characteristic) resulting from environmental effect(s).

Acrylamide: A compound that forms in some foods during the cooking process (e.g., frying, roasting, or baking), due to heat interacting with sugars and an amino acid naturally present in some foods.

Active immunity: The production of antibodies against a specific disease by the immune system. Active immunity can be acquired in two ways, either through contraction of the disease or through vaccination. Active immunity usually is permanent, meaning that an individual is protected from the disease for the duration of his or her life.

Acute: A short-term, intense health effect.

Acyl carrier protein (ACP): A class of molecules that bind acyl intermediates during the formation of long-chain fatty acids. ACPs are important because of their involvement in many of the reactions necessary for *in vivo* fatty acid synthesis.

Adaptation: Adjustment of a population to changes in environment over generations, associated (at least in part) with genetic changes resulting from selection imposed by the changed environment.

Additive genes: Genes whose net effect is the sum of their individual allelic effects, that is, they show neither dominance nor epistasis.

GLOSSARY

Additive genetic variance: The net effect of the expression of additive genes and thus the chief cause of the resemblance between relatives. It represents the main determinant of the response of a population to selection. Formally, the variance of breeding values.

Adenovirus: One of a group of DNA-containing viruses found in rodents, fowl, cattle, monkeys, and humans. In humans, they are responsible for respiratory tract infections, but they have been exploited as a vector in gene therapy, especially for genes targeted at the lungs.

Adequate intake: A dosage recommendation that may be used on a product label where recommended daily dietary allowance information is lacking and that is labeled as daily values.

Adult stem cell: An undifferentiated cell found in a differentiated tissue that can renew itself and, with certain limitations, differentiate to yield all the specialized cell types of the tissue from which it originated. The name is confusing because we are born with the so-called *adult stem cells* in our tissue; therefore, many scientists prefer to use the term *somatic stem cell* instead.

Adventitious bacteria: Bacteria that originate from a source other than the organism or environment in which they are found and are not inherent to that organism or environment.

Aflatoxins: A group of toxic compounds, produced by *Aspergillus flavus*, that bind to DNA and prevent replication and transcription. Aflatoxins can cause acute liver damage and cancer, a health hazard in certain stored foods or feed.

Agent: A factor, such as a microorganism, a chemical substance, or a form of radiation, which causes a disease or medical condition.

Agricultural biotechnology: A range of tools, including traditional breeding techniques, which alter living organisms or parts of organisms to make or modify products, improve plants or animals, or develop microorganisms for specific agricultural uses. Modern biotechnology includes the tools of genetic engineering.

Agriculture: The science, art, and business of producing crops and raising livestock.

Agrobacterium rhizogenes: A bacterium that causes hairy root disease in some plants. Similar to the crown gall disease caused by *Agrobacterium tumefaciens*, this is achieved by the mobilization of the bacterial Ri plasmid with the transfer to the plant of some of the genetic material from the plasmid. This process has been used to insert foreign genes into plant cells, but to a lesser extent than the *Agrobacterium tumefaciens*-mediated transformation system, because regeneration of whole plants from hairy root cultures is problematic.

Agrobacterium tumefaciens: A bacterium that causes crown gall disease in some plants. The bacterium characteristically infects a wound and incorporates a segment of Ti plasmid DNA into the host genome. This DNA causes the host cell to grow into a tumor-like structure that synthesizes specific opines that only the pathogen can metabolize. This DNA transfer mechanism is exploited in the genetic engineering of plants.

Agrobacterium tumefaciens-**mediated transformation:** The process of DNA transfer from *Agrobacterium tumefaciens* to plants that occurs naturally during crown gall disease and can be used as a method of transformation.

Agrobacterium: A naturally occurring pathogenic bacterium of plants that can incorporate a portion of a plasmid deoxyribonucleic acid (DNA) into plant cells.

Alien species: A species living in an area outside its historically known natural range as a result of intentional or accidental dispersal by human activities.

Allele frequency: The relative number of copies of an allele in a population, expressed as a proportion of the total number of copies of all alleles at a given locus in a population.

Allele: A variant form of a gene. In a diploid cell, there are two alleles of every gene (one inherited from each parent, although they could be identical). Within a population, there may be many alleles of a gene. Alleles are symbolized with a capital letter to denote dominance, and lower case for recessive. In heterozygotes with codominant alleles, both are expressed.

Allele-specific amplification (ASA): The use of the polymerase chain reaction at a sufficiently high stringency that only one allele is amplified. A powerful means of genotyping for single-locus disorders that have been characterized at the molecular level.

Allelic exclusion: A phenomenon whereby only one functional allele of an antibody gene can be assembled in a given B lymphocyte.

Allelopathy: The secretion of chemicals, such as phenolic and terpenoid compounds, by a plant's roots, which inhibit the growth or reproduction of competitor plants.

Allergen: Any substance, usually a protein, that can cause an allergy (specific hypersensitivity) or allergic reaction in the body. An antigen that provokes an immune response when an individual is exposed to it by consumption or contact.

Allergenic: Having the properties and activities of an allergen.

Allergic reaction: A reaction by the body's immune system after exposure to an allergen, often a protein. Food can contain proteins that induce an immune response. Allergic symptoms may include rash, hives, and in extreme cases, shortness or loss of breath or unconsciousness.

Allergies: Occur when the body detects something that is not as harmful as causing the body to attack it which can result in watery eyes, a rash, or if severe anaphylaxis shock.

Allogamy: Cross-fertilization in plants.

Alternative mRNA splicing: The inclusion or exclusion of different exons to form different mRNA transcripts from a single transcription unit.

Amplification: (1) Creation of many copies of a segment of DNA by the polymerase chain reaction. (2) Treatment (e.g., use of chloramphenicol) designed to increase the proportion of plasmid DNA relative to that of bacterial (host) DNA. (3) Evolutionary expansion in copy number of a repetitive DNA sequence through a process of repeated duplication.

Amplify: To increase the number of copies of a DNA sequence, either *in vivo* by inserting into a cloning vector that replicates within a host cell or *in vitro* by polymerase chain reaction.

Antagonism: An interaction between two organisms (e.g., molds or bacteria) in which the growth of one is inhibited by the other. *Opposite*: synergism.

Antagonist: A compound that inhibits the effect of an agonist in such a way that the combined biological effect of the two becomes smaller than the sum of their individual effects.

Antibiosis: The prevention of growth or development of an organism by a substance or another organism.

Antibiotic resistance marker gene (ARMG): Marker genes such as antibiotic resistance genes are used during the development of GM foods. The use of these antibiotic resistance genes has raised concerns that clinical effectiveness of antibiotics will be compromised. The possibility of transferring these genes to recipient cells is considered remote; nevertheless, the industry has been advised to use alternative methods.

Antibiotic resistance: The ability of a microorganism to disable an antibiotic or prevent its transport into the cell.

Antibiotic: A class of natural and synthetic compounds that inhibit the growth of or kill some microorganisms. Antibiotics are widely used medicinally to control bacterial pathogens, but resistance in bacteria to particular antibiotics is often rapidly acquired through mutation.

Antibody: A protein found in the blood that is produced in response to foreign substances, such as bacteria or viruses, invading the body. Antibodies protect the body from disease by binding to those organisms and destroying them.

Antidote: A substance that can counteract a form of poisoning. Sometimes, the antidote for a particular toxin is manufactured by injecting the toxin into an animal in small doses and the resulting antibodies are extracted from the animals' blood. However, some toxins have no known antidote. For example, the poison ricin, which is produced from the waste byproduct of castor oil manufacture, has no antidote, and as a result is often fatal if it enters the human body in sufficient quantities.

Antigen (Ag): A macromolecule (usually a protein foreign to the organism), which elicits an immune response on first exposure to the immune system by stimulating the production of antibodies specific to its various antigenic determinants. During subsequent exposures, the antigen is bound and inactivated by these antibodies. Examples include pollen grains, dust, bacteria, or viruses and most proteins.

Antinutrient: A compound (in food) that inhibits the normal uptake or utilization of nutrients or that is toxic in itself.

Antioxidant: A substance that blocks, slows, or inhibits the actions of free radicals, molecules that speed up the aging process and contribute to illness. Free radicals are found in rancid fats and oils and environmental hazards.

Antitoxins: Antibodies capable of neutralizing a poisonous substance, or toxin. The symptoms of certain diseases, such as botulism, tetanus, and diphtheria, actually are caused by the toxins produced by the infecting bacteria.

Antiviral: Any medicine capable of destroying or weakening a virus. The term literally means "against virus."

Apex: The portion of a root or shoot containing the primary or apical meristem.

Apical cell: A meristematic initial in the apical meristem of shoots or roots of plants.

Apical dominance: The phenomenon where growth of lateral (axillary) buds in a plant is inhibited by the presence of the terminal (apical) bud on the branch. Explained by the export of auxins from the apical bud.

Apical meristem: A region of the tip of each shoot and root of a plant in which cell division is continually occurring to produce new stem and root tissue, respectively. Two regions are visible in the apical meristem: (i) an outer 1–4-cell-layered region (the *tunica*), where cell divisions are anticlinal; and below the tunica, (ii) the *corpus*, where the cells divide in all directions, and increase in volume.

Apoptosis: The process of cell death, which occurs naturally as a part of normal development, maintenance, and renewal of tissue in an organism.

Aquaculture: Farming of aquatic organisms, including fish, mollusks, crustaceans, and aquatic plants.

Aquatic plants: The wide variety of aquatic biomass resources such as algae, giant kelp, other seaweed, and water hyacinth. Certain microalgae can produce hydrogen and oxygen, while others manufacture hydrocarbons and a host of other products. Examples of microalgae include chlorella, dunaliella, and euglena.

Artificial selection: The practice of choosing individuals from a population for reproduction, usually because these individuals possess one or more desirable traits.

Asexual reproduction: Reproduction that does not involve the formation and union of gametes from the different sexes or mating types. It occurs mainly in lower animals, microorganisms, and plants. In plants, asexual reproduction is by vegetative propagation (e.g., bulbs, tubers, corms) and by formation of spores.

Association genetics: A means of establishing an association between a gene and an observable trait by comparing frequencies of alleles of the gene in groups of individuals who differ in the expression of that trait.

Asymmetric hybrid: A hybrid formed, usually via protoplast fusion, between two donors, where the chromosome complement of one of the donors is incomplete. This chromosome loss can be induced by irradiation or chemical treatment, or can occur naturally.

Autosomal dominant mutation: A dominant mutation in a gene that is carried on an autosome.
Autosomal gene: Any gene that is located on an autosome.
Autosomal recessive mutation: A recessive mutation in a gene that is carried on an autosome.
Autosomal traits: Traits carried on the chromosomes other than the sex chromosomes (X and Y).
Autosome: Any chromosome that is not a sex chromosome (not an X or Y chromosome).

B

B chromosome: A supernumerary chromosome present in some individuals (both plants and animals). They are smaller than the normal chromosomes, behave abnormally in both mitosis and meiosis, can vary in number between somatic cells, and are not thought to have any significant gene content.
Bacillus thuringiensis **(Bt):** A common soil bacterium species used in genetically modified crops that kills insects when they eat it. It produces a protein (Bt toxins) toxic to certain insects that cause significant crop damage. Some organic farmers use this bacterium as an alternative to using chemicals to control pest insects. The genes for Bt toxins have been genetically engineered into cotton plants, so the plants produce the insecticides.
Bacillus: Large family of bacteria that have a rod-like shape. They include the bacteria that cause food to spoil as well as those responsible for some types of diseases. Helpful members of the *Bacillus* family are used to make antibiotics or to colonize the human intestinal tract and aid with digestion.
Back mutation: A second mutation at the same site in a gene as the original mutation. The second mutation restores the wild-type protein sequence.
Backcross: Crossing an individual with one of its parents or with the genetically equivalent organism. The offspring of such a cross is referred to as the backcross generation or backcross progeny.
Bacteria: Tiny, one-celled organisms present throughout the environment that require a microscope to be seen. Bacteria can exist either as independent (free-living) organisms or as parasites (dependent on another organism for life). While not all bacteria are harmful, some cause disease. Examples of bacterial disease include

diphtheria, pertussis, tetanus, *Haemophilus influenza*, and pneumococcus (pneumonia).

Bacterial toxin: A toxin produced by a bacterium, such as Bt toxin of *Bacillus thuringiensis*.

Bacteriocide: A chemical or drug that kills bacterial cells.

Bacteriophage (phage): A virus that infects bacteria. Altered forms are used as cloning vectors.

Bacteriostat: A substance that inhibits or slows down growth and reproduction of bacteria.

Baculovirus: A class of insect virus used to make DNA cloning vectors for gene expression in eukaryotic cells. Production of a target protein can be up to 50% of the cells' protein content, and several proteins can be made simultaneously, so that multi-subunit enzymes can be made by this system.

Baculovirus expression vector (BEV): A method for the *in vitro* production of complex recombinant eukaryotic proteins. A genetically engineered baculovirus (a virus that infects certain types of insects) is introduced into appropriate cultured insect cells, which then express the recombinant protein.

Base sequence: The order of nucleotide bases in a DNA molecule.

Base: Part of four types of simple molecules or nucleotides (adenine, cytosine, thymine, and guanine) that are the subunits (building blocks) of DNA and RNA.

Bio: A prefix used in scientific words to associate the concept of "living organisms." Usually written with a hyphen before vowels, for emphasis or in neologisms.

Bioaccumulation: A problem that can arise when a stable chemical such as a heavy metal or DDT is introduced into a natural environment. Where there are no agents present able to biodegrade it, its concentration can increase as it passes up the food chain and higher organisms may suffer toxic effects. This phenomenon may be employed beneficially for the removal of toxic metals from wastewater, and for bioremediation.

Bioassay: (1) The assessment of the activity of a substance on living cells or on organisms. Animals have been used extensively in drug research in bioassays in the pharmaceutical and cosmetics industries. Current trends are to develop bioassays using bacteria or animal or plant cells, as these are easier to handle than whole animals or plants, are cheaper to make and keep, and avoid the ethical problems associated with testing of animals. (2) An indirect method to detect submeasurable amounts of a specific substance by observing a sample's influence on the growth of live material.

Bioavailability: The proportion of a nutrient or administered drug, and so on that can be taken up by an organism in a biologically effective form. For example, some soils high in phosphorus have a low level of P availability because the pH of the soil renders much of the P insoluble.

Biochemical conversion: The use of fermentation or anaerobic digestion to produce fuels and chemicals from organic sources.

Biocontrol: Pest control by biological means. Any process using deliberately introduced living organisms to restrain the growth and development of other organisms, such as the introduction of predatory insects to control an insect pest. *Synonym*: biological control.

Bioconversion: Conversion of one chemical into another by living organisms, as opposed to their conversion by isolated enzymes or fixed cells, or by chemical processes. Particularly useful for introducing chemical changes at specific points in large and complex molecules.

Biodegrade: The breakdown by microorganisms of a compound to simpler chemicals. Materials that are easily biodegraded are colloquially termed biodegradable.

Biodiesel: Fuel derived from vegetable oils or animal fats. It is produced when a vegetable oil or animal fat is chemically reacted with an alcohol.

Biodiversity: The variability among living organisms from all sources, including, *inter alia*, terrestrial, marine, and other ecosystems and the ecological complexes of which they are part; this includes diversity within species, between species, and of ecosystems. *Synonyms*: biological diversity, ecological diversity. The more variety there is the higher amount of biodiversity and the less variety there is the lower the amount of biodiversity. The totality of genes, species, and ecosystems in a region or the world.

Bioenergetics: The study of the flow and the transformation of energy that occur in living organisms.

Bioenergy: (1) Useful, renewable energy produced from organic matter. (2) The conversion of the complex carbohydrates in organic matter to energy. Organic matter may either be used directly as a fuel, processed into liquids and gases or be a residual of processing and conversion.

Bioengineering: The use of artificial tissues, organs, and organ components to replace parts of the body that are damaged, lost or malfunctioning.

Bioethics: The branch of ethics that deals with the life sciences and their potential impact on society.

Biofuel: A gaseous, liquid, or solid fuel derived from a biological source, for example, ethanol, biodiesel, methanol, rapeseed oil, or fish liver oil.

Biogas: A mixture of methane and carbon dioxide resulting from the anaerobic decomposition of waste such as domestic, industrial, and agricultural sewage.

Bioinformatics: The science that uses advanced computing techniques for management and analysis of biological data. Bioinformatics particularly is important as an adjunct to genomic research, which generates a large amount of complex data involving DNA sequences and hundreds of thousands of genes.

Bioinsecticide: Any material used in insect control that is derived from living organisms, such as bacteria, plant cells, or animal cells. Examples include *Bacillus thuringiensis* (Bt) protein (from bacteria), and *Pyrethrum* (made from dried flower heads of certain chrysanthemum varieties), both used to control insects.

Biolistic device: A device that bombards (shoots) target cells with microscopic DNA-coated particles. Familiarly known as the Gene Gun, it was first developed in the early 1980s.

Biolistics: A technique to generate transgenic cells, in which DNA-coated small metal particles (tungsten or gold) are propelled by various means fast enough to puncture target cells. Provided that the cell is not irretrievably damaged, the DNA is frequently taken up by the cell. The technique has been successfully used to transform animal, plant, and fungal cells, and even mitochondria inside cells. *Synonym*: microprojectile bombardment.

Biological attack: The deliberate release of germs or other biological substances that can make people sick.

Biological containment: Restricting the movement of organisms from the laboratory. Can take two forms: making the organism unable to survive in the outside environment, or making the outside environment inhospitable to the organism. For microorganisms, the favored approach is to engineer organisms to require a supply of a specific nutrient that is usually available only in the laboratory. For higher organisms (plants and animals), it is more possible to ensure that the outside environment is unsuited to growth, spread, and reproduction.

Biological resources: The components of biodiversity that are of direct, indirect, or potential use to humanity.

Biomarker: An indicator, usually of a disease or a risk for a disease. For example, blood cholesterol is a biomarker for a risk for heart disease.

Biomass: Any plant-derived organic matter. Biomass available for energy on a sustainable basis includes herbaceous and woody energy crops, agricultural food and feed crops, agricultural crop wastes and residues, wood wastes and residues, aquatic plants and other waste materials, including some municipal wastes. Biomass is a heterogeneous and chemically complex renewable resource. Biomass covers a very wide range of headings, including phytomass, dendromass, zoomass, waste (domestic waste and industrial waste), biodegradable materials, residue-sourced materials, recycled materials, food production residues, agricultural residues, animal residues, vegetable residues, biomass materials, and innovative waste materials (poultry litter, coffee residues, mustard husks, and spice waste).

Biomass processing residues: Byproducts from processing all forms of biomass that have significant energy potential. An example is that solid wood products and pulp from logs produces bark, shavings, sawdust, and spent pulping liquors. Because these residues already are collected at the point of processing, they can be convenient and relatively inexpensive sources of biomass for energy.

Biome: A major ecological community or complex of communities, extending over a large geographical area and characterized by a dominant type of vegetation.

Biometry: The application of statistical methods to the analysis of continuous variation in biological systems. *Synonym*: biometrics.

Biopesticide: A compound that kills organisms by virtue of specific biological effects rather than as a broader chemical poison. Differs from biocontrol agents in being passive agents, whereas biocontrol agents actively seek the pest. The rationale behind replacing conventional pesticides with biopesticides is that the latter are more likely to be selective and biodegradable.

Biopharming: The use of genetically transformed crop plants and livestock animals to produce valuable compounds, especially pharmaceuticals. *Synonym*: molecular pharming.

Biopiracy: The patenting of genetic stocks, and the subsequent privatization of genetic resources collections. The term implies a lack of consent on the part of the originator.

Biopower: The use of biomass feedstock to produce electric power or heat through direct combustion of the feedstock, through gasification and then combustion of the resultant gas or through other thermal conversion processes. Power is generated with engines, turbines, fuel cells, or other equipment.

Biorefinery: A facility that processes and converts biomass into value-added products, which can range from biomaterials to fuels such as ethanol or important feedstocks for the production of chemicals and other materials.

Bioregion: A territory defined by a combination of biological, social, and geographic criteria, rather than by geopolitical considerations.

Bioremediation: (1) The use of plants and microorganisms to consume or otherwise help remove materials (such as toxic chemical wastes and metals) from contaminated sites (especially from soil and water). (2) A natural process in which environmental problems are treated by the use of bacteria or other microorganisms that break down a problem substance, such as oil, into less harmful molecules.

Biosafety protocol: An internationally agreed protocol setup to protect biological diversity from the potential risks posed by the release of genetically modified organisms. It establishes a procedure for ensuring that countries are provided with the information necessary to make informed decisions before agreeing to the import of such organisms into their territory. *Synonym*: Cartagena protocol.

Biosafety: Referring to the avoidance of risk to human health and safety and to the conservation of the environment, as a result of the use for research and commerce of infectious or genetically modified organisms.

Biosensor: A device that uses an immobilized biologically related agent (such as an enzyme, antibiotic, organelle, or whole cell) to detect or measure a chemical compound. Reactions between the immobilized agent and the molecule being analyzed are converted into an electric signal.

Biota: All the organisms, including animals, plants, fungi, and microorganisms, found in a given area.

Biotech foods, gene foods, bio-engineered foods, gene-altered foods, transgenic foods, GM foods: Terms used to describe foods that have been created through genetic engineering.

Biotechnology: (1) Any technological application that uses biological systems, living organisms, or derivatives thereof to make or modify products or processes for specific use (convention on biological diversity). (2) Interpreted in a narrow sense, ... a range of different molecular technologies such as gene manipulation and gene transfer, DNA typing, and cloning of plants and animals.

Biotechnology-derived: The use of molecular biology and/or recombinant DNA technology, or *in vitro* gene transfer, to develop products or to impart specific capabilities to plants or other living organisms.

Bioterrorism: Terrorism using biologic agents. Biological diseases and the agents that might be used for terrorism have been listed by the Centers for Disease Control and Prevention, and the list includes a sizable number of select agents—potential weapons whose transfer in the scientific and medical communities are regulated to keep them out of unfriendly hands.

Biotic factor: Other living organisms that are a component of an organism's environment and form the biotic environment, affecting the organism in many ways.

Biotic stress: Stress resulting from attack by pathogenic organisms.

Biotoxin: A naturally produced compound which shows pronounced biological activity, toxic to some or many organisms.

Botanical: A plant-based product.

Botulism: An uncommon but potentially very serious illness, it is a type of food poisoning that produces paralysis of muscles via the nerve toxin botulinum (botox), which in turn is manufactured by the bacteria *Clostridium botulinum*.

Bovine somatotropin (bST): A metabolic protein hormone used to increase milk production in dairy cows for commercial use. Scientists determined which gene in cattle controls or codes for the production of bST. They removed this gene from cattle and inserted it into a bacterium *Escherichia coli*. This bacterium produces large amounts of bST in controlled laboratory conditions. The bST produced by the bacteria is purified and then injected into cattle. *Synonyms*: bovine growth hormone.

Bovine spongiform encephalopathy (BSE): A disease of cattle, related to scrapie of sheep, also known as *mad cow disease*. It is thought to be caused by a prion, or small protein, that alters the structure of a normal brain protein, which in turn results in destruction of brain neural tissue.

Breed: (1) A sub-specific group of domestic livestock with definable and identifiable external characteristics that enable it to be separated by visual appraisal from other similarly defined groups within the same species. (2) A group of domestic livestock for which geographical and/or cultural separation from phenotypically similar groups has led to acceptance of its separate identity.

Breeding (traditional or selective): Making deliberate crosses or matings of plants or animals, so the offspring will have particular desired characteristics derived from one or both of the parents. Practices used in traditional plant breeding may include aspects of biotechnology such as tissue culture, mutational breeding, and marker-assisted breeding.

Breeding value: A quantitative genetics term, describing that part of the deviation of an individual phenotype from the population mean that is due to the additive effects of alleles. Thus, if an individual is mated with a random sample of individuals from a population, its breeding value for a given trait is twice the average deviation of its offspring from the population mean for that trait.

Bt crops: Crops that are genetically engineered to carry a gene from the soil bacterium *Bacillus thuringiensis* (Bt). The bacterium produces proteins that are toxic to some pests but nontoxic to humans and other mammals. Crops containing the Bt gene are able to produce this toxin, thereby providing protection for the plant. Bt corn and Bt cotton are examples of commercially available Bt crops.

Bt protein: A protein used as a natural pesticide to kill pests. It is produced by bacterium *Bacillus thuringiensis* (Bt).

Bt toxins: Insecticidal proteins produced by the soil microorganism *Bacillus thuringiensis*.

C

Callus (pl.: calli): (1) A protective tissue, consisting of parenchyma cells, that develops over a cut or damaged plant surface. (2) Mass of undifferentiated, thin-walled parenchyma cells induced by hormone treatment. (3) Actively dividing nonorganized masses of undifferentiated and differentiated cells often developing from injury (wounding) or in tissue culture in the presence of growth regulators.

Candidate gene: A gene whose function or location suggests it might be responsible for a disease or a trait in a population of individuals.

Canola: A specific subgroup of oilseed rape cultivars; canola oil is the highly mono-unsaturated fatty acid and low in erucic acid product produced in the seed of these cultivars.

Carbon footprint: Amount of greenhouse gases, specifically carbon dioxide or other carbon compounds, emitted by individuals, companies, or countries (i.e., a person's activities or a product's manufacture and transport) during a given period of time. Indicator of air quality often used to measure an entity's environmental impact.

Carcinogen: A cancer-producing agent or substance. A variety of chemical agents have been shown to induce malignancy in animals, but not all of them show the same capability in humans. It may or may not be a mutagen.

Carotenoids: A group of chemically similar red-to-yellow pigments responsible for the characteristic color of many plant organs or fruits, such as tomatoes and carrots. Carotenoids serve as light-harvesting molecules in photosynthetic assemblies and also play a role in protecting prokaryotes from the deleterious effects of light.

Carrier: An individual who has a recessive, disease-causing gene mutation at a particular locus on one chromosome of a pair and a normal allele at that locus on the other chromosome. It also may refer to an individual with a balanced chromosome rearrangement. Examples include a carrier for cystic fibrosis or for sickle cell anemia.

Carrier DNA: DNA of undefined sequence which is added to the transforming (plasmid) DNA used in physical DNA-transfer procedures. This additional DNA increases the efficiency of transformation in electroporation and chemically mediated DNA delivery systems. The mechanism responsible is not known.

Carrier of a mutated gene: Every cell contains two copies of each gene. One gene copy may be mutated (altered) and the other may be "correct." If the mutated gene is not expressed in the cells (resulting in a particular characteristic or a disorder), the mutated gene is said to be recessive to the other, correct copy of the gene. An individual who has one correct gene copy and one faulty (recessive) gene copy is said to be a carrier for the mutation leading to a specific condition. The carriers of a recessive mutation in a gene usually are not affected, but they are at risk for passing on the mutant gene to their offspring.

Carrier rate or frequency: The proportion of individuals in a population who have a single copy of a specific recessive gene mutation.

Carrier screening: Testing to determine whether individuals are carriers of a mutated or faulty gene for a particular disorder.

Carrier testing (also known as *carrier detection* **or** *heterozygote testing*): Testing used to identify usually asymptomatic individuals who have a gene mutation for an autosomal recessive or X-linked recessive disorder or who have a chromosome rearrangement (translocation or inversion, for example).

Carrying capacity: The maximum number of people or individuals of a particular species that a given part of the environment can maintain indefinitely.

Catalyst: A substance that increases the rate of a chemical reaction by reducing the activation energy. The substance is left unchanged by the reaction.

Cauliflower mosaic virus (CaMV): A DNA virus affecting cauliflower and many other dicot species. Its importance is due to the promoter of its 35S ribosomal DNA, which is constitutively active in most plant tissues, and has therefore been widely used as a promoter for the expression of transgenes.

Causative gene: A gene that in a variant form is known to be the reason for developing a specific genetic disease. Monogenic genetic diseases are due to causative genes. Examples include cystic fibrosis, sickle cell anemia, and Huntington's disease.

Cell: The lowest denomination of life thought to be possible and the fundamental unit of an animal body. Most organisms consist of more than one cell, which becomes specialized into particular functions to enable the whole organism to function properly. In humans, each body organ contains different types of cells, and at the heart of a cell is the nucleus, which contains chromosomes, or long coils of DNA. DNA provides not only a blueprint from which a cell can produce proteins to perform its function, but also a design for the entire body.

Cell culture: The *in vitro* growth of cells isolated from multicellular organisms.

Cell division: The mechanism by which cells multiply during the growth of tissues or organs.

Cell selection: The process of selecting cells that exhibit specific traits within a group of genetically different cells. Selected cells are often subcultured onto fresh medium for continued selection.

Cell-based therapies: Treatments in which stem cells are induced to differentiate into the specific cell type required to repair damaged or depleted adult cell populations or tissues.

Center of diversity: Geographic region with high levels of genetic or species diversity.

Characteristic diversity: The pattern of distribution and abundance of populations, species, and habitats under conditions where humanity's influence on the ecosystem is no greater than that of any other biotic factor.

Chromosome: A strand of coiled DNA that is the self-replicating genetic structure of cells. The nucleus of each animal cell (except red blood cells) contains at least one chromosome, and the number of chromosomes in each cell differs from animal to animal. Humans have 23 pairs of chromosomes, including the pair of sex chromosomes (two X chromosomes for females versus an X and a Y chromosome for males). Chromosomes are supported by proteins called *chromatin*.

Clone: (1) A group of cells or individuals that are genetically identical as a result of asexual reproduction, breeding of completely inbred organisms, or forming genetically identical organisms by nuclear transplantation. (2) Group of plants genetically identical in which all are derived from one selected individual by vegetative propagation. (3) Verb: to clone. To insert a DNA segment into a vector or host chromosome.

Cloning: The propagation of genetically exact duplicates of an organism by a means other than sexual reproduction, for example, the vegetative production of new plants or the propagation of DNA molecules by insertion into plasmids. Often, but inaccurately, used to refer to the propagation of animals by nuclear transfer.

***Clostridium perfringens*:** A type of bacteria that is the most common agent of gas gangrene and that also can cause food poisoning and a fulminant form of bowel disease called necrotizing colitis.

***Clostridium*:** A group of anaerobic bacteria that thrive in the absence of oxygen. It includes species that cause the diseases botulism, gas gangrene, and tetanus.

Cocultivation: Growth of cultured cells together.

Codex Alimentarius Commission (CAC): The Codex Alimentarius is a collection of internationally adopted food standards presented in a uniform manner. The food standards are developed to protecting consumers' health and ensuring fair practices in the food trade. The Codex Alimentarius Commission is the international body responsible for the execution of the food standard program.

Coding region: The part of a gene that directly specifies the amino acid sequence of its protein product.

Colony: (1) A group of genetically identical cells or individuals derived from a single progenitor. (2) A group of interdependent cells or organisms.

Commodity: A product of agriculture. Examples of agricultural commodities include wheat, rice, beets, corn, beef, soybeans, and coffee.

Communicable: Capable of spreading disease. Also known as *infectious*.

Community: An integrated group of species inhabiting a given area. The organisms within a community influence one another's distribution, abundance, and evolution.

Comparator: A product that is compared to another product (e.g., a genetically engineered food and a nongenetically engineered food).

Competent cells: Cells (e.g., bacteria, plant, or yeast) that can take up DNA and become genetically transformed.

Conditioning: (1) The effects on phenotypic characters of external agents during critical developmental stages. (2) The undefined interaction between tissues and culture medium resulting in the growth of single cells or small aggregates. Conditioning may be accomplished by immersing cells or callus contained within a porous material (such as dialysis tubing) into fresh medium for a period dependent on cell density and a volume related to the amount of fresh medium.

Conservation of biodiversity: The management of human interactions with genes, species, and ecosystems so as to provide the maximum benefit to the present generation while maintaining their potential to meet the needs and aspirations of future generations.

Control elements: DNA sequences in genes that interact with regulatory proteins (such as transcription factors) to determine the rate and timing of expression of the genes, as well as the beginning and end of the transcript.

Convention on Biological Diversity (CBD): The international treaty governing the conservation and use of biological resources around the world, which has also called for the establishment of rules to govern the international movement of nonindigenous living organisms and genetically modified organisms.

Correlation: A statistical association between variables.

CpG methylation: A heritable chemical modification of DNA (replacement of cytosine by 5-methyl cytosine) that, when present in a control region, usually suppresses expression of the corresponding gene.

Cross hybridization: The annealing of a single-stranded DNA sequence to a single-stranded target DNA to which it is only partially complementary. Often, this refers to the use of a DNA probe to detect homologous sequences in species other than the origin of the probe.

Cross pollination: When genes from one organism or crop carry over to another organism or crop that was not scientifically alter to have those genes.

Cross-breeding: Mating between members of different populations (lines, breeds, races, or species).

Cross-fertilization (cross-pollination): Introduction of pollen from one plant into the flower of another plant in order to breed plants with specific traits.

Crossing over: The process by which homologous chromosomes exchange material at meiosis through the breakage and reunion of nonsister chromatids.

Cultivar (cv): An internationally accepted term denoting a variety of a cultivated plant. Must be distinguishable from other varieties by stated characteristics and must retain their distinguishing characters when reproduced under specific conditions.

Culture medium: Any nutrient system for the cultivation of cells, bacteria, or other organisms; usually a complex mixture of organic and inorganic nutrients.

Culture: A population of plant or animal cells or microorganisms grown under controlled conditions.

Cybrid: A hybrid, originating from the fusion of a cytoplast (the cytoplasm without nucleus) with a whole cell derived from a different species.

Cytogenetic map: A map that illustrates where genes are located on each chromosome.

Cytogenetics: The biology of chromosomes and their relation to the transmission and recombination of genes. The microscopic study of chromosomes and how changes in chromosome structure and number affect individuals.

Cytology: The study of the structure and function of cells.

Cytolysis: Cell disintegration.

Cytoplasmic inheritance: Hereditary transmission dependent on cytoplasmic genes (genes located on DNA outside the nucleus).
Cytotoxicity: Poisoning of the cell.

D
Deoxyribonucleic acid (DNA): The material that makes up genes and carries information that tells cells what substances to produce. It carries the genetic information for most living systems. The DNA molecule consists of four base proteins (adenine, cytosine, guanine, and thymine) and a sugar phosphate backbone, arranged in two connected strands to form its characteristic double helix. The genome (all of the genetic information in a living organism), rather than single DNA molecules, determines the organism's characteristics.
Deregulation: The process or act of removing government restrictions or regulations on planting, import, and/or export. Plant commodities are deregulated upon the government receiving and evaluating scientific research demonstrating food, feed, and human safety and minimal impact on the environment.
Diagnostic testing: Testing designed to confirm or exclude a known or suspected genetic disorder in a symptomatic individual, or prenatally, in a fetus at risk for a certain genetic condition.
Dichlorodiphenyltrichloroethane (DDT): A colorless, crystalline, tasteless, and almost odorless organochloride, which has insecticidal properties. It has been used as an agricultural insecticide.
Dietary reference intakes: An umbrella term for groups of values that specify recommended dosages.
Dietary Supplement Health and Education Act (DSHEA): Passed in 1994, this law amended the Federal Food, Drug, and Cosmetic Act of the United States. It created a new regulatory framework for the safety and labeling of dietary supplements, placing them in a special category under the general umbrella of foods and requiring them to be labeled as dietary supplements.
Dietary supplement: A product taken by mouth that contains a dietary ingredient intended to supplement the diet. The dietary ingredients may include vitamins, minerals, herbs or other botanicals, amino acids, or dietary substances to supplement the diet by increasing the total dietary intake. Dietary supplements can be concentrates, metabolites, constituents, or extracts. They may be found in tablets, capsules, softgels, gelcaps, liquids, or powders. They also can be in other forms, such as a bar; in this case,

information on the label must not represent the product as a conventional food or a sole item of a meal or diet.

Differentiated cells: Cells that are specialized for a particular function (heart muscle or a blood cell, for example) and do not maintain the ability to generate other kinds of cells or to revert back to a less specialized type of cell (such as stem cells).

Differentiation: The process whereby an unspecialized early embryonic cell acquires the features of a specialized cell such as a heart, liver, or muscle cell.

Disease surveillance: The ongoing systematic collection and analysis of data and the provision of information, which leads to action being taken to prevent and control a disease, usually one of an infectious nature.

Disintegration: The drop in potency of a dietary supplement while in storage as a function of time and storage conditions (light, heat, moisture, and air). Stable supplements have a low rate of disintegration, allowing for a later expiration date, while others lose potency comparatively quickly.

DNA chip: A purpose-built microchip used to identify mutations or alterations in a gene's DNA.

DNA fingerprinting: The derivation of unique patterns of DNA fragments obtained using a number of marker techniques. *Synonym*: genetic fingerprinting.

DNA sequencing: Procedures for determining the nucleotide sequence of a DNA fragment in a gene, in a chromosome or in the entire genome. Two common methods available: (1) The Maxam Gilbert technique, which uses chemicals to cleave DNA into fragments at specific bases; or, most commonly. (2) The Sanger technique (also called the di-deoxy or chain-terminating method) which uses DNA polymerase to make new DNA chains, in the presence of di-deoxynucleotides (chain terminators) to stop the chain randomly as it grows. In both cases, the DNA fragments are separated according to length by polyacrylamide gel electrophoresis, enabling the sequence to be read directly from the gel. The procedure has become increasingly automated and large-scale in recent years.

Dolly: The first mammal (a sheep) to be created (via nuclear transfer) by the cloning of an adult cell (from the mammary tissue of a ewe). This showed that the process of differentiation into adult tissue is not, as previously thought, irreversible.

Dominant trait: A characteristic determined by an allele that is expressed over any other alleles for a given trait.

Dominant: Every cell contains two copies of each gene. When only one of the gene copies, or alleles, is mutated and the other allele is "correct," yet the person is affected by a disorder due to that mutation, the mutation is described as dominant. The mutated gene is said to be dominant over the other, correct copy of the gene. A disorder or characteristic caused by a dominant gene mutation only requires one of the genes to be mutated for the person to be affected.

Double helix: The structural arrangement of DNA, which resembles a twisted ladder. It describes the coiling of the two strands of the double-stranded DNA molecule, resembling a spiral staircase in which the base pairs form the steps and the sugar-phosphate backbones form the rails on each side. One strand runs 3′–5′, while the complementary one runs 5′–3′.

Dysplasia: Abnormal development, or growth, of tissues or cells.

E

Ebola virus: A notoriously deadly virus that causes fearsome symptoms, the most prominent being high fever and massive internal bleeding. The Ebola virus kills as many as 90% of the people it infects and is one of the viruses that is capable of causing hemorrhagic (bloody) fever.

Ecosystem: A network of organisms from many different species living together in a region and their connections through the flow of energy, nutrients, and matter. These connections occur as the organisms of different species interact with one another. The ultimate source of energy in almost every ecosystem is the sun.

Ectopic gene expression: Expression of a (trans) gene in a tissue or developmental stage when such expression is not expected.

Electroporation: The induction of transient pores in bacterial cells or protoplasts by the application of a pulse of electricity. These pores allow the entry of exogenous DNA into the cell. Widely used for the transformation of bacteria.

Electroscanning microscope (ESM): Used for the study of surface morphology (the study of the form or shape of an organism or part thereof) and the determination of the thickness of molecular beam epitaxy (MBE)-grown films.

Elemental analysis: The determination of carbon, hydrogen, nitrogen, oxygen, sulfur, chlorine, and ash in a sample.

ELISA: Abbreviation for enzyme-linked immunosorbent assay. An immunoassay, that is, an antibody-based technique for the diagnosis of the presence and quantity of specific molecules in a mixed sample. It combines the specificity of an immunoglobulin with the detectability of an enzyme-generated colored product. In one form, the primary antibody (specific to the test protein) is adsorbed onto a solid substrate, and a known amount of the sample is added; all the antigens in the sample are bound by the antibody. A second antibody (conjugated with an enzyme) specific for a second site on the test protein is added; and the enzyme generates a color change in the presence of a substrate reagent.

Embryo cloning: The creation of identical copies of an embryo by embryo splitting or by nuclear transfer from undifferentiated embryonic cells.

Embryo rescue: A sequence of tissue culture techniques utilized to enable a fertilized immature embryo resulting from an interspecific cross to continue growth and development, until it can be regenerated into an adult plant.

Embryo: An animal in the early stage of development before birth. In humans, the embryo stage lasts from the time of fertilization until the end of the eighth week of gestation, when it becomes known as a fetus.

Embryonic stem (ES) cell: A primitive (undifferentiated) cell from the embryo that has the potential to become a wide variety of specialized cell types.

Embryonic stem cell line: Embryonic stem cells that have been cultured under *in vitro* conditions, allowing proliferation without differentiation for months to years.

Endemic: (1) Restricted to a specified region or locality. (2) The continual, low-level presence of disease in a community.

Endogenous: Derived from within; from the same cell type or organism.

Energy balance: A state where energy intake is equivalent to energy expenditure, resulting in no net weight gain or weight loss. Energy balance in children is used to indicate equality between energy intake and energy expenditure that supports normal growth without promoting excess weight gain.

Energy density: The amount of energy stored in a given food per unit of volume or mass. Fat stores 9 kcal/g, alcohol stores 7 kcal/g, carbohydrate and protein each store 4 kcal/g, fiber stores 1.5–2.5 kcal/g, and water has no energy.

Enhancer: (1) A substance or object that increases a chemical activity or a physiological process. (2) A eukaryotic DNA sequence (also found in some eukaryotic viruses) which increases the transcription of a gene. Located up to several kbp, usually (but not exclusively) upstream of the gene in question. In some cases, can activate transcription of a gene with no (known) promoter. *Synonyms:* enhancer element; enhancer sequence. (3) A major or modifier gene that increases the rate of a physiological process.

Environmental Protection Agency (EPA): A U.S. governmental agency whose mission is to protect human health and safeguard the natural environment—air, water, and land—upon which life depends. EPA is one of the three agencies that review new products of agricultural biotechnology that express plant-incorporated pesticides (Bt), as well as the use of pesticides with a new crop variety developed through biotechnology.

Epidemic: The occurrence of disease within a specific geographical area or population that is in excess of what normally is expected.

Epithelial cell: The tissue that forms the superficial layer of skin and some organs. It also forms the inner lining of blood vessels, ducts, body cavities, and the interior of the respiratory, digestive, urinary, and reproductive systems.

Epitope: The part of an antigen that stimulates an immune response.

Ethnobotanist: A person who studies the relationship between plants and people. An ethnobotanist examines how plants have been or are used, managed, and perceived in human societies—especially how plants are used for food, medicine, cosmetics, dyeing, textiles, building, tools, currency, clothing, rituals, social life, and music.

Etiology: The cause of.

Evolution: Any gradual change. Organic evolution is any genetic change in organisms from generation to generation.

***Ex situ* conservation:** A conservation method that keeps components of biodiversity alive outside their original habitat or natural environment.

Exposure: Contact with infectious agents (bacteria or viruses) in a manner that promotes transmission and increases the likelihood of disease.

Extinction: The evolutionary termination of a species caused by the failure to reproduce and the death of all remaining members of the species.

F

F_1: Abbreviation for filial generation. The initial hybrid generation resulting from a cross between two parents.

F_2: The second hybrid generation, produced either by intercrossing two F_1 individuals, or by self-fertilizing an F_1 individual.

Fertilization: The union of two gametes from opposite sexes to form a zygote. Typically, each gamete contains a haploid set of chromosomes. Hence the zygotic nucleus contains a diploid set of chromosomes. Several categories can be distinguished: (1) Self-fertilization (selfing): fusion of male and female gametes from the same individual. (2) Cross-fertilization (crossing): fusion of male and female gametes from different individuals. (3) Double fertilization; restricted to flowering plants, in which the fusion of one male gamete with the ovum occurs at about the same time as the second male gamete nucleus fuses with the female polar nuclei (or secondary nucleus) to form the endosperm.

Fertilizer: Any substance that is added to soil in order to increase its productivity. Fertilizers can be of biological origin (e.g., composts), or they can be synthetic (artificial fertilizer).

Field test or trial: A test of a new technique or variety, including biotech-derived varieties, done outside the laboratory but with specific requirements on location, plot size, methodology, and more.

Fitness: The ability to survive to reproductive age and produce viable offspring. Fitness also describes the frequency distribution of reproductive success for a population of mature adults.

Food and Drug Administration (FDA): The U.S. regulatory agency responsible for ensuring the safety and wholesomeness of all foods sold in interstate commerce except meat, poultry, and eggs (which are under the jurisdiction of the U.S. Department of Agriculture). One of the three agencies that review new products of agricultural biotechnology that are intended for the food supply.

Food processing enzyme: Enzyme used to control food texture, flavor, appearance, or nutritional value. Amylases break down complex polysaccharides to simpler sugars; proteases tenderize meat proteins. A prominent target of food biotechnology is to develop novel food enzymes which can improve the quality of processed foods.

Food security: Availability of and access to sufficient, nutritious food on a consistent basis, as well as the knowledge and ability to select and prepare foods to ensure safety and adequate nutrition. *Antonym*: food insecurity.

Fortify: To add strengthening components or beneficial ingredients to a nutrient medium.

Frankenfoods: Another term for genetically modified foods (GMF) referring to the story of Frankenstein and science gone bad, usually used by GMF opponents and critics.

Freeze-dry: The removal of water as vapor from frozen material under vacuum. Used to measure water content and to preserve samples, particularly spores. Unlike oven-drying, bound water remains associated with the specimen. *Synonym*: lyophilize.

Functional food: A foodstuff that provides a health benefit beyond basic nutrition, demonstrating specific health or medical benefits, including the prevention and treatment of disease.

Functional genomics: The field of research that aims to determine patterns of gene expression and interaction in the genome, based on the knowledge of extensive or complete genomic sequence of an organism.

Fungicide: A chemical agent toxic to fungi.

Fungus (pl.: fungi): Multinucleate single-celled or multicellular heterotrophic microorganisms, including yeasts, moulds, and mushrooms. They live as parasites, symbionts, or saprophytes. Lacking any vascular tissues (unlike plants), their cell walls are made of chitin or other noncellulose compounds.

G

Gamete: A mature reproductive cell capable of fusing with a cell of similar origin but of opposite sex to form a zygote from which a new organism can develop. Gametes normally have a haploid chromosome content.

GE: Genetically engineered or genetic engineering.

Gene addition: The addition of a functional copy of a gene to the genome of an organism.

Gene amplification: The selective production of multiple copies of one gene without a proportional increase in others.

Gene bank: (1) The physical location where collections of genetic material in the form of seeds, tissues, or reproductive cells of plants or animals are stored. (2) Field gene bank: A facility established for the *ex situ* storage and maintenance, using horticultural techniques, of individual plants. Used for species whose seeds are recalcitrant, or for clonally propagated species of agricultural importance, for example, apple varieties. (3) A collection of cloned DNA fragments from a single genome. Ideally the bank should contain cloned representatives of all the DNA sequences in the genome.

Gene cloning: The synthesis of multiple copies of a chosen DNA sequence using a bacterial cell or another organism as a host. The gene of interest is inserted into a vector, and the resulting recombinant DNA molecule is amplified in an appropriate host cell. *Synonym*: DNA cloning.

Gene conversion: A process, often associated with recombination, during which one allele is replicated at the expense of another, leading to non-Mendelian segregation ratios.

Gene expression: The conversion of the gene's nucleotide sequence into an actual process or structure in the cell. Some genes are expressed only at certain times during an organism's life and not at others. The result of the activity of a gene or genes that influences the biochemistry and physiology of an organism and may change its outward appearance.

Gene families: Groups of closely related genes that make similar products, such as proteins.

Gene flow: The spread of genes from one breeding population to another (usually) related population by migration, thereby generating changes in allele frequency.

Gene insertion: The incorporation of one or more copies of a gene into a chromosome.

Gene introgression: Introduction of new genes into a population by crossing between two populations, followed by repeated backcrossing to that population while retaining the new genes.

Gene mapping: Determining the relative physical locations of genes on a chromosome. It is useful for plant and animal breeding.

Gene modification: Chemical change to a gene's DNA sequence.

Gene pool: (1) The sum of all genetic information in a breeding population at a given time. (2) In plant genetic resources, its usage is made of the terms "primary," "secondary," and "tertiary" gene pools. In general, members of the primary gene pool are interfertile; those of the secondary can be crossed with those in the primary gene pool under special circumstances; but to introgress variation from the tertiary gene pool, special techniques are required to achieve crossing.

Gene replacement: The incorporation of a transgene into a chromosome at its normal location by homologous recombination, thus replacing the copy of the gene originally present at the locus.

Gene sequencing (also known as *DNA sequencing***):** Determining the exact sequence of nucleotide bases in a strand of DNA to better understand the behavior of a gene.

Gene splicing: The isolation of a gene from one organism followed by the introduction of that gene into another organism using techniques of biotechnology.

Gene therapy: The addition of a functional gene or groups of genes to a cell using recombinant DNA techniques to correct a hereditary disease.

Gene: The fundamental physical and functional unit of heredity. It is a portion of a chromosome (DNA) that contains the hereditary information necessary for the production of a protein and for coding of a particular trait. A gene consists of three parts: promoter, coding region, and terminator. The promoter specifies the starting of a protein production process. The coding region specifies the structure of a protein and the terminator stops the production process.

Generally regarded as safe (GRAS): Designation given to foods, drugs, and other materials with a long-term history of not causing illness to humans, even though formal toxicity testing may not been conducted. Certain host organisms for recombinant DNA experimentation have recently been given this status.

Genetic code: The correspondence between the set of 64 possible nucleotide triplets and the amino acids and stop codons that they specify.

Genetic complementation: When two DNA molecules that are in the same cell together produce a function that neither DNA molecule can supply on its own.

Genetic counseling: A process involving an individual or family that comprises evaluation to confirm, diagnose, or exclude a genetic condition, malformation syndrome, or isolated birth defect; discussion of natural history and the role of heredity; identification of medical management issues; calculation and communication of genetic risks; and provision of or referral for psychosocial support.

Genetic diversity: Variation in the genetic composition of individuals within or among species.

Genetic engineering (GE): A science that involves taking a cell or gene out of one organism and replacing that cell or gene with one from another organism. It involves changes in the genetic constitution of cells resulting from the introduction or elimination of specific genes via molecular biology (i.e., recombinant DNA) techniques. By altering the genetic information, genetic engineering changes the type or amount of proteins an organism is capable of producing, thus enabling it to make new substances or to perform new functions. It is done to eliminate undesirable characteristics or to produce desirable new ones.

Genetic modification (GM): (1) The production of heritable improvements in plants or animals for specific uses, via either genetic engineering or other, more traditional methods. (2) Any process that alters the genetic material of living organism to express different characteristics.

Genetic pollution: Uncontrolled spread of genetic information (frequently referring to transgenes) into the genomes of organisms in which such genes are not present in nature.

Genetic predisposition: The presence of a gene or group of genes that might predispose a person to develop a particular health problem later in life.

Genetic testing: The analysis of human DNA, RNA, and/or chromosomes to detect inheritable or acquired disease-related genotypes, mutations, phenotypes, or karyotypes.

Genetically engineered (GE): The standard U.S. term for a process in which foreign genes are spliced into a nonrelated species, creating an entirely new organism.

Genetically engineered food: A food substance that has foreign genes inserted into its genetic code. Genetic engineering can be done with plants, animals, or microorganisms. Scientists can move desired genes from one plant into another and even from an animal to a plant, or vice versa.

Genetically engineered organism (GEO): An organism produced through genetic engineering.

Genetically modified (GM): The same as GE more widely used in Europe because it translates more easily among different languages.

Genetically modified organism (GMO): (1) The label GMO and the term *transgenic* often refer to organisms that have acquired novel genes from other organisms by laboratory gene transfer methods. GMOs have had genes from other species inserted into their genome.

Genetically modified: Refers to an organism whose genotype has been altered and includes alteration by genetic engineering and nongenetic engineering methods.

Genetically modified foods: Food plants that have been genetically altered by the addition of foreign genes to enhance a desired trait.

Genetics: The science of heredity. It studies how traits pass from parents to children and of the molecular basis of those traits.

Genome: (1) The entire complement of genetic material (genes plus non-coding sequences) present in each cell of an organism, virus, or organelle. (2) The complete set of chromosomes (hence of genes) inherited as a unit from one parent.

Genomic imprinting: Differing expression of genetic material depending on the sex of the transmitting parent.

Genomic library: A collection of biomolecules made from DNA fragments of a genome that represent the genetic information of an organism that can be propagated and then systematically screened for particular properties. The DNA may be derived from the genomic DNA of an organism or from DNA copies made from messenger RNA molecules.

Genomics: The mapping and sequencing of all the genetic material in the DNA of a particular organism as well as the use of information derived from genome sequence data to further reveal what genes do, how they are controlled and how they work together.

Genotype: The genetic identity of an individual, or the set of genes possessed by an individual organism. Genotype often is evident by outward characteristics. It also refers to the specific set of alleles inherited at a locus.

Genus (pl.: genera): A group of closely related species, whose perceived relationship is typically based on physical resemblance, now often supplemented with DNA sequence data.

GEO: Abbreviation for genetically engineered organism.

Germination: (1) The initial stages in the growth of a seed to form a seedling. (2) The growth of spores (fungal or algal) and pollen grains.

Germplasm: (1) An individual, group of individuals, or a clone representing a genotype, variety, species, or culture, held in an *in situ* or *ex situ* collection. (2) Original meaning, now no longer in use: The genetic material that forms the physical basis of inheritance and which is transmitted from one generation to the next by means of the germ cells.

Glyphosate: Glyphosate is a broad-spectrum herbicide used to kill crop weeds. Monsanto's trade name for this is Roundup. Roundup Ready crops are engineered to withstand exposure to glyphosate. This allows applications of the herbicide after crop emergence, killing weeds but not Roundup-resistant crop plants such as RRS (Roundup Ready Soybeans). Chemically, glyphosate is an organophosphate like many other pesticides but it does not affect the nervous system as other organophosphates do. It is a broad-spectrum, nonselective herbicide which kills all plants, including grasses, broad leaf, and woody plants. It is absorbed mainly through the leaves and is transported around the whole plant, killing all parts of it. It acts by inhibiting a biochemical pathway, the shikimic acid pathway. At low levels of application, it acts as a growth regulator. There are three forms of glyphosate used as weed killers; glyphosate-isopropylammonium and glyphosate-sesquiodium patented by Monsanto and glyphosate-trimesium, patented by ICI (now Zeneca). Other common brand names are Rodeo, Accord, and Vision. Glyphosate is technically extremely difficult to measure in environmental samples. Only a few laboratories have the sophisticated equipment and techniques necessary. This means that data are often lacking on residue levels in food and the environment and existing data may not be reliable. Because it is broad spectrum in action, it is used to control a great variety of annual, biennial, and perennial grasses, sedges, broad leafed weeds, and woody shrubs. It is used in fruit orchards, vineyards, conifer plantations, and many plantation crops (e.g., coffee, tea, bananas); precrop, postweed emergence in a wide range of crops (including soybean, cereals, vegetables, and cotton); on noncrop areas (e.g., road shoulders and rights of way); in cereal stubble; forestry; gardening and horticulture. Other uses of salts of glyphosate are in growth regulation in peanuts and in sugarcane to regulate growth and speed fruit ripening. Because the shikimic acid pathway does not exist in animals, the acute toxicity of glyphosate is very low. Glyphosate can interfere with some enzyme functions in animals but symptoms of poisoning are only seen at very high doses. However, products containing glyphosate also contain other compounds which can be toxic. In particular, most contain surfactants known as polyoxyethyleneamines (POEA). Some of these are much more toxic than glyphosate. These account for problems associated with worker exposure. They are serious irritants of the respiratory tract, eyes, and skin and are contaminated with dioxane (not dioxin) which is a suspected carcinogen. Some are toxic to fish. New formulations, with less irritating surfactants, have been developed by Monsanto (e.g., Roundup Biactive), but cheaper, older preparations are still available. Glyphosate is one of the most toxic herbicides, with many species of wild plants being damaged or killed by applications of less than 10 µm per plant. The U.S. EPA concluded that many endangered species of plants, as well as the Houston toad, may be at risk from glyphosate use. Fish and invertebrates are more sensitive to formulations of glyphosate. As with humans, the surfactants are responsible for much of the harm. Toxicity is increased with higher water temperatures, and pH. Of nine herbicides tested for their toxicity to soil microorganisms, glyphosate was found to be the second most toxic to a range of bacteria, fungi, actinomycetes, and yeasts. However, when glyphosate comes into contact with the soil, it rapidly binds to soil particles and is inactivated. Unbound glyphosate is degraded by bacteria. Low activity because of binding to soil particles suggests that glyphosate's effects on soil flora will be limited. However, some recent work shows that glyphosate can be readily released from certain types of soil particles, and therefore may leach into water or be taken up by plants.

GM food: Abbreviation for genetically modified food. Food that contains above a certain legal minimum content of raw material obtained from genetically modified organisms.

GMO: Abbreviation for genetically modified organism. The actual organism that is created through GE.

Golden rice: A biotechnology-derived rice, which contains large amounts of beta carotene (a precursor of vitamin A) in its seeds. Achieved by inserting two genes from daffodil and one from the bacterium *Erwinia uredovora*.

Good laboratory practice (GLP): Written codes of practice designed to reduce to a minimum the chance of procedural or instrument problems which could adversely affect a research project or other laboratory work.

Good manufacturing practice (GMP): Codes of practice designed to reduce to a minimum the chance of procedural or instrument/manufacturing plant problems which could adversely affect a manufactured product.

Graft: (1) Verb. To place a detached branch or bud (scion) in close cambial contact with a rooted stem (rootstock) in such a manner that scion and rootstock unite to form a single plant. (2) Noun.

GRAS: Generally recognized as safe—substances intentionally added to food that do not require a formal premarket review by FDA to assure their safety, because their safety has been established by a long history of use in food (from the FDA).

Green revolution: Name given to the dramatic increase in crop productivity during the third quarter of the twentieth century, as a result of integrated advances in genetics and plant breeding, agronomy, and pest and disease control.

H

Haploid: A cell or organism containing one of each of the pairs of homologous chromosomes found in the normal diploid cell.

Haplotype map: An effort to find DNA landmarks that identify specific DNA sequences shared by many individuals.

Hardening off: Adapting glasshouse or controlled environment-grown plants to outdoor conditions by reducing availability of water, lowering the temperature, increasing light intensity, or reducing the nutrient supply. The hardening-off process conditions plants for survival when transplanted outdoors.

Hazard: A substance or agent that, upon exposure, might result in a defined harm.

Helix: A spiral, staircase-like structure with a repeating pattern described by two simultaneous operations (rotation and translation).

Herbicide resistance: The ability of a plant to remain unaffected by the application of a herbicide.

Herbicide: A substance that is toxic to plants; the active ingredient in agrochemicals intended to kill specific unwanted plants, especially weeds.

Herbicide-tolerant crops: Crops that have been developed to survive (tolerate) exposure to particular herbicides by the incorporation of certain gene(s), either through genetic engineering or traditional breeding methods. The herbicide can therefore be applied to the field for weed control without damaging the crop.

Heredity: Resemblance among individuals related by descent; transmission of traits from parents to offspring.

Heritability: The degree to which a given trait is controlled by inheritance, as opposed to being controlled by nongenetic factors.

Heteroploid: Cells with nuclei containing chromosome numbers other than diploid.

Heterozygote: An individual who has two different alleles at a particular locus on the same pair of chromosomes.

Heterozygous insects: Contain only one copy of a resistance gene and are generally more susceptible than resistant homozygous individuals which contain two copies. When resistance is rare, resistant homozygous individuals are rare owing to crossing of resistant and nonresistant insects producing heterozygous individuals. Killing heterozygotes can therefore slow down the build-up of resistance in pest populations.

Heterozygous: The condition in which an organism has inherited two different alleles of a specific gene pair from its parents.

Homologous recombination: Rearrangement of DNA sequences on different molecules by crossing over in a region of identical sequence.

Homologs: In diploid organisms, a pair of matching chromosomes.

Homozygote: Refers to an individual in whom the two gene copies, or alleles, contain identical information. An individual can be homozygous for the correct copies of the gene or can be homozygous for the mutated copies of the gene.

Homozygous: The condition in which an organism has inherited two identical alleles of a specific gene pair from its parents.

Horizontal gene transfer: Transmission of DNA involving close contact between the donor's DNA and the recipient, uptake of DNA by the recipient, and stable incorporation of the DNA into the recipient's genome.

Human Genome Project: An extensive international research effort to determine the sequence in which human DNA is arranged. It ended in 2003 with a complete mapping of all the genes in the human body.

Hybrid seed: (1) Seed produced by crossing genetically dissimilar parents. (2) In plant breeding, it is used colloquially for seed produced by specific crosses of selected pure lines, such that the F_1 crop is genetically uniform and displays hybrid vigor. As the F_1 plants are heterozygous with respect to many genes, the crop does not breed true and so new seed must be purchased each season.

Hybrid selection: The process of choosing individuals possessing desired characteristics from among a hybrid population.

Hybrid vigor: The extent to which a hybrid individual outperforms both its parents with respect to one or many traits. The genetic basis of hybrid vigor is not well understood, but the phenomenon is widespread, particularly in inbreeding plant species. *Synonym*: heterosis.

Hybridization: (1) The process of forming a hybrid by cross-pollination of plants or by mating animals of different types. (2) The production of offspring of genetically different parents, normally from sexual reproduction, but also asexually by the fusion of protoplasts or by transformation. (3) The pairing of two DNA strands, often from different sources, by hydrogen bonding between complementary nucleotides.

Hybridoma: A fast-growing culture of cloned cells made by fusing a cancer cell to some other cell, such as an antibody-producing cell.

Hybrids: Are the result of crosses from two genetically distinct parents. Hybrids are often more vigorous and higher yielding than either of the parents. A hybrid contains one copy of each of the parents' chromosomes. Hybrid seed is not usually replanted because in the next generation the chromosomes will be inherited unevenly and the effects in the resulting crop may be difficult to predict. Overall, second-generation seed from a hybrid will be nonuniform and much lower yielding. Hybrid seeds are selected to have higher quality traits (yield or pest tolerance, for example).

I

Identity preservation: The segregation of one crop type from another at every stage from production and processing to distribution. This process usually is performed through audits and site visits and provides independent third-party verification of the segregation.

Immunoassay: Any detection system for a particular molecule, which exploits the specific binding of an antibody raised against it. For measurement, the antibody can incorporate a radioactive or fluorescent label, or be linked to an enzyme which catalyzes an easily monitored reaction such as a change in color. *Synonym*: immunodiagnostics.

In situ **conservation:** A conservation method that attempts to preserve the genetic integrity of gene resources by conserving them within the evolutionary-dynamic ecosystems of the original habitat or natural environment.

In situ: In the natural place or in the original place. (1) Experimental treatments performed on cells or tissue rather than on extracts from them. (2) Assays or manipulations performed with intact tissues.

In vitro: Outside the organism, or in an artificial environment. Performed in a test tube or other laboratory apparatus. The term literally means "in glass." It can be applied, for example, to cells, tissues, or organs cultured in glass or plastic containers.

In vivo: The natural conditions in which organisms reside. Refers to biological processes that take place within a living organism or cell under normal conditions.

Inbred line: The product of inbreeding, that is, the intercrossing of individuals that have ancestors in common. In plants and laboratory animals, it refers to populations resulting from at least six generations of selfing or 20 generations of brother–sister mating, so that they have become, for all practical purposes, completely homozygous. In farm animals, the term is sometimes used to describe populations that have resulted from several generations of the mating of close relatives, without having reached complete homozygosity.

Inbreeding depression: The reduction in vigor over generations of inbreeding. This affects species which are normally outbreeding and highly heterozygous.

Inbreeding: Matings between individuals that have one or more ancestors in common, the extreme condition being self-fertilization, which occurs naturally in many plants and some primitive animals. *Synonym*: endogamy.

Incubation period: The time elapsed between exposure to a pathogenic organism and when symptoms and signs are first apparent. Depending on the disease, the person may or may not be able to give the disease to others during the incubation period.

Indicator species: A species whose status provides information on the overall condition of the ecosystem and of other species in that ecosystem.

Informed consent: A process by which a subject voluntarily confirms his or her willingness to participate in a particular trial after having been informed of all aspects of the trial relevant to the subject's decision to participate. Informed consent typically is documented by means of a written, signed, and dated informed consent form.

Inoculate: To inject with a virus to create immunity.

Insect parasitoid: Species whose larvae develop on, or inside, individuals of other insect species, leading to the death of the host.

Insecticide resistance: The development or selection of heritable traits (genes) in an insect population that allow individuals expressing the trait to survive in the presence of levels of an insecticide (biological or chemical control agent) that otherwise would debilitate or kill the particular species of insect. The presence of such resistant insects makes the insecticide less useful for managing pest populations.

Insecticide: A class of crop protection and specialty chemicals used to control insects on farms and forests, as well as nonagricultural applications such as residential lawn care, golf courses, and public properties.

Insect-protected crops: Plants with the ability to withstand, deter, or repel insects, thereby preventing them from feeding on the plant. The traits (genes) determining resistance may be selected by plant breeders through cross-pollination with other varieties of this crop or through the introduction of genes such as *Bacillus thuringiensis* (Bt) through genetic engineering. See also: *Bacillus thuringiensis* (Bt).

Insect-resistance management: A strategy for delaying the development of pesticide resistance by maintaining a portion of the pest population in a refuge that is free from contact with the insecticide. For Bt crops, this allows the insects feeding on the Bt toxin to mate with insects not exposed to the toxin produced in the plants.

Insect-resistant crops: Plants with the ability to withstand, deter, or repel insects and thereby prevent them from feeding on the plant. The traits (genes) determining resistance may be selected by plant breeders through cross-pollination with other varieties of this crop or through the introduction of novel genes such as Bt genes through genetic engineering.

Integrated pest management (IPM): The coordinated, safe, and economical use of pest and environmental information along with available pest control methods (including cultural, biological, genetic, and chemical methods) to prevent unacceptable levels of pest damage.

Integration: The covalent joining of a piece of DNA into genomic DNA.

Intellectual property rights: The legal protection for inventions, including new technologies or new organisms (such as new plant varieties). The owner of these rights can control their use and earn the rewards for their use. This encourages further innovation and creativity. Intellectual property rights protection includes various types of patents, trademarks, and copyrights.

International Treaty on Plant Genetic Resources for Food and Agriculture: The international treaty resulting from the revision of the International Undertaking on Plant Genetic Resources was adopted by the 2001 FAO Conference as a binding international instrument to enter into force after ratification by 40 states. Its objectives are the conservation and sustainable use of plant genetic resources for food and agriculture and equitable sharing of the benefits of this use.

International Undertaking on Plant Genetic Resources: The first comprehensive voluntary, international agreement (adopted in 1983) dealing with plant genetic resources for food and agriculture. Designed as an instrument to promote international harmony in matters regarding access to plant genetic resources for food and agriculture. Following extensive negotiations to revise the Undertaking in harmony with the Convention on Biological Diversity, the binding International Treaty on Plant Genetic Resources for Food and Agriculture was adopted by the 2001 FAO Conference.

In situ **hybridization:** The visualization of *in vivo* location of macromolecules (particularly polynucleotides and polypeptides) by the histological staining of tissue sections or cytological preparations via labeled probes/antibodies.

K

Karyotype: A photomicrograph of an individual's chromosomes arranged in a standard format showing the number, size, and shape of each chromosome type. It is used in low-resolution physical mapping to correlate gross chromosomal abnormalities with the characteristics of specific diseases.

Keystone species: A species whose loss from an ecosystem would cause a greater-than-average change in other species' populations or ecosystem processes.

Knock in: Replacement of a gene by a mutant version of the same gene.

Knock out: The process of purposely removing a particular gene or trait from an organism.

L

Labeling of foods: The process of developing a list of ingredients contained in foods. Labels imply the list of ingredients can be verified. In the United States, the Food and Drug Administration has jurisdiction over what is stated on food labels.

Landrace: A primitive cultivar (in contrast to a named modern cultivar). Landraces of a particular crop are a collection of plants that were developed and maintained by traditional farmers. While they are genetically improved over wild versions of the species, they are not as well developed as modern commercial cultivars.

Lectins: A group of proteins which possess the unique ability to agglutinate erythrocytes and other types of cells. Widely distributed in the plant kingdom, particularly among the legumes.

Lipofection: A method of transfection in which DNA is incorporated into lipid vesicles (liposomes), which then are fused to the membrane of the target cells.

Living modified organism (LMO): Living organism that possess a novel combination of genetic material obtained through the use of modern biotechnology (convention on biological diversity). Synonym of GMO, but restricted to organisms that can endanger biological diversity.

Locus (pl.: loci): The location on a chromosome of a gene or other chromosome marker.

Locus-control regions: Segments of DNA important for the correct and coordinated expression of large regions (such as those encoding hemoglobins).

M

Macrophage: A large cell that helps the body defend itself against disease by surrounding and destroying foreign organisms (viruses or bacteria).

Marker gene: Genes with a phenotype that can be selected for gene transfer experiments. Marker genes are used to enable the selection/deletion of neighboring sequences in a gene construct.

Marker: An identifiable physical location on a chromosome whose inheritance can be monitored.

Marker-assisted selection (MAS): The use of DNA markers to improve response to selection in a population. The markers will be closely linked to one or more target loci, which may often be quantitative trait loci.

Meiosis: The special cell division process by which the chromosome number of a reproductive cell becomes reduced to half (n) the diploid ($2n$) or somatic number.

Mendelian population: A natural, interbreeding unit of sexually reproducing plants or animals sharing a common gene pool.

Mendelian segregation: Occurs when alleles are inherited according to Mendel's Laws.

Mendel's Laws: Two laws summarizing Gregor Mendel's theory of inheritance. The Law of Segregation states that each hereditary characteristic is controlled by two "factors" (now called alleles), which segregate and pass into separate germ cells. The Law of Independent Assortment states that pairs of "factors" segregate independent of each other when germ cells are formed.

Microbes: Tiny organisms (including viruses and bacteria) that only can be seen with a microscope.

Microinjection: A technique by which genes are directly inserted with a tiny syringe into the nucleus of a cell.

Microprojectile bombardment: Also known as particle acceleration, or biolistic bombardment (using the "Gene Gun"), this technique is used to transform cells using small gold or tungsten particles that are coated with DNA and literally shot into a cell.

Microsatellite DNA: A small segment of DNA with a repeated sequence. This segment is made up of short nucleotide sequences, which when tagged and amplified using polymerase chain reaction (PCR), can be used as markers for research purposes.

Minimum viable population: The smallest isolated population having a good chance of surviving for a given number of years despite the foreseeable effects of demographic, environmental, and genetic events and natural catastrophes.

Mitochondria: Compartments or organelles in the cell that are the cell's main energy source and often are called the powerhouse of the cell. The mitochondria also contain their own DNA and therefore genes. Mitochondrial genes follow maternal inheritance.

Mitosis: The process of cell division in all cells except reproductive cells. Mitosis results in "daughter" cells, which are identical genetically to the parent cells.

Mitotic division: A method of indirect division of a cell, consisting of a complex of various processes, by means of which the two daughter nuclei normally receive identical complements of the number of chromosomes characteristic of the somatic cells of the species.

Mobilization: The transfer of genes from one place to another (in the same or a different cell or organism) mediated by a retrovirus or transposable element.

Mode of inheritance: The manner in which a particular genetic trait or disorder is passed from one generation to the next. Autosomal dominant, autosomal recessive, X-linked dominant, X-linked recessive, multifactorial, and mitochondrial inheritance are examples.

Modern biotechnology: The application of: (a) *in vitro* nucleic acid techniques, including recombinant deoxyribonucleic acid (DNA) and direct injection of nucleic acid into cells or organelles, or (b) fusion of cells beyond the taxonomic family that overcome natural physiological reproductive or recombination barriers and that are not techniques used in traditional breeding and selection.

Modern farming practices: Farming practices that maximize the amount of production per unit (either per acre or per animal) while conserving soil and water resources. May include use of modern government-approved aids (e.g., fertilizers, insecticides, herbicides, and antibiotics), which undergo extensive safety testing before approval.

Molecular biology: The study of the structure and function of proteins and nucleic acids in biological systems.

Molecular genetic testing (also known as *DNA testing***):** Testing that involves the analysis of DNA.

Molecule: A group of atoms held together by chemical bonds; it is the typical unit manipulated by nanotechnology.

Monoclonal antibody: An antibody of a single type produced by a genetically identical group of cells (clone). Usually a fusion of an antibody-producing blood cell and a cancer cell.

Monogenic: A disease or trait caused by variation in a single gene.

Multipotent: Ability of a cell to develop into a small number of different cell types.

Mutagen: A physical or chemical agent (e.g., irradiation, alkylating agents) that causes a permanent change (mutation) in a gene. It may or may not be a carcinogen.

Mutagenesis (or mutation breeding): A process whereby the genetic information of an organism is changed in a stable, heritable manner, either in nature or induced experimentally via the use of chemicals or radiation. In agriculture, these genetic changes are used to improve agronomically useful traits.

Mutant: An organism or an allele bearing a mutation. Usually applied when a characteristic change in phenotype can be recognized.

Mutation: A permanent change in a gene, or the process by which a gene undergoes a change in the base DNA sequence. Some mutations result in the gene no longer coding for the correct protein or producing a reduced amount of the protein. If the mutation occurs in the germ line cells, it then is able to be inherited. Mutations in somatic cells cannot be inherited. Mutations can occur naturally and spontaneously, but they also can be due to exposure to mutagens. The identification and incorporation of useful mutations has been essential for traditional crop breeding, among other things.

Mycotoxins: Toxic substances of fungal origin, such as aflatoxins.

N

Nano: A prefix meaning one-billionth.

Nanotechnology: A science that involves the design and application of structures, devices, and systems on an extremely small scale, called the nanoscale, that is, billionths of a meter, or about 1-millionth the size of a pinhead. Potential applications related to food include food packaging and processing to improve food safety and quality, and better nutrient and ingredient profiles to improve health.

Natural selection: The differential survival and reproduction of organisms because of differences in characteristics that affect their ability to utilize environmental resources.

Necrotize: To undergo the process of necrosis, which is the death of tissue in the body.

Nematodes: Microscopic, slender worms, some of which, feed on plant roots.

No longer equivalent: A novel food or food ingredient is deemed to be "no longer equivalent" if scientific assessment, based on an appropriate analysis of exiting data, can demonstrate that the characteristics assessed are different in comparison with a conventional food or food ingredient, having regard to the accepted limits of natural variations for such characteristics.

Northern blot analysis: A technique used to analyze ribonucleic acid (RNA). RNA is separated by size, blotted (transferred onto a membrane), and then is detected by a special probe that allows information (e.g., size and abundance) about a particular species of RNA to be revealed.

Nuclear transfer (NT): The generation of a new animal nearly identical to another one by injection of the nucleus from a cell of the donor animal into an enucleated oocyte of the recipient.

Nuclear transfer technology (cloning): The process that involves the removal of the nucleus of a cell followed by the transfer of a nucleus from another cell into it.

Nuclei: The structure within the cell that contains the chromosomes.

Nucleotide: The subunit of DNA and RNA. It is composed of a nitrogenous base and a five-carbon sugar bonded to a phosphate group.

Nucleus: In biology, the part of a cell that controls growth and reproduction.

Nutraceutical: The term coined in the 1990s by Dr. Stephen DeFelice, who defined it as any substance that is a food or a part of a food and provides medical or health benefits, including the prevention and treatment of disease. Such products may range from isolated nutrients, dietary supplements, and specific diets to genetically engineered designer foods, herbal products, and processed foods such as cereals, soups, and beverages. Since the term was coined, its meaning has been modified. The term also has been defined as a product isolated or purified from foods and generally sold in medicinal forms not usually associated with food and demonstrated to have a physiological benefit or provide protection against chronic disease.

Nutrient density: The amount of nutrients that a food contains per unit volume or mass.

Nutrient: Any substance that can be metabolized by an organism to give energy and build tissue.

O

Obesity: An excess amount of subcutaneous body fat in proportion to lean body mass. Obesity in children and youth refers to the age and gender-specific BMI that are equal to or greater than the 95th percentile of the Centers for Disease Control and Prevention BMI charts. In most children, these values are known to indicate elevated body fat and to reflect the comorbidities associated with excessive body fatness.

Obesogenic: Environmental factors that may promote obesity and encourage the expression of a genetic predisposition to gain weight.

Organelles: Structures with special functions within cells such as the nucleus, mitochondria, and lysosomes.

Organic agriculture: A concept and practice of agricultural production that focuses on production without the use of synthetic inputs, for example, synthetic pesticides or fertilizers, and does not allow the use of transgenic organisms.

Organism: A living thing that contains DNA and is capable of cell replication by itself, from bacteria to fungi to plants to mammals.

Organoleptic: Relating to the senses (taste, color, odor, feel). For example, traditional meat and poultry inspection techniques are considered organoleptic because inspectors perform a variety of such procedures, involving visually examining, feeling, and smelling animal parts.

Outbreak: Sudden appearance of a disease in a specific geographic area, such as a neighborhood or a community, or in a population, such as adolescents.

Outbreeding: Crosses occur between distantly related parent plants.

Outcrossing: Mating between different populations or individuals of the same species that are not closely related. The term can be used to describe unintended pollination by an outside source of the same crop during hybrid seed production.

P

Pandemic: An epidemic occurring over a very large area.

Paradigm: An example, model, or pattern.

Parasite: A living organism that lives on, or in, an organism of another species, referred to as the host. The parasite draws its nutrients from the host. It also gets protection through the host.

Patent: A legal permission to hold exclusive right—for a defined period of time—to manufacture, use, or sell an invention.

Pathogen: A disease-causing organism (generally microbial: bacteria, fungi, viruses; but can extend to other organisms, e.g., nematodes, etc.). *Synonym:* infectious agent.

Pedigree: A diagram of the genetic relationships and medical history of a family using standard symbols and terminology.

Pesticide resistance: The development or selection of heritable traits (genes) in a pest population that allow individuals expressing the trait to survive in the presence of levels of a pesticide (biological or chemical control agent) that would otherwise debilitate or kill this pest. The presence of such resistant pests makes the pesticide less useful for managing pest populations.

Pesticide: A broad class of crop protection products, including four major types: insecticides used to control insects; herbicides used to control weeds; rodenticides used to control rodents; and fungicides used to control mold, mildew, and fungi. Both farmers and consumers use pesticides in the home or yard to control termites and roaches, clean mold from shower curtains, stave off crab grass on the lawn, kill fleas and ticks on pets, disinfect swimming pools, and so on.

Pest-resistant crops: Plants with the ability to withstand, deter, or repel pests and thereby prevent them from damaging the plants. Plant pests may include insects, nematodes, fungi, viruses, bacteria, weeds, and others.

Pharmacogenomics: The determination and analysis of the genome (DNA) and its products (RNAs) as they relate to drug response.

Pharming: (1) The process of farming genetically engineered plants or animals to be used as living pharmaceutical factories. The practice has used cows, sheep, pigs, goats, rabbits, and mice to produce large amounts of human proteins in their milk. Plants are being used to produce vaccines and diagnostic reagents. (2) The production of pharmaceuticals from genetically altered plants or animals.

Phenotype Mapping: A process used to identify landmarks on DNA through functional analysis of trait genes.

Phenotype: (1) The observable physical and/or biochemical characteristics of a person, animal, or other organism that are determined by that organism's genetic makeup and/or environment. (2) The clinical presentation of an individual with a particular genotype.

Phenotypic: Related to the observable characteristics of an organism produced by the interaction of genes and the environment; genotypic related to the genetic information contained within an organism.

Phytoalexins: Plant-produced substances that are toxic or inhibit the growth of microorganisms, especially phytopathogenic fungi (and can be harmful for the host plant itself). Phytoalexins are many chemically distinct chemical compounds (e.g., isoflavonids, furanocoumarins).

Phytoestrogens: Compounds that occur naturally in plants (phyto) and under certain circumstances can have actions like human estrogen.

Phytoremediation: The process of using plants for pollution clean-up of contaminated soils or water.

Plant breeding: The use of cross-pollination, selection, and certain other techniques involving crossing plants to produce varieties with particular desired characteristics (traits) that can be passed on to future plant generations.

Plant genetic resources (PGR): The reproductive or vegetative propagating material of: (1) Cultivated varieties (cultivars) in current use and newly developed varieties; (2) Obsolete cultivars; (3) Primitive cultivars (landraces); (4) Wild and weed species, near relatives of cultivated varieties; and (5) Special genetic stocks (including elite and current breeder's lines and mutants).

Plant pests: Organisms that might directly or indirectly cause disease, spoilage, or damage to plants, plant parts, or processed plant materials. Common examples include certain insects, mites, nematodes, fungi, molds, viruses, and bacteria.

Plant population: In a natural, or unmanaged, environment, a plant population typically is composed of many different species, some of which can be edible and desirable to humans, and others that might not. In the context of an agricultural, or managed, environment, a plant population could be a field of plants—typically of one variety—but invariably consisting of some plants of different varieties and some different species.

Plant stanols and sterols: Essential components of plant cell membranes that resemble cholesterol structurally. Plant sterols are present naturally in small quantities in many fruits, vegetables, nuts, seeds, cereals, legumes, vegetable oils, and other plant sources. Plant stanols occur in even smaller quantities than plant sterols in many of the same sources.

Plant-incorporated protectants (PIPs): Pesticidal substances introduced into plants by genetic engineering that are produced and used by the plant to protect it from pests. The protein toxins of Bt often are used as PIPs in the formation of Bt crops.

Plasmid: A circular DNA molecule capable of replication in host bacteria. Plasmids are the usual means of propagation of DNA for transfection or other purposes. Plasmids are also occasionally found in certain fungi and plants.

Plasticity: The ability of stem cells from one adult tissue to generate the differentiated cell types of another tissue.

Plastid: A general term for a number of plant cell organelles which carry nonnuclear DNA. Includes the pigment-carrying bodies: (1) chloroplasts in leaves, (2) chromoplasts in flowers, and (3) the starch-synthesizing amyloplasts in seeds.

Pleiotropic (of a gene): Having an effect simultaneously on more than one characteristic of the offspring.

Pleiotropy: The simultaneous effect of a given gene on more than one apparently unrelated trait.

Pluripotent: Ability of a single stem cell to develop into many different cell types of the body.

Polychlorinated biphenyl (PCB): A synthetic organic chemical compound of chlorine and two molecules of benzene rings. Polychlorinated biphenyls were widely used as dielectric and coolant fluids.

Polygenic: Refers to a trait or phenotype whose expression is the result of the interaction of numerous genes.

Polymerase chain reaction (PCR): A method for making multiple copies of fragments of DNA. It uses a heat-stable DNA polymerase enzyme and cycles of heating and cooling to successively split apart the strands of double-stranded DNA and use the single strands as templates for building new double-stranded DNA. One use of PCR is in the detection of DNA sequences that indicate the presence of a particular genetically engineered organism.

Polymorphism: A natural variation in a gene, DNA sequence, or chromosome, which may not have adverse effects on the individual and occurs with fairly high frequency in the general population.

Polyols: Chemical compounds containing multiple hydroxyl groups. Sugar alcohols, a class of polyols, commonly are added to foods because of their lower caloric content.

Postplanting surveillance: Monitoring system to detect any breaches of regulations of GM crop sites, and any adverse effects on health or the environment.

Prebiotics: Food substances that promote the growth of certain bacteria (generally beneficial) in the intestines.

Precautionary principle: A strategy for dealing with environmental risk and uncertainty, which guides us to act cautiously and embark on a systematic program of research to improve our understanding of the costs and benefits of particular actions.

Prevalence: Proportion of the whole population affected.

Prion-related protein (PrP): A normal protein, expressed in the nervous system of animals, whose structure when altered (by interaction with altered copies of itself) is the cause of scrapie in sheep, bovine spongiform encephalopathy in cattle, and Creutzfeldt–Jakob disease in humans.

Probiotics: Dietary supplements containing potentially beneficial bacteria or yeast.

Promoter: A region of DNA that regulates the level of function of other genes.

Propagation: The duplication of a whole plant from a range of vegetative materials; adapted for *in vitro* culture as micropropagation.

Proteomics: The analysis of complete complements of proteins. Proteomics includes not only the identification and quantification of proteins, but also the determination of their localization, modifications, interactions, activities, and, ultimately, their function.

Protoplast fusion: A technique in which protoplasts (plant cells from which the cell wall has been removed by mechanical or enzymatic means) are fused into a single cell.

Provirus: The integrated DNA form of a retrovirus.

Proximate analysis: The analytical determination of the major classes of food components, usually including total protein, fat, and carbohydrate, dietary fiber, water, and ash (minerals).

Psoralens: Mutagenic, carcinogenic agents that can act as phytoalexins. Psoralens are found in a variety of fruits and vegetables, such as

GLOSSARY

celery and carrots, and they are known to make human skin sensitive to long-wave ultraviolet radiation.

Public health: The approach to medicine that is concerned with the health of the community as a whole.

Pure line: A strain in which all members are genetically nearly identical and are indistinguishable by phenotype. Usually created by repeated generations of self-fertilization or close inbreeding.

R

Radioallergosorbent (test): Allergen testing using blood samples to identify an allergen capable of causing an allergic response.

Recessive trait: A characteristic determined by an allele that requires the presence of two identical alleles to be expressed.

Recessive: Every cell contains two copies of each gene. Each gene contains the information for a particular gene product, such as a protein. If a gene is mutated, the gene no longer codes for the gene product. When an individual has one gene copy, or allele, mutated and the other copy "correct," the cell only will be producing half the amount of gene product. If that does not result in any disorder for the individual, the mutation is described as being hidden or "recessive" to the correct copy of the gene. An individual with this genetic constitution is said to be a carrier of a recessive gene mutation. For a recessive gene mutation to result in a particular characteristic or a disorder, both copies of the genes must be mutated.

Recombinant Bovine Growth Hormone (rBGH): A synthetic form of growth hormone injected into cows to increase growth rates and milk production.

Recombinant bovine somatotropin (rbST): A protein produced through biotechnology that has the same genetic make-up as bovine somatotropin (BST), a naturally occurring protein hormone produced in cows. Somatotropin is also produced by humans and most animals to support tissue health, maintenance, and growth. FDA has approved the effectiveness and safety of rbST. All milk, regardless of production method, is safe and provides the same nutritional benefits.

Recombinant DNA technology (rDNA): Procedure used to join together DNA segments in a cell-free system (an environment outside a cell or organism). Under appropriate conditions, a recombinant DNA molecule can enter a cell and replicate there, either autonomously or after it has become integrated into a cellular chromosome.

Recombinant: (1) Of or resulting from new combinations of genetic material or cells. (2) The genetic material produced when segments of DNA from different sources are joined to produce recombinant DNA.

Recombination: The process by which progeny derive a combination of genes different from that of either parent.

Remediation: The cleanup or containment of a hazardous waste disposal site to the satisfaction of the applicable regulatory agency. This can sometimes be accomplished with naturally occurring or engineered microorganisms or plants.

Replication: The in vivo synthesis of double-stranded DNA by copying from a single-stranded template.

Residues of biomass: By-products from processing all forms of biomass that have significant energy potential. For example, making solid wood products and pulp from logs produces bark, shavings, sawdust, and spent pulping liquors. Because these residues already are collected at the point of processing, they can be convenient and relatively inexpensive sources of biomass for energy.

Restoration: The return of an ecosystem or habitat to its original community structure, natural complement of species, and natural functions.

Retroviral vectors: Vector constructs in which the internal genes of a retrovirus are replaced by the gene of interest, flanked by the viral long terminal repeats and packaging signals. After transfection of helper cells, the vector is packaged into virus particles. Infection of target cells with these particles leads to integration of the gene into cellular DNA as part of a provirus.

Retrovirus: An enveloped virus that replicates by reverse transcription of its RNA genome into DNA, followed by integration of the DNA into the cell genome to form a provirus. Expression of the provirus (as though it were a cellular gene) leads to the production of progeny virus particles.

Ribonucleic acid (RNA): An organic acid polymer composed of adenosine, guanosine, cytidine, and uridine ribonucleotides. The genetic material of some viruses, but more generally is the molecule, derived from DNA by transcription, that either carries information (messenger RNA), provides subcellular structure (ribosomal RNA), transports amino acids (transfer RNA), or facilitates the biochemical modification of itself or other RNA molecules.

Ribosomal DNA: The coding locus for ribosomal RNA. This is generally a large and complex locus, typically composed of a large number of repeat units, separated from one another by the intergenic spacer. A repeat unit comprises a gene copy for each individual ribosomal RNA component, separated from one another by the internal transcribed spacer.

Ribosomal RNA (rRNA): The RNA molecules that are essential structural and functional components of ribosomes, where protein synthesis occurs. Different classes of rRNA molecules are identified by their sedimentation (S) values. *E. coli* ribosomes contain one 16S rRNA molecule (1541 nucleotides long) in one (small) ribosomal subunit, and a 23S rRNA (2904 nucleotides) and a 5S rRNA (120 nucleotides) in the other (large) subunit. These three rRNA molecules are synthesized as part of a large precursor molecule which also contains the sequences of a number of tRNAs. Special processing enzymes cleave this large precursor to generate the functional molecules. Constitutes about 80% of total cellular RNA.

Ribosome: The subcellular structure that contains both RNA and protein molecules and is the site for the translation of mRNA into protein. Ribosomes comprise large and small subunits.

Ricin: A potent protein toxin made from the leftover waste after processing castor beans to make castor oil.

Risk analysis: A process consisting of three components: risk assessment, risk management, and risk communication performed to understand the nature of unwanted, negative consequences to human and animal health, or the environment.

Risk assessment: A scientifically based process consisting of the following steps: (i) hazard identification; (ii) hazard characterization; (iii) exposure assessment; and (iv) risk characterization.

Risk communication: The interactive exchange of information and opinions throughout the risk analysis process concerning hazards and risks, risk-related factors and risk perceptions, among risk assessors, risk managers, consumers, industry, the academic community and other interested parties, including the explanation of risk assessment findings and the basis of risk management decisions.

Risk management: The process, distinct from risk assessment, of weighing policy alternatives, in consultation with all interested parties, considering risk assessment and other factors relevant for the health protection of consumers and for the promotion of fair trade practices, and, if needed, selecting appropriate prevention and control options.

Roundup-ready: Describing transgenic crop varieties that carry the bacterial gene which detoxifies the herbicide glyphosate, thereby making them resistant to its application.

S

Saprophytes: Living organisms, which feed on dead organic matter.

Seedbank: A facility designed for the *ex situ* conservation of individual plan varieties through seed preservation and storage.

Selectable marker: A gene, often encoding resistance to an antibiotic or an herbicide, introduced into a group of cells to allow identification of those cells that contain the gene of interest from the cells that do not. Selectable markers are used in genetic engineering to facilitate identification of cells that have incorporated another desirable trait that is not easy to identify in individual cells.

Selection: Differential survival and reproduction phenotypes. Also, a system for either isolating or identifying specific organisms in a mixed culture; observing the characteristics of plants and choosing (selecting) to use only those organisms that have desired or superior characteristics.

Selective breeding: Choosing specific plants or animals with desirable traits to breed so that subsequent generations will have those traits.

Sequencing: Determination of the order of nucleotides (base sequences) in a DNA or RNA molecule or of the order of amino acids in a protein.

Sex chromosomes: The X or Y chromosome in human beings that determines the sex of an individual. Females have two X chromosomes; males have an X and a Y chromosome.

Sexual selection: The type of selection in which there is competition among males for mates, and characteristics enhancing the reproductive success of the carrier are perpetuated irrespective of their survival value.

Shelf life: The period of time during which a dietary supplement remains sufficiently potent to be effective. The expiration date on a product label should indicate the end of this time period.

Silencing: Loss of gene expression either through an alteration in the DNA sequence of a structural gene, or its regulatory region; or because of interactions between its transcript and other mRNAs present in the cell. Shutdown of transcription of a gene, usually by methylation of C residues.

Single-gene disorder: A hereditary disorder caused by a mutant allele of a single gene. An example is Huntington's disease.

SNP (single-nucleotide polymorphism): An SNP (pronounced "snip") is a single-base variation that occurs about every 1000 bases along the three billion base pairs of the human genome.

SNP map: A collection of single-nucleotide polymorphisms that can be superimposed over the existing genome map, creating greater detail and facilitating further genetic studies and analysis.

Solanaceous: Pertaining to plants of the family Solanaceae which includes tomato, tobacco, potato, pepper, and many weeds.

Somaclonal variation: Epigenetic or genetic changes induced during the callus phase of plant cells cultured *in vitro*. Sometimes visible as changed phenotype in plants regenerated from culture.

Somatic cell: (1) A cell that contains two complete sets of chromosomes. (2) A cell of the body other than those of the gamete-forming germ line.

Somatic cell embryogenesis: The process of differentiation of somatic embryos either from explants cells (direct embryogenesis), or from callus generated from explants (indirect embryogenesis). *Synonym*: asexual embryogenesis.

Somatic cell gene therapy: The delivery of a transgene(s) to a somatic tissue in order to correct a physiological defect.

Somatic cell hybrid panel: A panel of cells created by cell fusion, typically involving a reference species (e.g., hamster) and the species of interest (e.g., sheep) with each member of the panel containing a different mixture of chromosomes from the two species. By relating the presence or absence of cloned fragments (via *in situ* hybridization) or PCR products to the presence or absence of particular chromosomes from the species of interest, such panels can be used for physical mapping.

Somatic cell nuclear transfer (SCNT): The transfer of a cell nucleus from a somatic cell into an enucleated egg (an egg from which the nucleus has been removed). In SCNT, a nucleus from a patient's body cell, such as a skin cell, is introduced into an unfertilized egg from which the original genetic material has been removed. The egg then is used to produce a blastocyst whose stem cells could be used to create tissue that would be compatible with that of the patient. This is the procedure used for therapeutic cloning.

Somatic cell variant: A somatic cell with unique characters not present in the other cells, and which could be selected for by an appropriate screen.

Somatic embryo: An organized embryo-like structure. Although morphologically similar to a zygotic embryo, it is initiated from somatic plant cells. Under *in vitro* conditions, somatic embryos go through developmental processes similar to embryos of zygotic origin. Each somatic embryo is potentially capable of developing into a normal plantlet.

Somatic hybridization: Naturally occurring or induced fusion of somatic protoplasts or cells of two genetically different parents. The difference may be as wide as interspecific. Wide synthetic hybrids formed in this way (i.e., not via gametic fusion) are known as cybrids. Not all cybrids contain the full genetic information (nuclear and nonnuclear) of both parents.

Somatic hypermutation: The high frequency of mutation that occurs in the gene segments encoding the variable regions of immunoglobulins during the differentiation of B lymphocytes into antibody producing plasma cells.

Somatic mutation: A change or mistake in a gene that is found in the cells of the body but not in the germ or sex cells. Somatic mutations therefore cannot be passed on to future generations.

Somatic reduction: Halving of the chromosomal number of somatic cells; a possible method of producing "haploids" from somatic cells and calli by artificial means.

Somatic: Referring to cell types, structures, and processes other than those associated with the germ line.

Southern blot analysis: A technique used to obtain information about identity, size, and abundance of a specimen of DNA. DNA fragments are transferred to membrane filters so that specific base sequences can be detected.

Species diversity: A function of the distribution and abundance of species.

Species: A group of organisms capable of interbreeding freely with each other but not with members of other species.

Splicing junction: The DNA sequence immediately surrounding the boundary between an exon and an intron. There is a degree of sequence conservation in these regions, allowing the identification of introns in newly sequenced genes.

Stability: A function of several characteristics of community or ecosystem dynamics, including the degree of population fluctuations, the community's resistance to disturbances, the speed of recovery from disturbances, and the persistence of the community's composition through time.

Stacked genes: Refers to the insertion of two or more genes into the genome of an organism. An example would be a plant carrying a *Bt* transgene giving insect resistance and a *bar* transgene giving resistance to a specific herbicide.

Stacked traits: The biotechnology process by which more than one gene can be transferred, resulting in a plant with two or more transgenic traits. Usually, a result of the crossing of two transgenic plants with different transgenes.

Stacked-trait plant: A plant in which more than one characteristic has been genetically modified.

***Staphylococcus*:** A group of bacteria that cause a multitude of diseases. Under a microscope, *Staphylococcus* bacteria are round and bunched together. They can cause illness directly by infection or indirectly through products they make, such as the toxins responsible for food poisoning and toxic shock syndrome. *Staphylococcus* are the main culprit in hospital-acquired infections and cause thousands of deaths every year.

Staple crops: The most common crops in people's diets, such as rice, wheat, and maize (corn), which provide 60% of the world's food energy intake. Typically, staple crops are well adapted to the climate in which they are grown, and many are tolerant of drought, pests, or soil low in nutrients.

StarLink: A commercial brand of transgenic maize approved for animal feed only, but which also had been found in the human food supply.

Stem cells: Cells with the ability to divide for indefinite periods in culture and give rise to specialized cells. Scientists work primarily with two kinds of stem cells from animals and humans: embryonic stem cells and adult stem cells, which have different functions and characteristics.

Substantial equivalence: A concept that has been proposed to measure whether a biotechnology-derived food or crop shares similar health and nutritional characteristics with its conventional counterpart. The Food and Agriculture Organization of the United Nations, and the World Health Organization have attempted to develop substantial equivalence as an internationally agreed upon principle.

Substantially equivalent: A food or food product that has the same intended use and characteristics as an earlier food or food product that has already passed the FDAs safety inspections or has different characteristics but data demonstrate that the new food or food additive is as safe as the previous food or food product and does not raise different issues of safety. No additional safety inspections are required for new foods or food products that are deemed substantially equivalent.

Super weed: Term used to describe the possible outcome of genetically modified crops cross-pollinating with wild relatives, producing hybrids containing new genes, which could result in highly competitive and uncontrollable new plant species.

Surrogate: One that acts in another's place. In Dolly the sheep's case (the first mammal cloned from an adult cell), the surrogate was another Scottish Blackface ewe.

Susceptibility gene: A gene that confers a risk to develop a disease but is not necessary or sufficient by itself to cause the disease.

Susceptible: Unprotected against disease.

Sustainable agriculture: An integrated system of plant and animal production practices that will, over the long term, satisfy human food and fiber needs; enhances environmental quality and the natural resource based upon which the agricultural economy depends; makes the most efficient use of nonrenewable resources and integrate natural biological cycles and controls; sustains the economic viability of farm operations; and enhances the quality of life for farmers and society.

Sustainable development: Development that meets the needs and aspirations of the current generation without compromising the ability to meet those of future generations.

Syndrome: A group of characteristics or symptoms that occur together in a recognizable pattern.

T

T-DNA: DNA encoded on a plasmid of *Agrobacterium* that integrates into the genome of a plant cell.

Telomeres: The simple repeated sequences at the ends of chromosomes that protect them from loss of coding sequence during replication. In the absence of telomerase, telomeres become progressively shorter with each cell division, and this shortening is the major cause of senescence of cells in culture.

Terminator gene: A gene used in crops to make the next generation of seeds sterile, forcing farmers to buy new seeds every year.

Tillage: Practice of preparing the ground for planting and controlling weeds between plantings by turning or aerating the soil. Conventional tillage can lead to increased risk of erosion; therefore, conservation tillage has been increasingly adopted to preserve soil, a nonrenewable resource.

Tissue culture: (1) A process of growing a plant in the laboratory from cells rather than from seeds. The technique is used in traditional plant breeding as well as when using techniques of agricultural biotechnology. (2) The growth of animal or plant cells *in vitro* in an artificial culture medium for experimental research. Also known as *cell culture*.

Tissue: A part of an organism consisting of a collection of cells having a similar structure and function (a piece of skin or bone, for example).

Totipotent: Ability of a single cell to develop into all the different types of cells in the body (its potential is total).

Toxicant: Any substance or material that can injure living organisms through physicochemical interactions.

Toxin: A poisonous substance produced by living cells or organisms. Toxins almost always are proteins that are capable of causing disease on contact or absorption with body tissues by interacting with biological macromolecules such as enzymes or cellular receptors. Toxins vary greatly in their severity, ranging from usually minor and acute (as in a bee sting) to almost immediately deadly (as in botulinum toxin).

Traditional breeding: Modification of plants and animals through selective breeding. Practices used in traditional plant breeding can include aspects of biotechnology such as tissue culture and mutation breeding.

Trait: One of the many characteristics that define an organism. The phenotype is a description of one or more traits. *Synonym:* character.

Transcription: The process of converting genetic instructions coded in a segment of DNA into messenger RNA.

Transfection: Alteration of the genome of a cell by direct introduction of DNA, a small portion of which becomes covalently associated with the host cell DNA.

Transgene: A gene from one organism inserted into another organism by recombinant DNA techniques.

Transgenic crops: Crops whose DNA has one or more genes from another crop changing the characteristics of the crop (ex-crop becoming herbicide-resistant).

Transgenic mouse: A mouse that has been genetically altered by injecting human or other foreign DNA from another animal into fertilized mouse eggs. This DNA becomes incorporated into the mouse DNA and the mouse will translate the information contained in the foreign gene. This has become a useful model for the study of various human disorders.

Transgenic organism: A plant, animal, or other organism with different traits from the parent organism, resulting from the use of recombinant DNA techniques to insert genetic material from another organism.

Transgenic plants: An organism in which a foreign DNA gene (a transgene) is incorporated into its genome early in development. The transgenic plant usually contains material from at least one unrelated organism, such as from a virus, animal, or other plant.

Transgenic: (1) Containing genes altered by insertion of DNA from an unrelated organism. (2) Taking genes from one species and inserting them into another species to get that trait expressed in the offspring.

Translation: The conversion of genetic information coded in a segment of mRNA into a sequence of amino acids.

Transposon: A DNA element capable of moving (transposing) from one location in a genome to another in the same cell through the action of transposase.

Trisomy: The presence of a single extra chromosome, yielding a total of three chromosomes of that particular type instead of a pair. *Partial trisomy* refers to the presence of an extra copy of a segment of a chromosome.

U

U.S. Department of Agriculture (USDA): U.S. government agency charged with agricultural oversight to ensure a safe, affordable, nutritious, and accessible food supply. The USDA works to enhance the quality of life for the U.S. population by supporting production of agricultural products; caring for agricultural, forest, and range lands; supporting sound development of our rural communities; providing economic opportunities for farm and rural residents; expanding global markets for agricultural and forest products and services; and working to reduce hunger in the United States and throughout the world.

Undifferentiated: A cell that has not changed to become a specialized cell type.

V

Variants: Alleles that are rare; they are found in fewer than 1% of a population. Some mutations are variants.

Variety, plant: A group of individual plants that is uniform, stable, and distinct genetically from other groups of individuals in the same species. Also referred to as a cultivar.

Variety: A subdivision of a species for taxonomic classification. A variety is a group of individual plants that is uniform, stable, and distinct genetically from other groups of individuals in the same species.

Vector: A type of DNA, such as a plasmid or phage, that is self-replicating and that can be used to transfer DNA segments among host cells. Also, an insect or other organism that provides a means of dispersal for a disease or parasite.

Vertical transmission: Inheritance of a gene from parent to offspring.

Virion: The extracellular form of a virus (i.e., a virus particle).

Virulence: The relative capacity of a pathogen to overcome body defenses.

Virus resistant (crops): Plants with the ability to withstand plant viral diseases. Developed through traditional breeding or through genetic engineering (e.g., papaya ringspot virus-resistant papaya).

Virus: An infectious agent composed of a single type of nucleic acid, DNA or RNA that is enclosed in a coat of protein. It does not have a cellular structure and therefore cannot replicate outside of a living, host cell. A virus invades living cells and uses their chemical machinery to keep itself alive and to replicate itself. It can reproduce with fidelity or with errors (mutations). The ability to mutate is responsible for the ability of some viruses to change slightly in each infected person, which makes treatment more difficult. Viruses are smaller than bacteria and are not affected by antibiotics, the drugs used to kill bacteria. Viruses cause diseases such as chicken pox, measles, mumps, rubella, pertussis, and hepatitis.

Volunteer plant: Plants that grow in a field in the year after the original crop was grown as a result of seed being shed from the crop and remaining dormant in the soil for some time. Some volunteer plants may germinate several years after the original seed was shed.

W

Waning immunity: The loss of protective antibodies over time.

Weed: A plant that is growing in an undesired area and is able to overtake other plants by overcrowding, depleting soil nutrients and moisture that would otherwise be available to preferred plants or crops.

Well-being: A view of health that takes into account a child's physical, social, and emotional health.

Wet nanotechnology: The study of biological systems that exist primarily in a water environment.

X

X chromosome: A sex chromosome. Normal females carry two X chromosomes; normal males, one.

Xenograft: The transplanted tissue in a xenotransplantation.

Xenotransplantation: (1) Transplantation of cells, tissues, or organs from one species to another. (2) The term used to describe the transfer of living cells, tissues, and organs from nonhuman animals into humans for medical purposes.

X-linked dominant mutation: A dominant mutation in a gene carried on the X chromosome.

X-linked gene: Any gene that is located on the X chromosome.

X-linked recessive mutation: A recessive mutation in a gene carried on the X chromosome.

X-linked trait: A trait that is passed on from mother to child or from father to daughter on the X chromosome.

Y

Y chromosome: A sex chromosome. Normal males carry one Y and one X chromosome, while females carry none.

Yield: The amount of an agricultural crop, such as a grain, fruit, or vegetable, produced in a season. It can be measured in pounds or bushels per acre, or kilograms or metric tons per hectare.

Z

Zygote: The single cell with 46 chromosomes resulting from the fertilization of an egg (23) by a sperm (23). Through cell division (mitosis), the zygote develops into a multicellular embryo and then into a fetus.

SUGGESTED REFERENCES

Centre for Food Safety. 2014. The Government of the Hong Kong Special Administrative Region 2007. http://www.cfs.gov.hk/english/programme/programme_gmf/programme_gmf_gi_glossary.html (accessed September 21, 2014).

Food Land. 2014. Glossary for genetically modified foods. 2014. http://foodland.wikispaces.com/Glossary+for+Genetically+Modified+Foods (accessed September 21, 2014).

IFIC Foundation. 2014. *Food Biotechnology: A Communicator's Guide to Improving Understanding*, Third Edition. http://www.foodinsight.org/education/food-biotechnology-communicator%E2%80%99s-guide-improving-understanding (accessed September 21, 2014).

National Research Council. 2004. *Safety of Genetically Engineered Foods Approaches to Assessing Unintended Health Effects.* National Academic Press, Washington, DC.

North Carolina Association for Biomedical Research (NCABR). 2014. http://www.ncabr.org/About Bioscience. Bioscience Glossary http://www.aboutbioscience.org/glossary (accessed September 21, 2014).

USDA National Agricultural Library. 2014. Inter-American Institute for Cooperation on Agriculture. Glossary: Browse A–Z. http://agclass.nal.usda.gov/glossary_az_ae.shtml#sym (accessed September 23, 2014).

Zaid, A., Hughes, H.G., Porceddu, E., and Nicholas, F. 2001. Glossary of biotechnology for food and agriculture—A revised and augmented edition of the glossary of biotechnology and genetic engineering. FAO Research and Technology Paper 9. Food and Agriculture Organization of the United Nations.

INDEX

A

AAAS, *see* American Association for Advancement of Science (AAAS)
AAEM, *see* American Academy of Environmental Medicine (AAEM)
AAFC, *see* Agriculture and Agri-Food Canada (AAFC)
ABC, *see* Agricultural Biotechnology Council (ABC); Australian Broadcasting Corporation (ABC)
Abiotic factors, 118
Abiotic stresses, 118
ABN, *see* African Biodiversity Network (ABN)
ACB, *see* African Centre for Biosafety (ACB)
ACNFP, *see* Advisory Committee on Novel Foods and Processes (ACNFP)
ADB, *see* Asian Development Bank (ADB)
Adenine (A), 27
Advisory Committee on Novel Foods and Processes (ACNFP), 233
Africa, 258–259
　Gaia Foundation, 259
　GM crops/foods, 130–134
　GSR project, 244
　hybrid seeds, 119
AfricaBio, 238
African Biodiversity Network (ABN), 258
African Centre for Biosafety (ACB), 258
Agricultural/agriculture, 2
　adoption of GM technology, 276
　in African continent, 131
　arguments against GM technologies, 251
　biotechnology applications, 47
　GMO-free, 168
　green biotechnology, 50
　modern biotechnology, 139
　organic, 263
　practices, 3
　techniques, 30
Agricultural biotechnology, 5, 109, 120, 134; *see also* Plant biotechnology
　application, 243
　effects and benefits in GM food production, 246
Agricultural Biotechnology Council (ABC), 239
Agriculture and Agri-Food Canada (AAFC), 197
Agrobacterium, 40
Agrobacterium-mediated gene transfer, 76
Agrobacterium-mediated transformation, 64
Agrobacterium tumefacien (*A. tumefacien*), 76
Agrobiotechnology, 4, 5, 6
AI, *see* Artificial insemination (AI)
AIAB, *see* Italian Association for Organic Agriculture (AIAB)
Alliance for Better Foods, 241
AMA, *see* American Medical Association (AMA)
Amaranth plant, 141
American Academy of Environmental Medicine (AAEM), 252, 256, 266

INDEX

American Association for Advancement of Science (AAAS), 237, 250
American Chemical Society, 47
American Medical Association (AMA), 52, 187, 193
American Society for Microbiology (ASM), 52
American Society of Plant Biologists (ASPB), 52
American Soybean Association (ASA), 237
Amflora, 45
Animal biotechnology, 135; *see also* Food biotechnology; Plant biotechnology
   animal genetic modification techniques, 135–138
   biotechnology for animal feed production, 138
Animal feed, 150
   biotech crops in, 246
   Regulation EC 1829/2003, 201
Animal genetic modification techniques, 135
   non-usable techniques in GE, 135
   usable techniques in GE, 136–138
Animal health and welfare, 231
Anther culture, 67–68
Anti-GMO activists, vandalism and threats by, 261–262
Anti-labeling arguments, 190–191
Apples, 141–142
Aquatic invertebrates, 137
Aquatic organisms, 148
Artificial insemination (AI), 135
Artificial selection, 135
ASA, *see* American Soybean Association (ASA)
Asian Development Bank (ADB), 52
Asia Pacific, 260
ASM, *see* American Society for Microbiology (ASM)
ASPB, *see* American Society of Plant Biologists (ASPB)

Assisted reproductive procedures, 135
AusBiotech Ltd, 238–239
Australia, 258
Australian Broadcasting Corporation (ABC), 262
Australia–New Zealand, 195
   FSANZ, 197
   safety assessment in, 196
Australia New Zealand Food Authority (ANZFA), *see* Food Standards Australia New Zealand (FSANZ)
Azospirillum, 125
Azotobacter, 125

B

Bacillus thuringeinsis (Bt), 115, 116, 142; *see also* Genetically modified foods (GM foods)
   corn, 142
   cotton, 132, 142
   maize, 132
   toxin, 115
Backcross breeding, 86–88
Bacterial carriers, 76–77
Bacterial mating, *see* Conjugation
Baker's yeast 164
Basic food biotechnology, 140
bEcon, 133–134
Beet necrotic yellow vein virus (BNYVV), 117–118
BGA, *see* Blue green algae (BGA)
BGH, *see* Bovine growth hormone (BGH)
BIO, *see* Biotechnology Industry Organization (BIO)
Biobased economy, *see* Bioeconomy
Biochip, *see* Microarrays technology
Biocolonization, 276
Biocolors, 166
Bioeconomy, 51
Bioengineered method, 7
Bioethanol, 110, 170

INDEX

Biofertilizer, 50
  production, 122, 124–126
Biofuels production, 15, 109–110, 170–172
Biolistics, see Microprojectile bombardment
Biological pesticides, see Biopesticides
Biopesticides, 50, 115–116
Biopharm, 32
Bioremediation, 249
Biotech, see Biotechnology
Biotech, agrochemical, and associated companies, 233
  Advanta Company, 235
  Bayer Company, 234–235
  Cargill Inc., 235
  Dow Company, 235
  DuPont Company, 235
  Monsanto Company, 234
  Pioneer Hi-Bred Company, 235
  Syngenta Company, 234
Biotech century, 7
Biotechnology, 23, 25, 242
  for animal feed production, 138
  for biofuel production, 170–172
  branches, 26
  DNA, 26–29
  in functional foods and nutraceuticals, 156–166
  future, 45–49
  GM and GE, 29–32
  GMCs, 32
  GM foods, 33
  GMOs, 32
  historical developments in, 33–36
  position statements, 51–53
  timeline, 36–45
  types, 49–51
Biotechnology Industry Organization (BIO), 237
Biotechonomy, see Bioeconomy
Biowatch South Africa, 259
Blue biotechnology, 51
Blue green algae (BGA), 125

BMA, see British Medical Association (BMA)
BNYVV, see Beet necrotic yellow vein virus (BNYVV)
Boveri–Sutton Chromosome Theory, 38
Bovine growth hormone (BGH), 150
Bovine somatotropin (bST), 42
Bovine spongiform encephalopathy (BSE), 136, 199
British Medical Association (BMA), 245, 246
Broad-spectrum herbicides, see Nonselective herbicides
Browning process, 141–142
BSE, see Bovine spongiform encephalopathy (BSE)
bST, see Bovine somatotropin (bST)
Bt, see *Bacillus thuringeinsis* (Bt)

C

CAC, see Codex Alimentarius Commission (CAC)
Calcium phosphate precipitation, 83
"California Proposition 37", 209
"Callus", 67
Canada, 258
  labeling, 198–199
  regulations, 197–198
Canadian Biotechnology Advisory Committee (CBAC), 52
Canadian Food Inspection Agency (CFIA), 197, 198
Canola, 121, 143
Cargill Inc., 235
CBAC, see Canadian Biotechnology Advisory Committee (CBAC)
CBD, see Convention on Biological Diversity (CBD)
Cell fusion, see Somatic hybridization
Cell selection process, 69
CEN, see European Committee for Standardization (CEN)
Center for Environmental Risk Assessment (CERA), 130

INDEX

Center for Food Safety (CFS), 191, 253
Central and South America, 260
"Central Dogma" theory, 39
CERA, see Center for Environmental Risk Assessment (CERA)
Certified reference material (CRM), 96, 97
CFIA, see Canadian Food Inspection Agency (CFIA)
CFS, see Center for Food Safety (CFS)
Chain termination DNA sequencing, 40
"Charter of Florence", 44
Cheese, 149
Chemical fertilizers, 122
Chromosome maps, 46
Chromosome theory of inheritance, see Boveri–Sutton Chromosome Theory
Chymax chymosin, 41
Chymosin, 149
Cisgenesis, 71
CJD, see Creutzfeldt-Jakob disease (CJD)
Classical crossbreeding, 34–35
Cloning, see Somatic cell nuclear transfer (SCNT)
Codex Alimentarius Commission (CAC), 13, 185
Composition of GE foods, 140
  intended changes, 140–141
  unintended changes, 141
Conjugation, 164
Consumer choice, 190, 231
Consumer issues, 295; see also Genetically modified foods (GM foods)
  consumer perceptions, 296
  consumer rights, 298–300
  mass media, 313–314
  impact of moving from GM components to non-GM components, 312–313
Consumer perspectives on GM food labeling, 210–211

Consumer rights, 298
  GM foods, 299–300
  GM technology, 299
  PLU system, 298
Consumers, benefits for, 243
  BMA, 245–246
  food quality, 244
  GM foods, 243
  healthy foods, 245
  Super Rice, 244
Conventional biotechnology, see Nongenetic engineering techniques
Convention on Biological Diversity (CBD), 212
Corn (Maize), 142
Corn flour, 142
Corn meal, 142
Cotton (Seed Oil), 143
Council for Responsible Genetics (CRG), 253
CPA, see Crop Protection Association (CPA)
Creutzfeldt-Jakob disease (CJD), 136
CRG, see Council for Responsible Genetics (CRG)
CRM, see Certified reference material (CRM)
CropGen, 239
Crop Protection Association (CPA), 239–240
Crop protection strategies, 114–115
"Cutting-copying-pasting" approach, 71
Cyanophyceae, 125
Cytosine (C), 27

D

D-psicose, 165
D-tagatose, 165
DDT, see Dichlorodiphenyltrichloroethane (DDT)
Delayed ripening, 111
Delta endotoxin, 115

386

Deoxyribonucleic acid technique
  (DNA technique), 25, 26
  DNA fingerprinting, 29
  living organism's, 28
  recombinant DNA technology, 27
  structure, 28
Department of Fisheries and Oceans
  (DFO), 197
Detection of GMO, 91
DFO, *see* Department of Fisheries and
  Oceans (DFO)
Dichlorodiphenyltrichloroethane
  (DDT), 234
Disease resistance, 111
DNA chip technology, *see* Microarrays
  technology
DNA fingerprinting, 29
DNA microarray, 96
DNA microinjection, *see*
  Microinjection
DNA profiling, *see* DNA
  fingerprinting
DNA technique, *see* Deoxyribonucleic
  acid technique (DNA
  technique)
Dolly (The sheep, 1st cloned animal), 137
Domestication, 135
"Double haploid" techniques, 39
Double helix, 27
Dow Company, 235
Drought-tolerant plants, 118
DuPont Pioneer Company, 235

E

Earth Liberation Front (ELF), 253
Earthsave Canada, 258
EC, *see* Environment Canada (EC);
  European Commission (EC)
Edible oils and fats, 121
EFSA, *see* European Food Safety
  Authority (EFSA)
Electropermeabilization, *see*
  Electroporation
Electroporation, 77–78, 164

Electroporator, 78, 79
ELF, *see* Earth Liberation Front (ELF)
ELISA, *see* Enzyme-linked
  immunosorbent analysis
  (ELISA)
Elves, *see* Earth Liberation Front (ELF)
Embryo culture, *see* Embryo rescue
Embryo rescue, 62–63
Endocytosis, 83
Enhancers, 86
Enterologics, Inc., 162
Enucleated oocyte, 137
Environmental concerns, 228
  biodiversity, 230
  genetic pollution, 229
Environment Canada (EC), 197
Environment, GM food benefits for, 246
  bioremediation, 249
  pest-resistant crops development,
    247
  soil erosion reduction, 248
Enviropig™, 43
Enzymatically active molecules, 51
Enzyme-linked immunosorbent
  analysis (ELISA), 92–94
EPA, *see* U.S. Environmental Protection
  Agency (EPA)
Erythritol, 165
Escherichia coli bacteria, 40, 162
EU, *see* European Union (EU)
European Association for Bioindustries
  (EuropaBio), 238
European Commission (EC), 52, 199
European Committee for
  Standardization (CEN), 99
European Food Safety Authority
  (EFSA), 200
European GMO-free regions'
  network, 44
European Union (EU), 199, 256–258, 303
  future of GM food, 167–168
  GMOs, 200
  labeling regulation, 201
  Regulation EC 1829/2003, 201–202,
    203

INDEX

Expression constructs, *see* Expression vectors
Expression vectors, 74

F

FAO, *see* Food and Agriculture Organization (FAO)
FAO-BioDeC, *see* FAO Biotechnology in Developing Countries Database (FAO-BioDeC)
FAO Biotechnology in Developing Countries Database (FAO-BioDeC), 133
Farmers
    agricultural practices, 34
    agricultural techniques, 30
    DroughtGard Hybrids production, 119
    Farmer to Farmer Campaign on Genetic Engineering, 253
    GM crop benefits for, 242–243
    modern biotechnological techniques and tools, 131
    using simple selection method, 60
Farming, negative impact on, 275
FAS, *see* French Academy of Sciences (FAS)
FASS, *see* Federation of Animal Science Societies (FASS)
Fatty acid profile of soybean, maize, rapeseed, and other oil crops, 140–141
FD&C Act, *see* Federal Food and Drug Cosmetic Act (FD&C Act)
FDA, *see* U. S. Food and Drug Administration (USFDA)
FDF, *see* Food and Drink Federation (FDF)
Federal Food and Drug Cosmetic Act (FD&C Act), 205
Federation of Animal Science Societies (FASS), 52
Fermentation-produced chymosin (FPC), 149

Fermentation, 35, 162–163
Fermented foods, 160, 162–164
Feulgen stain, 38
Field food crops, 111
FIRAB, *see* Foundation for Organic and Biodynamic Research (FIRAB)
First-generation GM crops, 6
First GM labeling rules, 43
FlavrSavr tomato, 8, 145
FoE, *see* Friends of the Earth (FoE)
Food, 1
    color, 166
    companies using GMO ingredients, 145, 146
    GM to non-GM component moving impact on food sales, 312–313
Food and Agriculture Organization (FAO), 223
Food and Drink Federation (FDF), 240
Food biotechnology, 138; *see also* Animal biotechnology; Plant biotechnology
    advances, 5–6
    biotechnology in functional foods and nutraceuticals, 156–166
    food companies using GMO ingredients, 145, 146–148
    foods from animals, 150–151
    future of GM food, 166–169
    genetically modified foods, 138–140
    GM crops as food, 141–145
    non-GMO companies and food products, 151–155
    non-GMO Project verified, 156, 157
    non-GMO seed companies, 156
    processed foods and ingredients, 149–150
    recognition, 3–5
Food, Drug, and Cosmetic Act, 206
Food First, 254
Food safety
    CAC guideline principles, 13
    concerns about, 226–228
    establishment, 185
    issue of GM, 190

Food Safety Authority of Ireland (FSAI), 52
Food Standards Agency (FSA), 233
Food Standards Australia New Zealand (FSANZ), 52, 195
FOS, *see* Fructo-oligosaccharides (FOS)
Foundation for Organic and Biodynamic Research (FIRAB), 256
FPC, *see* Fermentation-produced chymosin (FPC)
Frankenstein foods, 11
French Academy of Sciences (FAS), 52
Friends of the Earth (FoE), 253–254
Fructo-oligosaccharides (FOS), 160
FSA, *see* Food Standards Agency (FSA)
FSAI, *see* Food Safety Authority of Ireland (FSAI)
FSANZ, *see* Food Standards Australia New Zealand (FSANZ)
Functional foods, 120, 156, 157–158
　animal-based functional food production, 159
　biotechnology application for foods, 160–166
　plant-based functional food production, 159
Fungi-based biopesticides, 116
Fungi, 116, 117, 149
"Future For All" website, 47

G

GA21 GM maize, 45
Gaia Foundation, 259
Galactose, 165
Gastrointestinal tract (GI tract), 160
GE, *see* Genetic engineering (GE)
GEAN, *see* Genetic Engineering Action Network (GEAN)
GE crops, *see* Genetically engineered crops (GE crops)
"GE era", 3
GE method, *see* Genetically engineered method (GE method)

Gene
　biotechnology, 26
　cloning, 84–85
　design and packaging, 85–86
　Gun, 41, 72
　insertion, 87
　knock-out technology, 136
　silencing, 81–83
　splicing, 80–81
　technology, 26
　therapy, 27, 46
　transfer process, 82
Generally recognized as safe (GRAS), 165, 206
Genetically engineered method (GE method), 7
　corn in field, 8
Genetically engineered crops (GE crops), 225
Genetically modified crops (GMCs), 32, 111
　global area of Biotech crops, 127–129, 130
　global status, 126
　plant breeding, 130
　in United States, 129, 130
Genetically modified foods (GM foods), 1, 5, 14–16, 23, 29–32, 33, 138–140, 145, 148–149, 221, 295; *see also* Opponents of GM technologies; Proponents of GM technologies
　American consumers, 302
　animal health and welfare, 231
　composition, 140–141
　concerns about bias in scientific publishing, 232
　consumer choice, 231
　consumer purchasing behavior, 306
　consumers' perceptions, attitudes, and preferences, 300
　controversy, 223
　controversy, 8–14
　cultural factors, 301
　debates, 277–278

INDEX

Genetically modified foods (*Continued*)
  economic concerns, 230–231
  environmental concerns, 228–230
  ethical concerns, 230
  European consumer, 303, 307
  food production, regulations, and labeling, 9
  food safety and human health, concerns about, 226–228
  foods in grocery stores, 12
  fruit and vegetables, 33
  future of, 166–169
  genetic modification, 222
  GM food labeling, 309
  GMO, 225
  "Horizon 2020" program, 311
  IFIC, 301
  invisible flag, 305
  issue of GM foods, 11, 14
  issues of concern and controversy, 226
  legal concerns, 230
  market dynamics, 224
  non-GM foods, 308
  regulatory concerns, 230
  safety of GM crops, 10
  sustainable agriculture company, 13
  technology, 6–8
Genetically modified organisms (GMOs), 4, 23, 31, 32, 71, 88, 225, 297
  assessment, 91
  Detection Method Database, 90
  future, 97–99
  GM components, 89
  high technological methods, 94–96
  ISO standards for detecting, 98–100
  low technological methods, 92–94
  reference material for, 97
  sampling, 90
  standardization of methods, 96
  testing methods, 91–96
  validation of methods, 96
Genetically modified organisms, 7

Genetic engineering (GE), 24, 29–32, 59, 72, 117
  calcium phosphate precipitation, 83
  electroporation, 77–78
  fermentation development and improvements, 164
  general principles, 70
  gene silencing, 81–83
  gene splicing, 80–81
  gene transfer process, 82
  Helios PDS 1000/He Biolistic Particle Delivery System, 80
  from laboratory to greenhouse, 72
  lipofection, 83
  microinjection, 78–79
  microprojectile bombardment, 79–80
  non-usable techniques in, 135
  plants growing in greenhouse, 75
  recombinant gene technology, 71
  seedlings, 74, 75
  steps, 83–87, 89
  transgenic inspection of GE plants, 73
  transgenic plants in petri dishes, 73
  unintended effects, 100–103
  usable techniques in, 136–138
  vitamin profile improvement, 120
Genetic Engineering Action Network (GEAN), 254
Genetic fingerprinting, *see* DNA fingerprinting
Genetic linkage, 61
Genetic modified/modification (GM), 5, 7, 24, 30, 222; *see also* GM hazards and risks
  alternatives to GM technology, 263
  animal biotechnology, 135–138
  biotechnology for biofuel production, 170–172
  corn, 142
  food biotechnology, 138–169
  GM soybean, 144
  GM Watch, 119
  plant biotechnology, 110–134
  of plants, 59

Genetic pollution, 224 229
   causing by vector-mediated HGT, 270–271
Genuity® DroughtGard™ Hybrids, 119
GI tract, see Gastrointestinal tract (GI tract)
Global warming effect, 119
   resistant to bacterial diseases, 117
   resistant to fungal diseases, 117
   resistant to viral diseases, 117–118
Glyphosate, 113
GM, see Genetic modified/modification (GM)
GM crop varieties development
   with improved nutritional quality traits, 119–121
   with improved resistance or tolerance to abiotic factors, 118–119
   with improved resistance to bacterial, fungal, and viral diseases, 116–118
   with improved resistance to herbicides, 112–113
   with improved resistance to pests, 114–116
GMCs, see Genetically modified crops (GMCs)
GM foods, see Genetically modified foods (GM foods)
GM hazards and risks, 264
   ethical and moral objections, 275
   GM technology, 275–276
   hazards to human and animal health and safety, 265–270
   impact on ecosystem, 270–275
   impact on environment, 270–275
   impact on farming, 270–275
   interconnected genetic network, 265
   natural boundaries, 264
   transgenic lines instability, 265
   unnatural technique, 264
   unpredictability and unknown, 264–265
"GMO-free" agriculture, 168, 206, 208
GMOs, see Genetically modified organisms (GMOs)
Golden rice, 120, 144, 244
Golden rice enriched with provitamin A, 140
GRAS, see Generally recognized as safe (GRAS)
Gray biotechnology, 50
Green biotechnology, 50, 109
Greenpeace, 254, 261, 267, 269–270, 274
Green Revolution, 2, 3, 133
Green Super Rice (GSR), 244
Grocery Manufacturers Association, 240
GSR, see Green Super Rice (GSR)
Guanine (G), 27

H

Haploid plants, 67
HC, see Health Canada (HC)
Health Canada (HC), 197
Heat-stable DNA polymerase, 94
Helios PDS 1000/He Biolistic Particle Delivery System, 80
Herbicide-tolerant crop (HT crop), 112, 132, 243
Herbicide bioassays, see Phenotypic characterization
Herbicide resistance, 111, 112, 113
Herbicide tolerance, 112, 113
HGT, see Horizontal gene transfer (HGT)
"High-oleic" soybean varieties, 45
High pro-vitamin A transgenic rice, see Golden rice
High technological methods, 94; see also Low technological methods
   microarrays, 96
   PCR, 94–95
High technology methods, 92
Honey, 145
"Horizon 2020" program, 311
Horizontal gene transfer (HGT), 229
   vector-mediated, 270–271

HT crop, *see* Herbicide-tolerant crop (HT crop)
Human and animal health and safety, hazards to, 265–270
Human genome, 27
Human health, concerns about, 226–228
Hybrid plant, 61

I

ICSU, *see* International Council for Science Union (ICSU)
Identification of GMOs, 91
IFIC, *see* International Food Information Council (IFIC)
IFICF, *see* International Food Information Council Foundation (IFICF)
IFPS, *see* International Federation for Produce Standards (IFPS)
IFST, *see* Institute of Food Science and Technology (IFST)
ILSI, *see* International Life Science Institute (ILSI)
Immunoassays, 93
  ELISA, 93–94
  lateral flow sticks, 94
"Individualized nutrition" approach, 49
Industrial biotechnology, *see* White biotechnology
Infertile aquatic species production, 137–138
Input characteristics, *see* First-generation GM crops characteristics
Insect pest species, 114
Institute for Food and Development Policy, *see* Food First
Institute for Responsible Technologies, 254
Institute of Food Science and Technology (IFST), 113, 139
Institute of Science in Society (ISIS), 256

Integrated pest management programs (IPM programs), 116
Intellectual Property (IP), 211
  biopiracy, 214–215
  biotechnology, 214
  seed treaty, 212
  technology protection system, 212–213
Interior toxins in GM foods, 266
International bodies and governments, 232–233
International Council for Science Union (ICSU), 51
International Federation for Produce Standards (IFPS), 298
International Food Information Council (IFIC), 236, 301
International Food Information Council Foundation (IFICF), 139, 160
International Life Science Institute (ILSI), 52
International Organization for Standardization (ISO), 97
International Service for Acquisition of Agri-biotech Applications (ISAAA), 51, 126, 236
International Society of African Scientists (ISAS), 52
International Treaty on Plant Genetic Resources for Food and Agriculture, 212
Interspecies crossing, 62
Investigational New Drug application, 162
In vitro anther culture, 68
In vitro embryo rescue techniques, 63
In vitro nucleic acid techniques, 25
IP, *see* Intellectual Property (IP)
IPM programs, *see* Integrated pest management programs (IPM programs)
ISAAA, *see* International Service for Acquisition of Agri-biotech Applications (ISAAA)

ISAS, *see* International Society of African Scientists (ISAS)
ISIS, *see* Institute of Science in Society (ISIS)
ISO, *see* International Organization for Standardization (ISO)
Italian Association for Organic Agriculture (AIAB), 257

J

Japan
   documentation, 204
   MAFF and MHLW, 202
   "zero tolerance" policy, 204
Japanese Agricultural Standards (JAS), 202

L

Labeling, GM food, 187
   anti-labeling arguments, 190–191
   Australia–New Zealand, 195–197
   Canada, 197–199
   consumer perspectives on, 210–211
   countries with mandatory GE foods labeling, 192
   European Union, 199–202
   global, 191
   issues with, 187–189
   Japan, 202–204
   labeling requirements in countries, 195
   pro-labeling arguments, 189–190
   South Africa, 204–205
   United States, 205–209
Lateral flow sticks, 92, 94
Laurical™, 121
Legal concerns, 230
Lepidoptera, 114
Lipofection, 83
Liposomes, 48
Liposome transfection, *see* Lipofection
Low technological methods, 92; *see also* High technological methods
   immunoassays, 93–94
   phenotypic characterization, 92–93
Low technology methods, 92

M

Mad cow disease, 199
MAFF, *see* Ministry of Agriculture, Forestry, and Fisheries (MAFF)
Marker-aided selection, *see* Marker-assisted selection (MAS)
Marker-assisted selection (MAS), 60–61, 263
MAS, *see* Marker-assisted selection (MAS)
Mass media, 313–314
messenger RNA (mRNA), 39
MHLW, *see* Ministry of Health, Labor, and Welfare (MHLW)
Microarray technology, 96
Microbial inoculants, 125
Microbial vectors, 74
   bacterial carriers, 76–77
   bacterial colony picking for DNA cloning, 76
   viral carriers, 77
Microinjection, 78–79
Microinjector, 79
Micronutrient, 120, 169
   biofortification, 159
   malnutrition, 120
Microorganisms, 50, 74, 162, 163
   agricultural practices, 34
   in biofertilizers, 125
   in food fermentation and production, 35
   plant disease-causing, 117
   probiotics, 160
Microprojectile bombardment, 64, 79–80
Micropropagation, 65, 66, 68–69
Ministry of Agriculture, Forestry, and Fisheries (MAFF), 202, 204
Ministry of Health, Labor, and Welfare (MHLW), 202, 204

INDEX

Modern agricultural biotechnology, 110
MON810 corn, 168
Monsanto corporation, 129
Monsanto Protection Act, 265
mRNA, *see* messenger RNA (mRNA)
Mutations, 64, 70
    breeding, 64–65
Mutation theory of evolution, 38
Mycorrhiza, 125

N

National Academy of Sciences (NAS), 52
National Center of Food and Agricultural Policy (NCFAP), 247
National Corn Growers Association (NCGA), 237–238
National Research Council (NRC), 52
Nature Institute, 103
NCFAP, *see* National Center of Food and Agricultural Policy (NCFAP)
NCGA, *see* National Corn Growers Association (NCGA)
Nematodes, 114
New Zealand, 258
NGO, *see* Nongovernmental organization (NGO)
Nitrogen bases, 27
Non-GM biotechnology, 263
Non-GMO companies and food products, 151–155
"Non-GMO Project Verified", 156, 157
Non-GMO seed companies, 156
Nonbrowning Arctic apple, 142
Nongene biotechnology, 26
Nongenetic engineering techniques, 30, 60
    crossing, 61–62
    embryo rescue, 62–63
    interspecies crossing, 62
    MAS, 60–61
    mutation breeding, 64–65

    plant TC and micropropagation, 65–69
    simple selection, 60
    somaclonal variation, 63–64
    somatic hybridization, 63
    unintended effects, 69–70
Nongovernmental organization (NGO), 223
Nonselective herbicides, 113
Nontarget effects, *see* Unintended effects of GE
Northwest Resistance Against Genetic Engineering (NW RAGE), 255
NRC, *see* National Research Council (NRC)
Nucleic acid extraction, 83–84
Nuclein, 37
Nucleotides, 27, 28
Nutraceutical biotechnology, 157
Nutritional quality traits, 119–121
NW RAGE, *see* Northwest Resistance Against Genetic Engineering (NW RAGE)

O

OCA, *see* Organic Consumers Association (OCA)
Occupy the World Food Prize, 5
OECD, *see* Organization for Economic Cooperation and Development (OECD)
Office of Gene Technology Regulator (OGTR), 196
Office of International Affairs National Research Council, 164
OGTR, *see* Office of Gene Technology Regulator (OGTR)
Old biotechnology techniques, 30
"Omics" era, 43
OP crop varieties, *see* Open-pollinated crop varieties (OP crop varieties)
Open-pollinated crop varieties (OP crop varieties), 35

INDEX

Opponents of GM technologies, 251; *see also* Proponents of GM technologies
   Africa, 258–259
   alternatives to GM technology, 263
   arguments against GM foods, 262
   Asia Pacific, 260
   Australia, 258
   Canada, 258
   Central and South America, 260
   Europe, 256–258
   GM hazards and risks, 264–270
   New Zealand, 258
   protests against GMOs and GM foods, 260–261
   United States, 252–256
   vandalism and threats by anti-GMO activists, 261–262
Organic agriculture, 263
Organic Consumers Association (OCA), 255
Organization for Economic Cooperation and Development (OECD), 48, 185
Output characteristics, *see* First-generation GM crops characteristics

P

Papaya, 143
Patents, 211
   biopiracy, 214–215
   biotechnology, 214
   in United States, 213
Patho-biotechnology, 161
PBO, *see* Plant Biosafety Office (PBO)
PCB, *see* Polychlorinated biphenyl (PCB)
PCR, *see* Polymerase chain reaction (PCR)
Peas, 143
PEM, *see* Protein-energy malnutrition (PEM)
Penicillium bilaii (*P. bilaii*), 122

Personalized medicine, 49
Peruvian Environmental Law Society, 260
Pew Initiative on Food and Biotechnology, 157
PGPR, *see* Plant-growth promoting rhizobacteria (PGPR)
Pharmacogenomics, 49
"Pharma crop" production, 6
Phenotypic characterization, 92–93
Pioneer Hi-Bred Company, *see* DuPont Pioneer Company
PIPs, *see* Plant-incorporated protectants (PIPs)
Plant-growth promoting rhizobacteria (PGPR), 125
Plant-incorporated protectants (PIPs), 116
Plant Biosafety Office (PBO), 198
Plant biotechnology, 110; *see also* Animal biotechnology; Food biotechnology
   biofertilizer production, 122, 124–126
   field food crops, 111
   GM crops, 111, 126–130
   GM crops/foods in developing countries, 130–134
   GM crop varieties development, 112–121
   weighing risks and benefits, 121–122, 123–124
Plant breeding, 34, 47, 69, 87, 130
Plant disease-causing microorganisms, *see* Viruses
Plant protection products (PPP), 116
Plants as factories, *see* Third-generation GM crops
Plant TC, 65
   advantages, 69
   multiple benefits, 67
   process, 66
   types, 67
Pleiotropic effects, 101–102
PLU Code, *see* Price LookUp Code (PLU Code)

Polychlorinated biphenyl (PCB), 234
Polymerase chain reaction (PCR), 40, 86, 94–95
"Polyploidization" process, 39
Potato, 42, 143
Poultry production, 242
PPP, see Plant protection products (PPP)
Prebiotics, 160, 161
Predictable unintended effects, 102–103
Price LookUp Code (PLU Code), 298
Primers, 94, 95
Prion-related peptide gene (PRP gene), 136
Pro-corporate activists global network, 241
Pro-GM Lobby Groups, 236
   ABC, 239
   AfricaBio, 238
   Alliance for Better Foods, 241
   ASA, 237
   AusBiotech Ltd, 238–239
   BIO, 237
   CPA, 239–240
   CropGen, 239
   EuropaBio, 238
   Grocery Manufacturers Association, 240
   IFIC, 236
   ISAAA, 236
   NCGA, 237–238
   SAS, 240
Pro-labeling arguments, 189–190
Probiotic-containing foods, 160
Probiotics, 160, 161
   benefits, 161
   goals of biotechnology, 161–162
Promoter, 85, 86
Pronuclear microinjection, see Microinjection
Proponents of GM technologies, 232; see also Opponents of GM technologies
   AAAS, 250
   arguments in GM foods favor, 241
   biotech, agrochemical, and associated companies, 233–235
   consumers, benefits for, 243–246
   economy, benefits for, 249–250
   environment, benefits for, 246–249
   farmers, benefits for, 242–243
   international bodies and governments, 232–233
   pro-corporate activists global network, 241
   Pro-GM Lobby Groups, 236–241
   UNIDO, 251
Protein-energy malnutrition (PEM), 120
Protein methods, 92
PRP gene, see Prion-related peptide gene (PRP gene)
Pure lines, 61
Purines, 27
Pyrimidines, 27

Q

Quantification of GMOs, 91
Quantitative polymerase chain reaction (Q-PCR), 95

R

RAFI, see Rural Advancement Foundation International (RAFI)
Rapeseed, 143
Rare sugar, see D-psicose
rBGH, see recombinant bovine growth hormone (rBGH)
rDNA, see recombinant DNA (rDNA)
Real-time PCR, 95
recombinant bovine growth hormone (rBGH), 267
recombinant bovine somatotropin (rBST), see Bovine growth hormone (BGH)
recombinant DNA (rDNA), 74, 184
   molecules, 40
   technology, 27

Recombinant gene technology, 71
Red biotechnology, 49
Regulations, GM food, 183
    Australia–New Zealand, 195–197
    Canada, 197–199
    characteristics in country groups, 194
    in countries, 193
    European Union, 199–202
    Japan, 202–204
    SE principles, 185–187
    South Africa, 204–205
    United States, 205–209
Regulatory concerns, 230
Rennet, 149
Rennin, 149
Reporter genes, 85
Restriction enzymes, 81
Reverse transcriptase enzyme, 39
Rhizobium, 125
Rice, 143–144; *see also* Golden rice
Rod-shaped *Baculoviruses*, 116
Roundup herbicide, 113
RoundupReady, 113
Royal Society of London (RSL), 52
Rural Advancement Foundation International (RAFI), 255

S

Saccharomyces cerevisiae strains, 163
Safeguard clause, 43, 44
SAGENE, *see* South African Committee for Genetic Experimentation (SAGENE)
Sampling, 90
SAS, *see* Sense About Science (SAS)
Save our Seeds (SOS), 257
SCNT, *see* Somatic cell nuclear transfer (SCNT)
SE, *see* Substantial equivalence (SE)
Second-generation GM crops characteristics, 6
Seed proteins, 120
Seed treaty, *see* International Treaty on Plant Genetic Resources for Food and Agriculture
Selectable marker genes, 85
Sense About Science (SAS), 240
SE principle, *see* Substantial equivalence principle (SE principle)
Sexual reproduction process, 65
Sierra Club, 255
Simple selection, 60
"Single seed descent" technique (SSD technique), 39
Sociedad Peruana de Derecho Ambiental (SPDA), 260
Society supported biotechnology, 48
Soil Association, 257
Somaclonal variation, 40, 63–64
Somatic cell nuclear transfer (SCNT), 136–137
Somatic hybridization, 63
SOS, *see* Save our Seeds (SOS)
South Africa, 204–205
    GM crops in, 132
South African Committee for Genetic Experimentation (SAGENE), 132
Soybean, 12, 111, 144
SPDA, *see* Sociedad Peruana de Derecho Ambiental (SPDA)
Squash, 144
SSD technique, *see* "Single seed descent" technique (SSD technique)
Stacked-trait organism, 71
Standardization of methods, 96
Substantial equivalence (SE), 185
    GM foods, 185–186
    OECD, 186
    principles, 185, 186–187
Substantial equivalence principle (SE principle), 185–187, 206, 228
Sugar alcohols, 164, 165
Sugar beets, 45, 149
Sugar cane, 145, 170

INDEX

"Superpests", 272–273
Super rice, 244
"Supertrees", 273–274
"Superviruses", 268
"Superweeds", 229, 271
Sutton–Boveri theory, *see* Boveri–Sutton Chromosome Theory
Sweet corn, 145
Sweeteners, 164–165
Switching off technique, 81
Synbiotics, 161

T

Taq polymerase, 94
TC, *see* Tissue culture (TC)
Technology protection system, 212–213
Terminator technology, 11, 43, 85, 86, 275
Terminator trees, 273
Thermocycler, 94
Third-generation GM crops, 6, 7
Third World Academy of Sciences (TWAS), 51
Thymine (T), 27
Tissue culture (TC), 63
Tomato, 145
Trade-related aspects of intellectual property rights (TRIPS), 213
Traditional biotechnology, 30, 31, 34, 163
Transcription, 82
  vectors, 74
Transduction, 76, 77, 164
Transformation, 86, 164
Transgenesis, 5, 162
Transgenic(s), 136
  GMO, 71
  plants, 86
  process, 27
Translation, 82
TRIPS, *see* Trade-related aspects of intellectual property rights (TRIPS)

Turning off technique, *see* Switching off technique
TWAS, *see* Third World Academy of Sciences (TWAS)

U

Uganda, GM crops in, 132–133
U.K. Bodies, 233
UN and International bodies, 232–233
UNECA, *see* United Nations Commission for Africa (UNECA)
UNIDO, *see* United Nations Industrial Development Organization (UNIDO)
Unintended effects of GE, 100
Unintended effects of genetic manipulation, 103
United Nations Commission for Africa (UNECA), 52
United Nations Industrial Development Organization (UNIDO), 251
United States, 205, 252
  AAEM, 256
  "California Proposition 37", 209
  CFS, 253
  FD&C Act, 205
  FDA guidance, 207
  FDA policy, 206
  FoE, 253–254
  Food, Drug, and Cosmetic Act, 206
  Food First, 254
  GM food production and consumption, 167
  "GMO-free", 208
  NW RAGE, 255
  OCA, 255
  Sierra Club, 255
Unpredictable unintended effects, 103

INDEX

U.S. Department of Agriculture (USDA), 39, 205, 233
U.S. Environmental Protection Agency (EPA), 41, 205
U.S. Food and Drug Administration (USFDA), 42, 149, 162, 205, 233
U.S. Government Departments, 233

V

Validation of methods, 96
Vandalism, 261–262
Vector-mediated HGT, genetic pollution causing by, 270–271
Vector recombination, 271
Vegetarian Society, 257
Viral carriers, 77
Virus-resistant papaya, 43
Viruses, 117
Vitamin profile, 120

W

Water conservation, 248
WDM, *see* World Development Movement (WDM)
Weeds, 112, 113
Weed Science Society of America (WSSA), 112
White biotechnology, 49–50
WHO, *see* World Health Organization (WHO)
Wide crosses, 63
Willingness-to-pay (WTP), 305
Witchweed, 112
World Development Movement (WDM), 257
World Food Prize, 3
World Food Summit (1996), 1
World Health Organization (WHO), 223
World Trade Organization (WTO), 213
WSSA, *see* Weed Science Society of America (WSSA)
WTO, *see* World Trade Organization (WTO)
WTP, *see* Willingness-to-pay (WTP)

X

Xenotransplantation, 136

Z

"Zero tolerance" policy, 204